A HISTORY OF VECTOR ANALYSIS

The Evolution of the Idea of a Vectorial System

MICHAEL J. CROWE

Dover Publications, Inc.
New York

To Mary Ellen

Copyright

Bibliographical Note

This Dover edition, first published in 1994, is a slightly corrected reprint of the Dover edition of 1985, which was an unabridged and corrected republication of the work originally published by the University of Notre Dame Press in 1967, and which contained a new Preface by the author. This 1994 edition adds a Publisher's Note.

Library of Congress Cataloging-in-Publication Data

Crowe, Michael J.
A history of vector analysis : the evolution of the idea of a vectorial system / Michael J. Crowe.
p. cm.
Reprint. Originally published: Notre Dame : Notre Dame University Press, 1967.
Includes bibliographical references and index.
ISBN-13: 978-0-486-67910-5
ISBN-10: 0-486-67910-1
1. Vector analysis—History. I. Title.
QA433.C76 1993
515′.63—dc20 93–6116
CIP

Manufactured in the United States by RR Donnelley
67910108 2015
www.doverpublications.com

Publisher's Note to the 1994 Edition

In deciding to bring this book back into print, Dover was influenced not only by the substantial interest in the book on the occasion of the first Dover republication in 1985, but also by the fact that in late 1992 the exceptional merits of the book were recognized by its being awarded in international competition a Jean Scott prize of $4000 by the Maison des Sciences de l'Homme in Paris. Charles Morazé, writing on behalf of the Maison des Sciences de l'Homme and the awards committee, praised the book in the following terms:

> The Jean Scott Foundation, the purpose of which is to contribute to research related to cultural and social inequalities, decided in October 1991 to grant an award to the best of the works attempting to clarify how—as early as the nineteenth century—Europe and then the entire Western world found themselves in a position to foresee that the world was made of forces rather than things, and to analyze their functions. Thus, the subject was the so-called imaginary numbers that a succession of algebraic discoveries would finally identify with vectors capable of analyzing and calculating systems of forces.
>
> Taking into account the conclusions which previous discussions, mostly on an international level, had established, in June 1992 the Jury unanimously decided to establish a "special prize" to honor Michael J. Crowe's work, thus recommending the reading of his *History of Vector Analysis*. With regard to the nineteenth century and its immediate extensions, this pioneer book remains unequaled. This is due mostly to its remarkable amount of information, given with a perfect and accurate scientific density.
>
> The President of the Jury was Professor André Lichnerowicz, of the French Academy of Sciences. The other members were Professor Pierre Chaunu, historian, member of the Institute; Professor Marc Barbut, mathematician, University of Paris, Hautes Etudes; Norbert Beyrard, doctor of physics and economics; and Professor Charles Morazé, University of Paris, Ecole Polytechnique, Haute Etudes, founding member of the Maison des Sciences de l'Homme.

We concur with the judgment of this distinguished committee that this volume deserves reading by those interested in this important subject matter.

Dover Publications, Inc.

Preface

Shortly before becoming President of Harvard in 1862, the mathematician Thomas Hill made the following statement concerning the best known vectorial system of his day: "The discoveries of Newton have done more for England and for the race than whole dynasties of British monarchs; and we doubt not that in the great mathematical birth of 1853, the Quaternions of Hamilton, there is as much real promise of benefit to mankind as in any event of Victoria's reign." Lord Kelvin, writing when Victoria was very old and the modern vector system very new, took a very different view. "Quaternions came from Hamilton after his really good work had been done; and, though beautifully ingenious, have been an unmixed evil to those who have touched them in any way . . . vectors . . . have never been of the slightest use to any creature." Though Kelvin's barbs in this attack of the 1890's were directed against the quaternion system, he had been waging war against all vectorial methods since the 1860's.

If a scientist of the present day were forced to take sides in this dispute on the value of vectorial methods, he might view Hill as overly enthusiastic, but he would not side with Kelvin. The view of vectorial methods championed by this great physicist has been refuted by the thousands of uses that have been found for vectorial methods. Nearly all branches of classical physics and many areas of modern physics are now presented in the language of vectors, and the benefits derived thereby are many. Vector analysis has likewise proved a valuable aid for many problems in engineering, astronomy, and geometry.

Despite the importance of vector analysis, its history has been little studied. Not a single book and not more than a handful of scholarly papers have up to now been written on its history. Consequently many historical errors may be found in the relevant literature. The present study was not written in the expectation that all or even most historical questions about vector analysis would be answered; rather it was written in the hope of presenting an essentially correct outline of the history of this important area. In undertaking this study I have frequently been hindered by the scarcity of scholarly studies of the history of such related areas as complex numbers, linear algebra, tensors, theoretical electricity, and nineteenth-century mechanics. This is of course the common plight of historians of science, and I have been consoled by the hope that the present study may shed light on the history of other areas of science, such as those mentioned above.

In this study I have concentrated on the more fundamental aspects of vectorial analysis; the history of the following topics is treated in detail: vector addition and subtraction, the forms of vector multiplication, vector division (in those systems where it occurs), and the specification of vector types. Less attention has been given to the history of vector differentiation and integration, and the operator ∇ and the associated transformation theorems, since these were for the most part developed originally in a Cartesian framework. No detailed presentation of the complicated history of the linear vector function has been attempted.

Though the above statement indicates the materials included, it does not sufficiently specify the approach taken in this study. For a number of reasons I have chosen to focus (as the subtitle indicates) on the history of the idea of a vectorial system. It should not be forgotten that the modern system of vector analysis is but one of the many vectorial systems created in the course of history. Each of these systems embodied an idea or conception of the form that a vectorial system can have and should have. And it is the history of these ideas that I have tried to describe. To do this I have discussed each of the major vectorial systems created before 1900 and attempted to determine what ideas (mathematical and motivational) led to the creation, development, and acceptance or rejection of these systems.

The history of vectorial analysis may in one sense be viewed as the history of systems of abbreviation, since any problem that can be solved by vectorial methods can also be solved (though usually less conveniently) by the older Cartesian methods. The history of vectorial analysis may equally well be viewed as the history of a way of looking at physical and geometrical entities. Consideration of these two aspects of the history will help explain why I have chosen to focus on the evolution of the idea of a vectorial system, rather than on the history of the major theorems in vectorial analysis, many of which were in any case discovered before and outside of the vectorial traditions.

Concerning the references. The reader will find that a simple and not uncommon system of reference has been employed in the text. The notes for each chapter are located at the end of that chapter; within each chapter ordinary note numbers will be found as well as references to these notes of the form (3,II,I; 27). The latter are read as follows: the first number always refers to a note at the end of the chapter; the numbers to the right of the semicolon always refer to the page numbers in the publication indicated in that note. In some cases (as above) one or two other numbers are included to the left of the semicolon; these numbers (when they occur) refer to the volume number and part number of the publication indicated. Thus the reference above is read: see volume II, part I, of item 3 in the notes; consult page 27. Through this method it has been possible to provide the reader with many references that otherwise could be included only through a substantial increase in the size of the book.

Concerning quotations and translations. Since many of the sources for this study were books and journals of limited circulation I have used quotations rather liberally. All quotations from documents written in foreign languages (French, German, Italian, Russian, and Danish) have been translated into English. In the few cases where previously published translations were available, I have used these after checking them against the original and noting deviations. The sole exception to this statement occurs in the case of Wessel's Danish; here I have checked Nordgaard's English translation against the French translation of Zeuthen. The remaining translations (the majority) are my own.

Concerning bibliography. No formal bibliographical section has been included in this book. The reader will find however that the sections of notes at the end of each chapter will serve rather well as a bibliography for that chapter. Moreover the need for a bibliography is greatly diminished by the existence of a book that lists nearly all relevant primary documents published to about 1912; this is Alexander Macfarlane's *Bibliography of Quaternions and Allied Systems of Mathematics* (Dublin, 1904). Supplements to this uncommonly accurate bibliography were published up to 1913 in the *Bulletin of the International Society for Promoting the Study of Quaternions and Allied Systems of Mathematics.*

The author wishes to express his gratitude to those who have aided him in preparing this study. Published and unpublished materials have been obtained from libraries too numerous to mention, and this through the kindness of the librarians of the universities of

Notre Dame, Wisconsin, Cambridge, and Yale. Assistance at important points has come from Professor Stephen J. Rogers of Notre Dame University and from Professor Derek J. Price of Yale University. Sincere thanks are extended to Professors C. H. Blanchard and William D. Stahlman of the University of Wisconsin and to Professor James W. Bond of Pennsylvania State University. These three scholars (a physicist, an historian of science, and a mathematician) gave generously of their time (in reading the entire manuscript) and of their wisdom (in saving the manuscript from a number of errors). To Professor Erwin N. Hiebert, of the University of Wisconsin, my most sincere thanks for his numerous, detailed, and perceptive comments on the entire manuscript. Portions of the research for this book were carried out with financial assistance provided by funds administered by Committee on Grants for the Arts and Humanities of the University of Notre Dame.

Michael J. Crowe

Notre Dame, Indiana
March, 1967

Acknowledgments

Grateful acknowledgment is hereby made to the following publishers and libraries for permission to quote from books and unpublished materials:

B. G. Teubner Verlag, Stuttgart, for permission to quote from Friedrich Engel, *Grassmanns Leben*, contained in Vol. III of *Hermann Grassmanns Gesammelte mathematische und physikalische Werke.*

Cambridge University Library for permission to quote from unpublished material in the correspondence of James Clerk Maxwell and Peter Guthrie Tait.

Cambridge University Press for permission to quote from Cargill Gilston Knott, *Life and Scientific Work of Peter Guthrie Tait.*

Ernst Benn Limited, London, for permission to quote from Oliver Heaviside, *Electromagnetic Theory*, Vols. I and III.

Macmillan & Co., Ltd., London, for permission to quote from Oliver Heaviside, *Electrical Papers.*

Thomas Nelson and Sons, Ltd., London, for permission to quote from Sir Edmund Whittaker, *A History of the Theories of Aether and Electricity*, Vol. I.

Yale University Library for permission to quote from the unpublished material in the correspondence of Josiah Willard Gibbs.

Yale University Press for permission to quote from Lynde Phelps Wheeler, *Josiah Willard Gibbs: The History of a Great Mind.*

List of Graphs and Tables

Contents

Preface to the Dover Edition

It is very gratifying that interest in the materials presented in this volume is sufficient to justify a second edition. This has permitted the correction of a number of small errors and, more importantly, provides an opportunity to bring to readers' attention some of the relevant studies of specific areas which have appeared since the book's first publication in 1967.

Recent researches have shed light particularly on the history of algebra during the nineteenth century. The most broadly conceived of such works is Luboš Nový's *Origins of Modern Algebra.*[1] British developments in algebra have received most attention, important studies having been published by Harvey W. Becher, J. M. Dubbey, Philip C. Enros, Elaine Koppelman, Luis M. Laita, and Joan L. Richards.[2] Interest in Sir William Rowan Hamilton's achievements in algebra has been especially intense. Research in this area has been aided by the appearance in 1967 under the editorship of H. Halberstam and R. E. Ingram of the third volume of Hamilton's *Mathematical Papers,* that volume being devoted to his publications in algebra.[3] Thomas L. Hankins has enriched Hamiltonian scholarship by various publications, most notably his engaging biography of the great Irish mathematician and scientist.[4] The scholar most actively engaged in assessing Hamilton's place in the history of British algebra is Helena M. Pycior, whose doctoral dissertation in this area has been followed by a number of studies of the algebraic ideas of Hamilton and his British contemporaries.[5] Jerold Mathews has published a paper on Hamilton's algebraic/analytic researches during the 1830's,[6] while B. L. van der Waerden has provided a new analysis of Hamilton's 1843 discovery of quaternions.[7] David Bloor has broadly considered Hamilton's algebraic approach in relation to the social, political, and philosophical context of his times,[8] whereas T. L. Hankins and John Hendry have focused studies on the genesis and importance of Hamilton's conception of algebra as the "Science of Pure Time."[9] Arnold R. Naiman in his doctoral dissertation surveyed the role of quaternions in the overall development of mathematics.[10]

The fascination I felt for Hamilton while researching this book was rivaled, if not surpassed, as I learned more of his remarkable contemporary Hermann Grassmann. Many issues I encountered in studying his mathematical creations have been treated in depth by Albert C. Lewis whose doctoral dissertation is a careful analysis of Grassmann's *Ausdehnungslehre* of 1844 and its sources. Dr. Lewis has now published some of his results in papers on the influence of Grassmann's father and of Schleiermacher on his mathematical system.[11] Moreover, Jean Dieudonné and Desmond Fearnley-Sander have each published essays on Grassmann's place in the history of linear algebra.[12]

Recent researches have also developed new perspectives on figures less central than Hamilton and Grassmann in this history. Helena M. Pycior has presented a fresh analysis of Benjamin Peirce's pioneering *Linear Associative Algebra*,[13] and Hubert Kennedy has investigated James Mills Peirce's place in the "cult of quaternions" that arose in late nineteeth-century America.[14] G. C. Smith in a recent paper has urged that Matthew O'Brien deserves significantly more credit as a pioneer of the modern vectorial system than has traditionally been accorded him,[15] and the late B. R. Gossick has provided new insights on the contrasting views of vectorial methods espoused by Oliver Heaviside and Lord Kelvin.[16] The first systematic study of the history of Stokes' Theorem and of the associated theorems named after Gauss and Green is due to Victor J. Katz.[17] The involvement of Russian mathematicians with vectorial methods has been treated by W. Dobrovolskij,[18] and studies of the history of vector analysis in general have been undertaken by Adalbert Apolin and James W. Joiner, both of whom seem to have written without knowledge of my book.[19]

For the sake of completeness, mention should be made of three publications which appeared before my book, but which escaped my bibliographic searches. Two of these, both from the 1930's, treat the history of complex numbers; the earlier was written by J. Budon whereas the second is Ernest Nagel's "Impossible Numbers."[20] The latter omission is especially regrettable because that essay provides a historiographic perspective which would have enriched my presentation. This is even more true of the third essay, that published in 1963–64 by the late Imre Lakatos.[21] Although containing essentially nothing relevant to the history of vector analysis, Lakatos' now famous essay is rich in philosophic and historiographic insights which, were I to rewrite this book, would certainly be included. Some hints as to the direction I would take are provided in a historiographic essay I published in 1975.[22]

Persons who may acquire from this book an interest in further readings in the history of mathematics may wish to consult the excellent general histories of mathematics written by Carl B. Boyer and Morris Kline,[23] or for articles on individual mathematicians, the many volumes of the *Dictionary of Scientific Biography*.[24] Bibliographic searches which required days in the early 1960's can now be accomplished by spending a few hours with the late Kenneth O. May's *Bibliography and Research Manual of the History of Mathematics* and with the *ISIS Cumulative Bibliography.*[25]

In concluding this updated preface, I extend warmest thanks to the two persons who have made this new edition possible: John W. Grafton, Assistant to the President of Dover Publications, and James R. Langford, Director of the University of Notre Dame Press. The encouragement of the former and the cooperation of the latter are chiefly responsible for this book being once again available to readers.

Michael J. Crowe

University of Notre Dame
January, 1985

Notes

[1] Luboš Nový, *Origins of Modern Algebra,* trans. by Jaroslav Tauer (Leyden, 1973).

[2] Harvey W. Becher, "Woodhouse, Babbage, Peacock, and Modern Algebra," *Historia Mathematica, 7* (1980), 389–400; J. M. Dubbey, "Babbage, Peacock and Modern Algebra," *Historia Mathematica, 4* (1977), 295–302; Philip C. Enros, "The Analytical Society (1812–1813): Precursor of the Renewal of Cambridge Mathematics," *Historia Mathematica, 10* (1983), 24–47; Elaine Koppelman, "The Calculus of Operations and the Rise of Abstract Algebra," *Archive for History of Exact Sciences, 8* (1971), 155–242; Luis M. Laita, "The Influence of Boole's Search for a Universal Method in Analysis on the Creation of His Logic," *Annals of Science, 34* (1977), 163–176; Luis M. Laita, "Influences on Boole's Logic: The Controversy between William Hamilton and Augustus De Morgan," *Annals of Science, 36* (1979), 45–65; Luis M. Laita, "Boolean Algebra and Its Extra-logical Sources: The Testimony of Mary Everest Boole," *History and Philosophy of Logic, 1* (1980), 37–60; Joan L. Richards, "The Art and the Science of British Algebra: A Study in the Perception of Mathematical Truth," *Historia Mathematica, 7* (1980), 343–365.

[3] H. Halberstam and R. E. Ingram, *The Mathematical Papers of Sir William Rowan Hamilton,* vol. III: *Algebra* (Cambridge, England, 1967).

[4] Thomas L. Hankins, *Sir William Rowan Hamilton* (Baltimore, 1980). For a shorter biography which discusses Hamilton's mathematics on a more elementary level, see: Sean O'Donnell, *William Rowan Hamilton: Portrait of a Prodigy* (Dublin, 1983).

[5] Helena M. Pycior, *The Role of Sir William Hamilton in the Development of British Modern Algebra* (A 1976 Cornell University doctoral dissertation); H. M. Pycior, "George Peacock and the British Origins of Symbolical Algebra," *Historia Mathematica, 8* (1981), 23–45; H. M. Pycior, "Early Criticisms of the Symbolical Approach to Algebra," *Historia Mathematica, 9* (1982), 392–412; H. M. Pycior, "Augustus De Morgan's Algebraic Work: The Three Stages," *Isis, 74* (1983), 211–226.

[6] Jerold Mathews, "William Rowan Hamilton's Paper of 1837 on the Arithmetization of Analysis," *Archive for History of Exact Sciences, 19* (1978), 177–200.

[7] B. L. van der Waerden, *Hamiltons Entdeckung der Quaternionen* (Göttingen, 1974). Now available in English as "Hamilton's Discovery of Quaternions," *Mathematics Magazine, 49* (1976), 227–234.

[8] David Bloor, "Hamilton and Peacock on the Essence of Algebra" in *Social History of Nineteenth Century Mathematics,* ed. by Herbert Mehrtens, Henk Bos and Ivo Schneider (Boston, 1981), pp. 202–232.

[9] Thomas L. Hankins, "Algebra as Pure Time: William Rowan Hamilton and the Foundations of Algebra" in *Motion and Time, Space and Matter: Interrelations in the History of*

Philosophy and Science, ed. by P. J. Machamer and R. G. Turnbull (Columbus, 1976), pp. 327–359.

[10] Arnold R. Naiman, *The Role of Quaternions in the History of Mathematics* (A 1974 New York University doctoral dissertation).

[11] Albert C. Lewis, *An Historical Analysis of Grassmann's Ausdehnungslehre of 1844* (A 1975 University of Texas at Austin doctoral dissertation); A. C. Lewis, "H. Grassmann's 1844 *Ausdehnungslehre* and Schleiermacher's *Dialektik*," *Annals of Science, 34* (1977), 103–162; A. C. Lewis, "Justus Grassmann's School Programs as Mathematical Antecedents of Hermann Grassmann's 1844 *Ausdehnungslehre*" in *Epistemological and Social Problems of the Sciences in the Early Nineteenth Century*, ed. by Hans Niels Jahnke and Michael Otte (Dordrecht, 1981), pp. 255–267.

[12] Jean Dieudonné, "The Tragedy of Grassmann," *Linear and Multilinear Algebra, 8* (1979), 1–14; Desmond Fearnley-Sander, "Hermann Grassmann and the Creation of Linear Algebra," *American Mathematical Monthly, 86* (1979), 809–817; see also D. Fearnley-Sander, "Hermann Grassmann and the Prehistory of Universal Algebra," *American Mathematical Monthly, 89* (1982), 161–166.

[13] Helena M. Pycior, "Benjamin Peirce's *Linear Associative Algebra*," *Isis, 70* (1979), 537–551.

[14] Hubert Kennedy, "James Mills Peirce and the Cult of Quaternions," *Historia Mathematica, 6* (1979), 423–429.

[15] G. C. Smith, "Matthew O'Brien's Anticipation of Vectorial Mathematics," *Historia Mathematica*, 9 (1982), 172–190.

[16] B. R. Gossick, "Heaviside and Kelvin: A Study in Contrasts," *Annals of Science, 33* (1976), 275–287.

[17] Victor J. Katz, "The History of Stokes' Theorem," *Mathematics Magazine*, 52 (1979), 146–156.

[18] W. Dobrovolskij, "Développement de la théorie des vecteurs et des quaternions dans les travaux des mathématiciens russes du XIXe siècle," *Revue d'histoire des sciences, 21* (1968), 345–349.

[19] Adalbert Apolin, "Die geschichtliche Entwicklung der Vektorrechnung," *Historia Naturalis, 12* (1970), 357–365; James Walter Joiner, *A History of Vector Analysis* (A 1971 doctoral dissertation at George Peabody College for Teachers).

[20] J. Budon, "Sur la représentation géométriques des nombres imaginaires (Analyse de quelques mémoires parus de 1795 à 1820)," *Bulletin des sciences mathématiques*, ser. 2, 57 (1933), 175–200; 220–232; Ernest Nagel, "Impossible Numbers," *Studies in the History of Ideas, 3* (1935), 427–474; reprinted in Ernest Nagel, *Teleology Revisited and Other Essays in the Philosophy and History of Science* (New York, 1979), pp. 166–194.

[21] Imre Lakatos, "Proofs and Refutations," *British Journal for the Philosophy of Science, 14* (1963–1964), 1–25; 120–139; 221–245; 296–342. These essays have now been republished as: Imre Lakatos, *Proof and Refutations: The Logic of Mathematical Discovery*, ed. by John Worrall and Elie Zahar (Cambridge, England, 1976).

[22] M. J. Crowe, "Ten 'Laws' Concerning Patterns of Change in the History of Mathematics," *Historia Mathematica, 2* (1975), 161–166.

[23] Carl B. Boyer, *A History of Mathematics* (New York, 1968) and Morris Kline, *Mathematical Thought from Ancient to Modern Times* (New York, 1972).

[24] *Dictionary of Scientific Biography*, 14 vols., ed. by Charles Coulston Gillispie (New York, 1970–1980).

[25] *ISIS Cumulative Bibliography. A Bibliography of the History of Science Formed from ISIS Critical Bibliographies 1–90: 1913–1965*, 5 vols., ed. by Magda Whitrow (London, 1971–1982); *ISIS Cumulative Bibliography 1966–1975*, vol. I, ed. by John Neu (London, 1980); and Kenneth O. May, *Bibliography and Research Manual of the History of Mathematics* (Toronto, 1973).

CHAPTER ONE

The Earliest Traditions

I. *Introduction*

The early history of vectorial analysis is most appropriately viewed within the context of two broad traditions in the history of science. One of these traditions relates to mathematics, the other to physical science.

The first tradition, that within the history of mathematics, extends from the time of the Egyptians and Babylonians to the present and consists in the progressive broadening of the concept of number. Throughout time the concept of number has been expanded so as to include not only positive integers, but negative numbers, fractions, and algebraic and transcendental irrationals. Eventually complex and higher complex numbers (including vectors) were introduced. The activities of some of the figures in the history of vector analysis may be viewed as belonging to this tradition.

The second tradition, that within the history of physical science, also extends back to ancient times and consists in the search for mathematical entities and operations that represent aspects of physical reality. This tradition played a part in the creation of Greek geometry, and the natural philosophers of the seventeenth century inherited from the Greeks the geometrical approach to physical problems. However in the course of the seventeenth century the physical entities to be represented passed through a transformation. This transformation consisted in the shift in emphasis from such scalar quantities as position and weight to such vectorial quantities as velocity, force, momentum, and acceleration. The transition was neither abrupt nor was it confined to the seventeenth century. Later developments in electricity, magnetism, and optics acted further to transform the space of mathematical physics into a space filled with vectors.

These two broad traditions converged at a number of periods in history; one such period was in the nineteenth century, and this convergence is marked by the creation and development of vectorial methods. The first major three-dimensional vectorial systems were created in the 1840's. Before this time, however, three important ideas were put forth which led to the major vectorial systems. These three ideas are the subject of the present chapter; they are the concept of a parallelogram of forces, Leibniz' concept of a geometry of situation, and the concept of the geometrical representation of imaginary numbers.

II. *The Concept of the Parallelogram of Velocities and Forces*

One of the most fundamental mathematical ideas in vector analysis is the idea of the addition of vectors. The sum of two vectors which have a common point of origin is defined as the vector originating at the same point and extending to the opposite corner of the parallelogram defined by the two original vectors. Certain physical entities, such as velocities and forces, may be compounded in a similar way, and from this correspondence stems much of the usefulness of vector analysis.

The idea of a parallelogram of velocities may be found in various ancient Greek authors,[8] [*] and the concept of a parallelogram of forces was not uncommon in the sixteenth and seventeenth centuries.[9] By the early nineteenth century parallelograms of physical entities frequently appeared in treatises, and this usage *indirectly* led to vector analysis, for this idea provided a striking example of how vectorial entities could be used for physical applications. It should not be inferred, however, that all of those who used the concept of a parallelogram of physical entities were aware of the idea of a vector or of vector addition. The essential idea in the parallelogram of physical entities is the construction of a diagram in terms of which the operations involved in determining the resultant become evident. The idea of *adding* the lines need not be introduced or was it (to my knowledge) ever introduced before the creation of vectors. Thus this idea alone could not and almost certainly did not *directly* stimulate anyone to the creation of a vectorial system. Its influence was *indirect* but important, for it was the first and most obvious case in which vectorial methods could be brought to the aid of physical science.

[*] The system used for numbering notes is described in the preface.

III. *Leibniz' Concept of a Geometry of Situation*

Gottfried Wilhelm Leibniz (1646–1716) made many contributions to mathematics; among the less well known is his concept of a geometry of situation. In this regard Leibniz discussed the possibility of creating a system which would serve as a direct method of space analysis. Although the details of this idea were never fully worked out by Leibniz, he advanced far enough to be ranked as a conceptual forerunner of the first vectorial analysts. Moreover his essay, which was first published in 1833, played a part in the later history of vectorial analysis.

Leibniz' main ideas were contained in a letter dated September 8, 1679, and written to Christian Huygens.[1] In this letter Leibniz wrote:

> I am still not satisfied with algebra, because it does not give the shortest methods or the most beautiful constructions in geometry. This is why I believe that, so far as geometry is concerned, we need still another analysis which is distinctly geometrical or linear and which will express *situation* [*situs*] directly as algebra expresses *magnitude* directly. And I believe that I have found the way and that we can represent figures and even machines and movements by characters, as algebra represents numbers or magnitudes. I am sending you an essay which seems to me to be important. (1; 382)

In his essay, which was contained in the letter, Leibniz described his system further:

> I have discovered certain elements of a new characteristic which is entirely different from algebra and which will have great advantages in representing to the mind, exactly and in a way faithful to its nature, even without figures, everything which depends on sense perception. Algebra is the characteristic for undetermined numbers or magnitudes only, but it does not express situation, angles, and motion directly. Hence it is often difficult to analyze the properties of a figure by calculation, and still more difficult to find very convenient geometrical demonstrations and constructions, even when the algebraic calculation is completed. But this new characteristic, which follows the visual figures, cannot fail to give the solution, the construction, and the geometric demonstration all at the same time, and in a natural way and in one analysis, that is, through determined procedure.
>
> . . . .
>
> But its chief value lies in the reasoning which can be done and the conclusions which can be drawn by operations with its characters, which could not be expressed in figures, and still less in models, without multiplying these too greatly or without confusing them with too many points and lines in the course of the many futile attempts one is forced to make. This method, by contrast, will guide us surely and without

> effort. I believe that by this method one could treat mechanics almost like geometry, and one could even test the qualities of materials, because this ordinarily depends on certain figures in their sensible parts. Finally, I have no hope that we can get very far in physics until we have found some such method of abridgment to lighten its burden of imagination. (1; 384–385)

His system as actually sketched by him shows that he by no means discovered a primitive vector analysis, though the above quotations show that he was searching for something akin to vector analysis.

Leibniz' system centered on the idea of the congruence of sets of points. He used *A*, *B*, . . . to represent fixed points and *X*, *Y*, . . . to represent unknown points. The symbol ♉ was used to express the relation congruence; thus he wrote *ABC* ♉ *DEF* to express that a set of three points *A*, *B*, *C*, each of which was a fixed distance from the other two points, could be made to coincide with another set of similarly fixed points *D*, *E*, *F*. He then discussed locus relations and stated that the locus of points congruent to a fixed point "will be a *space infinite* in all directions." (1; 387) If it is given that *AB* ♉ *AY*, the point values of *Y* will be points on a sphere with center at *A* and radius of length *AB*. The relation *AX* ♉ *BX* determines a plane whose points (*X*) are equidistant from *A* and *B*, and the relation *ABC* ♉ *ABY* determines a circle. Leibniz then discussed the locus of points *Y* satisfying the relation *AY* ♉ *BY* ♉ *CY* and concluded that "the locus of all *Y*'s will be a *straight line*." (1; 389) After showing that the relation *AY* ♉ *BY* ♉ *CY* ♉ *DY* determines a point, Leibniz applied his analysis to four simple problems. One of these may be discussed as typical. The problem is to show that the intersection of two planes is a straight line. The relation *AY* ♉ *BY* determines one plane, and the relation *AY* ♉ *CY* determines a second plane. By combining these we have *AY* ♉ *BY* ♉ *CY*, which, as it was shown before, determines a straight line.[10]

Proceeding from this summary of Leibniz' best-known exposition of his system,[11] we may discuss its relation to modern vector analysis. First, Leibniz deserves much credit for suggesting that a new algebra, wherein geometrical entities are symbolically represented and the symbols operated upon directly, was desirable. However, he failed to discover a system in which geometrical entities could be added, subtracted, and multiplied. Likewise he failed to see that *AB* and *BA* (for example) can be viewed as distinct entities and that −*AB* could have a significant meaning. Though his idea of directly representing a fixed point by a symbol makes him a partial forerunner of Möbius and Grassmann, he certainly did not introduce the concept of a vector. Despite the fact that angle considerations did not enter into his system, he still must be viewed as having constructed a system which allowed for the use of co-ordinates. Leibniz saw that a new algebra of the form sought would have numerous applications in mathematics and in the physical sciences, but he failed to develop practical methods for these tasks. The view of Leibniz' system taken by Couturat, though stated in relation to Grassmann's system, is also applicable in relation to modern vector analysis; Couturat wrote: "In summary, the calculus of Grassmann seems to bring fully into reality the geometrical characteristic conceived by Leibniz, and shows that Leibniz' idea was not simply a dream. But there is such a disproportion between Leibniz' conception of a system and the very defective essay which he actually produced that Grassmann felt a sharp distinction should be made between the ideal conceived and the sketch actually written." [12]

Shortly after 1833, when Leibniz' essay was first published, the Jablonowski Gesellschaft expressed their interest in and enthusiasm for the essay by offering a prize for the further development of Leibniz' system. One mathematician entered the competition and won the prize, even though he had created his system before hearing of Leibniz' ideas; this mathematician was Grassmann and this incident will be more fully discussed in the third chapter.

IV. *The Concept of the Geometrical Representation of Complex Numbers*

Though the term *vector analysis* is now used primarily in reference to systems of mathematics that may be applied in three-dimensional space, it should not be forgotten that the complex number system may legitimately be considered as a vectorial system. The two-dimensional vectorial system based on the geometrical representation of complex numbers is certainly not as useful as the three-dimensional vectorial systems which are the primary subject of this history. Nevertheless it is important to discuss briefly the early history of the geometrical representation of complex numbers, not only because the complex number system is (broadly speaking) a vectorial system but also because Hamilton discovered quaternions in the course of a search for a three-dimensional analogue to the complex number system.

At least six men are commonly credited with the discovery of the geometrical representation of complex numbers; they are Wessel, Gauss, Argand, Buée, Mourey, and Warren.[13] Since the systems created by these six men are very similar and are of limited relevance to the present study, they need not all be discussed in detail. In what follows, the system published by Wessel, which was the earliest and among the most impressive, will be analyzed in some depth; the ideas of the other five men will be treated less fully though with special attention to certain aspects of their development. Thus it will be shown that some of these mathematicians searched for three-dimensional vectorial systems and that one of them influenced Hamilton in an important manner.

Although Hero of Alexandria and Diophantus in ancient times had encountered the question of the meaning of the square root of a negative number, and although Cardan had in his 1545 *Ars Magna* used complex numbers in computation, nevertheless complex numbers were not accepted by most mathematicians as legitimate mathematical entities until well into the nineteenth century. This is hardly surprising since numbers such as $\sqrt{-1}$ seem to be neither less than, greater than, nor equal to zero.

In modern mathematics complex numbers are usually justified either by representing them in terms of couplets of real numbers or by representing them geometrically. The origin of the first method will be discussed in the next chapter. The first attempt (which was unsuccessful) to represent complex numbers geometrically was made in the seventeenth century by John Wallis.[14]

Where Wallis failed, a Norwegian surveyor succeeded; in 1799 Caspar Wessel (1745–1818) published the first explanation of the geometrical representation of complex numbers.[15] His ideas were presented before the Royal Academy of Denmark in 1797 and published two years later in the memoirs of that society.[2] Unfortunately, however, Wessel's publication went unnoticed by European mathematicians until 1897, when it was republished in a French translation.[3]

In the first paragraph of his memoir Wessel stated: "This present attempt deals with the question, how may we represent direction analytically; that is, how shall we express right lines so that in a single equation involving one unknown line and others known, both the length and the direction of the unknown line may be expressed." (2; 55) As this quotation suggests and later passages confirm, Wessel's chief interest was the creation of geometrical methods; his representation of complex numbers was subservient to this aim. Nonetheless the latter played a fundamental role as is indicated by the following statement: "The occasion for its being [his treatise] was my seeking a method whereby I could avoid the impossible operations. . . ." (2; 57)

After stating that previously only oppositely directed lines could be represented analytically, Wessel suggested that it should be possible to find methods to represent inclined lines. Wessel then gave a definition of the addition of straight lines: "Two right lines are added if we unite them in such a way that the second line begins where the first one ends, and then pass a right line from the first to the last point of the united lines. This line is the sum of the united lines." (2; 58) In the subsequent discussion of addition Wessel stated that the same definition can be used in adding more than two (not necessarily coplanar) lines and that the order of addition is immaterial. (2; 59) Hence Wessel had introduced three-dimensional vector addition and realized the importance of the commutative law for addition.

Though Wessel had up to this point only discussed what he called the positive unit (our 1 of $x \cdot 1 + y\sqrt{-1}$) and had not yet indicated how lines in general were to be represented in terms of complex numbers, nevertheless he proceeded to introduce the multiplication of lines. The product of two lines (coplanar with each other and with the positive unit) was to have a length equal to the product of the lengths of the two factors. The product line was to be coplanar with the two factor lines and was to have its inclination or direction angle (defined by reference to the inclination of the positive unit as 0°) equal to the sum of the inclinations of the factor lines. (2; 60) Wessel then added:

> Let $+1$ designate the positive rectilinear unit and $+\epsilon$ a certain other unit perpendicular to the positive unit and having the same origin; then the direction angle of $+1$ will be equal to 0°, that of -1 to 180°, that of $+\epsilon$ to 90°, and that of $-\epsilon$ to $-90°$ or 270°. By the rule that the direction angle of the product shall equal the sum of the angles of the factors, we have: $(+1)(+1) = +1$; $(+1)(-1) = -1$; $(-1)(-1) = +1$; $(+1)(+\epsilon) = +\epsilon$; $(+1)(-\epsilon) = -\epsilon$; $(-1)(+\epsilon) = -\epsilon$; $(-1)(-\epsilon) = +\epsilon$; $(+\epsilon)(+\epsilon) = -1$; $(+\epsilon)(-\epsilon) = +1$; $(-\epsilon)(-\epsilon) = -1$.
>
> From this it is seen that ϵ is equal to $\sqrt{-1}$; and the divergence of the product is determined such that not any of the common rules of operation are contravened. (2; 60)

Wessel stated that any straight line in a plane may be represented analytically by the expressions $a + \epsilon b$ and $r(\cos v + \epsilon \sin v)$ and showed how such expressions are to be multiplied, divided, and raised to powers. After giving two examples of the application of his methods, Wessel developed an elementary three-dimensional vector analysis. (3; 23–28)

Wessel began by constructing three mutually perpendicular lines which passed through the center of a sphere of radius r. Wessel specified that three radii of the sphere which were collinear with the three mutually perpendicular axes should be designated by r, ηr, and ϵr and that any point in space could be designated by a vector of the form $x + \eta y + \epsilon z$. (3; 23–24) By analogy with ordinary complex numbers Wessel defined $\eta\eta$ and $\epsilon\epsilon$ as equal to −1. The multiplication of vectors corresponded to the rotation and extension of one vector by another. Thus $(x + \eta y + \epsilon z)$, , $(\cos u + \epsilon \sin u)$ represented the rotation of the vector $x + \eta y + \epsilon z$ through the angle u around the η or y axis. Wessel stated that the component of the vector that lies on the axis of rotation should remain unchanged, and thus the product of the above is $\eta y + x \cos u - z \sin u + \epsilon x \sin u + \epsilon z \cos u$. (3; 25–26) The symbol , , was used to indicate multiplication. A rotation of v degrees around the ϵ or z axis was expressed in the following way: $(x + \eta y + \epsilon z)$, , $(\cos v + \eta \sin v) = \epsilon z + x \cos v - y \sin v + \eta x \sin v + \eta y \cos v$. (3; 26) Rotations around the η axis could be compounded with rotations around the ϵ axis and vice versa, but rotations around the axis of the positive unit (the x axis) were never discussed by Wessel. The reason for this is that serious mathematical difficulties were involved in determining how such rotations should be represented, for example, the products $\eta\epsilon$ and $\epsilon\eta$ would have had to be defined. Wessel presumably encountered these difficulties but could not solve them.[16] But even with this limitation on his methods Wessel was able to use them to derive a number of important results in spherical trigonometry.

Wessel's three-dimensional vectorial system exhibited an *ad hoc* character that makes it appear seriously deficient when compared to modern systems; nevertheless, if it is viewed as a creation of the late eighteenth century, it can only be viewed with awe. Wessel's treatment of ordinary complex numbers is equally impressive, and it was unfortunate for Wessel and for mathematics that his memoir lay buried for nearly a century.

In the early history of complex numbers a striking phenomenon occurred: on three separate occasions two men independently and simultaneously discovered the geometrical representation of complex numbers. In 1806 Argand and Buée both published independent treatments of imaginary numbers, and the same coincidence occurred in 1828 with Mourey and Warren. What is even more surprising is that Gauss probably discovered the geometrical representation of complex numbers at the same time as Wessel.

Gauss' first published treatment of the geometrical representation of complex numbers appeared in 1831;[17] herein Gauss commented that he had had this idea for many years and that traces of it could easily be found in his 1799 "Demonstratio Nova."[18] Julian Lowell Coolidge investigated this point and showed that Gauss' claim was amply supported by the fact that some methods used in the 1799 paper seem "blind and meaningless" unless the author already possessed this idea.[19] It was through Gauss' 1831 publication that most mathematicians came into contact with the geometrical representation of complex numbers, although Hamilton heard of Gauss' paper only in 1852 (4; 312) and Grassmann in 1844.[20] However Hamilton heard in 1845 that Gauss had been searching for a "triple algebra" corresponding to the double algebra of complex numbers. (4; 311–312) Felix Klein argued in an 1898 publication that Gauss had in fact discovered quaternions, but Tait and Knott vigorously denied this.[21] Ironically Gauss himself did not accept the geometrical representation of complex numbers as a sufficient justification for them.[22] In conclusion it may be noted that Gauss' publication was the shortest, the most precise, the last, and the most influential of the six independent presentations.

The next publication to be considered was the longest, the least precise, the earliest (except for Wessel's), and the least influential. On June 20, 1805, a long essay entitled "Mémoire sur les quantités imaginaires" was read before the Royal Society of London. The author was Abbé Buée and his paper was published (without translation) in the 1806 *Transactions of the Royal Society.*[23] Buée's treatment of complex numbers was not of high quality; Coolidge in fact has expressed surprise that it was published.[24] The well-founded consensus among those who have studied Buée's paper is that some ingenuity mixed with much obscurity is to be found there, as well as a near approach to the concept of the multiplication of directed lines. Hamilton asserted that Buée attempted to extend his methods to space (5; [57]), but if Buée did do this, he did it in a very unorthodox manner.

A far superior work also appeared in 1806; this was Jean Robert Argand's small book, *Essai sur une manière de représenter les quantités imaginaires dans les constructions géométriques.*[6] Herein Argand gave the modern geometrical representation of the addition and multiplication of complex numbers, and showed how this representation could be applied to deduce a number of theorems in trigonometry, elementary geometry, and algebra. At this time Argand did not attempt to expand his methods for application to three-dimensional space. For seven years Argand shared the fate of Wessel; however in 1813 attention was called to his book in a very unexpected way.

In 1813 J.-F. Français published a short memoir in volume IV of Gergonne's *Annales de mathématiques* (6; 63–74), in which Français presented the geometrical representation of complex numbers. At the conclusion of his paper Français stated that the fundamental ideas in his paper were not his own; he had found them in a letter written by Legendre to his (Français') brother who had died. In this letter Legendre discussed the ideas of an unnamed mathematician. Français added that he hoped that this mathematician would make himself known and publish his results. (6; 74)

The unnamed mathematician had in fact already published his ideas, for Legendre's friend was Jean Robert Argand. Hearing of Français' paper, Argand immediately sent a communication to Gergonne in which he identified himself as the mathematician of Legendre's letter, called attention to his book, summarized its contents, and finally presented an (unsuccessful) attempt to extend his system to three-dimensions. (6; 76–96) Before seeing Argand's publications, Français had written a letter to Gergonne containing his admittedly unsatisfactory attempts to extend the geometrical representation of complex numbers to space. (6; 96–101) And soon after Argand's publication, Servois published a paper criticizing Argand's attempt and outlining his own ideas on a method of space analysis. (6; 101–109) Hamilton attributed to Servois "the *nearest approach* to an *anticipation of the quaternions*, or at least to an anticipation of triplets. . . ." (5; [57]) In making this statement Hamilton had the following passage from Servois' paper in mind.

> Analogy would seem to indicate that the tri-nominal should be of the form $p \cos \alpha + q \cos \beta + r \cos \gamma$, α, β, and γ being the angles made by a right line with three rectangular axes, and that we should have $(p \cos \alpha + q \cos \beta + r \cos \gamma)(p' \cos \alpha + q' \cos \beta + r' \cos \gamma) = \cos^2 \alpha + \cos^2 \beta + \cos^2 \gamma = 1$. The values of p, q, r, p', q', r' satisfying this condition would be *absurd;* but would they be imaginaries, reducible to the general form $A + B\sqrt{-1}$? (7; 114–115)

Concerning this passage Hamilton wrote:

> The six NON-REALS which Servois thus with remarkable sagacity *foresaw,* without being able to *determine* them, may now be identified with the then unknown symbols $+i, +j, +k, -i, -j, -k$, of the quaternion theory: at least, these latter symbols fulfil precisely the *condition* proposed by him, and furnish an *answer* to his "singular question." It may be proper to state that my own theory had been constructed and published for a long time, before the lately cited passage happened to meet my eye. (5; [57])

The series of articles in Gergonne's *Annales* was concluded by a letter written by Argand in which he responded to a notice sent in by Lacroix (6; 111) calling attention to Buée's (1806) publication. Argand wrote that he had had no knowledge of Buée's work. (6; 123)

There is very strong evidence that all the men discussed up to this point were unknown to Hamilton when he discovered quaternions in 1843.[25] The ideas of the next man to be discussed were known to Hamilton as early as 1829 and moreover influenced his thinking, as he repeatedly acknowledged. (4; 190) John Warren published in 1828 a short book entitled *A Treatise on the Geometrical Representation of the Square Roots of Negative Quantities.* Warren's presentation of the geometrical representation of complex numbers exhibited great care and understanding; he, unlike Buée and Argand, was aware of the importance of the commutative, associative, and distributive laws, though he did not use these terms.[26] Warren discovered his ideas in complete independence of the other mathematicians who wrote on the geometrical representation of complex numbers, but he, unlike the majority of them, did not discuss the extension of his system to space.[27]

The final independent discoverer of the geometrical representation of complex numbers was the Frenchman C. V. Mourey, who in 1828 published an excellent treatise entitled *La vrai Théorie des quantités négatives et des quantités prétendues imaginaires.*[28] At the conclusion of his book Mourey stated that there exists an algebra surpassing not only ordinary algebra but also the two-dimensional algebra created by him. This algebra, he stated, extends to three-dimensions.[29] Presumably Mourey searched for such an algebra; if he found it, he did not publish his discovery.

V. *Summary and Conclusion*

Thus we can say that at least five men, working independently of each other, had by 1831 discovered and published the geometrical representation of complex numbers. These men were Wessel, Gauss, Argand, Warren, and Mourey. At least two others, Wallis and Buée, had come close to the same idea. Wessel, Gauss, Argand, and Mourey, as well as Servois and Français, and perhaps Buée, had attempted to find higher complex numbers for the analysis of space, and all had failed.

A number of conclusions may be drawn from what has been discussed. The first is that the idea of a graphical representation of complex numbers was certainly "in the air" at that time. However, the acceptance of this idea was very slow, and little attention was paid to these ideas until Gauss published his paper of 1831. The fact that the idea was neglected until Gauss entered the field should not, I think, be taken as surprising. Historians of science have repeatedly shown that radically new ideas presented only on their own merits are usually neglected. The men before Gauss were all little known; indeed they are now known only because of their one great discovery. But when Gauss wrote, he wrote with the authority of one who had already acquired fame through impressive work in traditional fields and through his widely known prediction of the position of the lost planetoid Ceres. It may be noted now and discussed later that the pattern exhibited in this instance will recur in the later history of vectorial analysis.

Second, it has been noted that most of those who worked on the geometrical representation of complex numbers attempted to construct analogous methods for three-dimensional space. That many embarked on this quest illustrates what is probably mathematically obvious: the search for a system of space analysis was a natural concomitant to the idea of the geometrical representation of complex numbers. Up to this point only those who made their attempts before 1831 have been discussed; many others also puzzled over this problem after 1831. Among them was Hamilton, who, working precisely in this tradition, discovered quaternions.

Notes

[1] Gottfried Wilhelm Leibniz, "Studies in a Geometry of Situation with a Letter to Christian Huygens" in Leibniz, *Philosophical Papers and Letters*, ed. and trans. Leroy E. Loemker, vol. I (Chicago, 1956), 381–396. Leibniz' essay and letter were first published in 1833; the citation as given in Hermann Grassmann, *Gesammelte mathematische und physikalische Werke*, vol. I, pt. I (Leipzig, 1894), 415–416, is "Christi. Huygenii aliorumque seculi XVIII. virorum celebrium exercitationes mathematicae et philosophiae. Ed. Uylenbroek. Hagae comitum 1833. fasc. II, p. 6." Loemker based his translation on the text as given in Leibniz, *Mathematische Schriften*, ed. C. I. Gerhardt, vol. II (Berlin, 1850), 17–27, and on Uylenbroek's text (which is superior) as given in Leibniz, *Hauptschriften zur Gründung der Philosophie*, ed. Ernst Cassirer, trans. A. Buchenau, 2nd. ed., 2 vols. (Leipzig, 1924). Quotations are from Loemker and have been checked with the Gerhardt's text (cited above) and Uylenbroek's text as given in Grassmann, *Werke*, vol. I, pt. I, 417–420.

[2] All quotations have been taken from Martin A. Norgaard's English translation of the first sixteen sections of Wessel's book; see Wessel, "On Complex Numbers" in *A Source Book in Mathematics*, vol. I, ed. David Eugene Smith (New York, 1959), 55–66. I have also used the French translation of Wessel's book which is cited in note (1) above. The title for Wessel's original publication is "Om Directionens analytiske Betegning," and it appeared in vol. V (1799) of *Nye Samling af det Kongelige Danske Videnskabernes Selskabs Skrifter*. Wessel's essay was rediscovered in 1895 by S. D. Christensen and C. Juel; it was republished without translation by Sophus Lie in the 1896 *Archiv for Mathematik og Naturvidenskab*. In this connection see Viggo Brun, "Caspar Wessel et l'introduction géométrique des nombres complexes" in *Revue d'histoire des sciences, 12* (1959), 20–21.

[3] Caspar Wessel, *Essai sur la représentation analytique de la direction*, ed. H. Valentiner and T. N. Thiele, trans. H. G. Zeuthen and others (Copenhagen, 1897).

[4] Robert Perceval Graves, *Life of Sir William Rowan Hamilton*, vol. III (Dublin, 1889).

[5] Sir William Rowan Hamilton, *Lectures on Quaternions* (Dublin, 1853). All references are to Hamilton's Preface, where Arabic numerals set in parentheses are used to indicate page numbers.

[6] Jean Robert Argand, *Essai sur une manière de représenter les quantités imaginaires dans les constructions géométriques*, 2nd ed., ed. J. Hoüel (Paris, 1874). This contains a reprint of the first edition (Paris, 1806) along with selections from the papers on complex numbers by Français, Argand, Gergonne, Lacroix, and Servois, papers which were originally published in Gergonne's *Annales des Mathematiques, 4* (1813–1814) and *5* (1814–1815). See the work cited in note (7) below for an English translation of Argand's book; Hardy included less material than Hoüel from the series of papers in Gergonne's *Annales* but supplied valuable commentary not found in Hoüel's edition.

[7] Jean Robert Argand, *Imaginary Quantities: Their Geometrical Representation*, trans. A. S. Hardy (New York, 1881).

[8] The three Greek authors who used this concept are (1) the author of the so-called "pseudo-Aristotelian *Mechanica*," (2) Archimedes, and (3) Hero of Alexandria. For

the first and the third see Marshall Clagett, *The Science of Mechanics in the Middle Ages* (Madison, 1959), 4–5, 41. On Archimedes see Joseph Louis Lagrange, *Mécanique Analytique* in Lagrange, *Œuvres*, vol. XI (Paris, 1888), 12, and Archimedes, "On Spirals" in *The Works of Archimedes*, trans. Thomas Heath (New York, n.d.), 165.

[9] The history of this concept is discussed by numerous authors; the following are among the most important: (1) René Dugas, *A History of Mechanics*, trans. J. R. Maddox (New York, 1955); (2) Ernst Mach, *The Science of Mechanics*, trans. Thomas J. McCormack (La Salle, Ill., 1960); (3) Max Jammer, *Concepts of Force* (New York, 1962); (4) A. Voss, "Grundlegung der Mechanik" in *Encyklopädie der mathematischen Wissenschaften*, vol. IV, pt. I (Leipzig, 1901–1908), 43–46.

[10] For this example see (1; 390) but note that Loemker wrote "*AB* ȣ *BY* for one plane . . . ," whereas the Uylenbroek text (see Grassmann, *Werke*, vol. I, pt. I, 420) has (correctly) "*AY* ȣ *BY*."

[11] There is a fuller but similar exposition in Leibniz, *Mathematische Schriften*, vol. V, ed. C. I. Gerhardt (Halle, 1858), 141–171. Many minor statements of Leibniz (for example, statements in letters) are referred to and discussed by Louis Couturat, *La Logique de Leibniz* (Paris, 1901), which includes a full discussion of Leibniz' ideas, particularly as they relate to Grassmann's system. Leibniz' system was also discussed by A. E. Heath, "The Geometrical Analysis of Grassmann and Its Connection with Leibniz's Characteristic" in *The Monist*, 27 (1917), 36–56, and by Grassmann in his *Geometrische Analyse* in *Werke*, vol. I, pt. I, 321–399.

[12] Louis Couturat, *La Logique de Leibniz* (Paris, 1901), 538.

[13] There have been a number of studies on the early history of complex numbers that have aided me in this study; among the most important are (1) Wooster Woodruff Beman, "A Chapter in the History of Mathematics" in *The Proceedings of the American Association for the Advancement of Science, 46* (1897), 33–50; (2) Florian Cajori, "Historical Note on the Graphic Representation of Imaginaries Before the Time of Wessel" in *American Mathematical Monthly, 19* (September-October, 1912), 167–171; (3) Julian Lowell Coolidge, *Geometry of the Complex Domain* (Oxford, 1924); (4) Hermann Hankel, *Theorie der complexen Zahlensysteme* (Leipzig, 1867); (5) P. S. Jones, "Complex Numbers: An Example of Recurring Themes in the Development of Mathematics" in *Mathematics Teacher, 47* (1954), 106–114, 257–263, 340–345; (6) George Peacock, "Report on the Recent Progress and Present State of Certain Branches of Analysis" in *Report of the British Association for the Advancement of Science* (1834), 185–352; (7) Angelo Romorino, "Gli Elementi imaginarii nella geometria" in Battaglini's *Giornale di matematica, 35* (1897), 242–258; and *36* (1898), 317–345; (8) G. Windred, "History of the Theory of Imaginary and Complex Quantities" in *Mathematical Gazette, 14* (1929), 533–541.

Unfortunately a recent excellent study came to my attention too late to take full advantage of it. This is F. D. Kramar's "Vektornoe ischislenie kontsa XVIII i nachala XIX vv" (in Russian) in *Istoriko-Matematicheskie Issledovaniia, 15* (1963), 225–290.

[14] The important passage from Wallis may be found in *A Source Book in Mathematics*, ed. David Eugene Smith, vol. I (New York, 1959), 46–54.

[15] The words "to publish" qualify this statement sufficiently that no mention need be made in the text of Leonard Euler, Charles Walmesley, and Dominique Truel. The basis for attributing the geometrical representation to the first two of these men is that it seems from reading their writings on relevant subjects that they probably had this representation. The sole basis for mentioning Truel is a statement by Cauchy that Truel had this representation as early as 1786. For fuller discussion see Florian Cajori, "Historical Notes on the Graphic Representation of Imaginaries before Wessel" in *American Mathematical Monthly, 19* (1912), 167–171.

[16] Some of these difficulties will be discussed more fully in Chapter II, where Hamilton's efforts to find a three-dimensional vectorial system are treated.

[17] Gauss' untitled publication, which was a discussion of his "Theoria residuorum biquadraticum, Commentatio secunda," was originally published in the *Göttingische gelehrte Anzeigen* of April 23, 1831. I have used the text as given in Carl Friedrich Gauss, *Werke,* vol. II (Göttingen, 1863), 169–178.

[18] *Ibid.,* 175.

[19] Julian Lowell Coolidge, *The Geometry of the Complex Domain* (Oxford, 1924), 28–29.

[20] Hermann Günther Grassmann, *Gesammelte mathematische und physikalische Werke,* vol. I, pt. II (Leipzig, 1896), 8–9, 397–398.

[21] Felix Klein, "Über den Stand der Herausgabe von Gauss' Werken" in *Mathematische Annalen, 51* (1898), 128–133; Peter Guthrie Tait, "On the Claim Recently Made for Gauss to the Invention (not the Discovery) of Quaternions" in *Proceedings of the Royal Society of Edinburgh, 23* (1900), 17–23; Cargill Gilston Knott, "Professor Klein's View of Quaternions: a criticism" in *Proceedings of the Royal Society of Edinburgh, 23* (1900), 24–34. For the document on which Klein based his claim, see Gauss, *Werke,* vol. VIII (Leipzig, 1900), 357–362.

[22] On this point see Ludwig Schlesinger, "Über Gauss Arbeiten zur Functionentheorie" in Gauss, *Werke,* vol. X, pt. 2 (Göttingen, 1922–1933), 55–57.

[23] Abbé Buée, "Mémoire sur les quantités imaginaires" in *Transactions of the Royal Society of London, 96* (1806), 23–88.

[24] Coolidge, *Geometry of the Complex Domain,* 24.

[25] This is strongly implied by the fact that in an 1844 paper (Hamilton, "On Quaternions" in *Philosophical Magazine,* 3rd Ser., *25* (1844), 489–495) Hamilton discussed the authors that had influenced him; of the six men who discovered the geometrical representation of complex numbers, only Warren was mentioned. The same conclusion is implied in Hamilton's richly historical preface to his *Lectures on Quaternions;* see especially (5; [31]–[57]) as well as the work listed in note (4) above, wherein many letters from Hamilton to De Morgan were published in which Hamilton discussed these men. Hamilton explicitly denied having seen (1) Gauss' paper (4; 312), (2) Mourey's book (4; 489), (3) Argand's book (4; 435), (4) Servois' paper (5; [57]), and (5) Français' papers (Robert Perceval Graves, *Life of Sir William Rowan Hamilton,* vol. II [Dublin, 1885], 606). From the fact that Hamilton had not read Servois' and Français' papers or Argand's book, it seems reasonable to conclude that he had not read any of the relevant papers in volumes IV and V of Gergonne's *Annales* before 1844. Hamilton did not explicitly deny knowledge of Wessel, since he never, even after 1843, heard of Wessel, and he did not explicitly deny knowledge of Buée's paper, since he had already denied that it had any merit. On the other hand, Hamilton attended the 1833 meeting of the British Association for the Advancement of Science, and at this meeting George Peacock presented his "Report on the Recent Progress and Present State of Certain Branches of Analysis" (*B.A.A.S. Report,* 185–352), in which Peacock briefly discussed (*ibid.,* page 228) Argand's book and the papers from Gergonne's *Annales.*

[26] See John Warren, *A Treatise on the Geometrical Representation of the Square Roots of Negative Quantities* (Cambridge, 1828), page 3 for commutative law of addition, page 9 for commutative law for multiplication, page 18 for a hint that he was aware of the associative law for multiplication, and page 13 for the distributive law. The importance of this is that the discovery of quaternions depended to some extent on the recognition of the importance of these laws. On the origin of the names "associative," "commutative," and "distributive" in a mathematical sense probably the first historical statement was made by Hermann Hankel, *Theorie der*

complexen Zahlensysteme (Leipzig, 1867), footnote on page 3, where he said, "These names have been adopted universally in England since 1840 and hence I have not hesitated to transplant them to German soil; 'distributive' and 'commutative' were introduced by Servois (GERGONNE'S Ann. vol. V. 1814, p. 93); 'associative' was it seems first introduced by Sir. W. R. Hamilton." This statement is repeated by both David Eugene Smith and Florian Cajori. The earliest recognition of the necessity of proving the commutative law for multiplication is in Euclid, Book VII, Proposition 16. The first publication, to my knowledge, in which Hamilton used the term "associative" is in the paper "On a New Species of Imaginary Quantities, Connected with a Theory of Quaternion," communicated November 13, 1843, published in the *Proceedings of the Royal Irish Academy*, 2 (1844) 424–434. He wrote: "However, in virtue of the same definitions, it will be found that another important property of the old multiplication is preserved, or extended to the new, namely, that which may be called the *associative* character of the operation. . . ." *Ibid.*, 429–430.

[27] At least no extension is suggested in his book or in the two subsequent papers which he published on this subject. His two later papers were both published in the *Philosophical Transactions of the Royal Society of London, 119* (1829); they were entitled "Considerations of the Objections Raised Against the Geometrical Representation of the Square Roots of Negative Quantities," pages 241–254, and "On the Geometrical Representation of the Power of Quantities Whose Indicies Involve the Square Roots of Negative Quantities," pages 339–359. In the first paper (*ibid.*, 251–254) Warren stated that he had written his book before he heard of Buée's or Mourey's publication; Argand was not mentioned, presumably because Warren still had not heard of Argand's book.

[28] C. V. Mourey, *La vrai Théorie des quantités négatives et des quantités prétendues imaginaires* (Paris, 1828). The second edition (Paris, 1861) was used; this was a reprint of his 1828 work. Nowhere in the work does Mourey mention Argand or Buée; in fact no mathematicians are ever mentioned in the book. Mourey could not have known Warren's book since it was published after his own. Mourey (*ibid.*, IX) made the interesting comment that his book was an abridgement of a longer treatise he had written but had not published.

[29] *Ibid.*, 95.

CHAPTER TWO

Sir William Rowan Hamilton and Quaternions

I. *Introduction: Hamiltonian Historiography*

The task of the historian who wishes to treat any aspect of the work of Sir William Rowan Hamilton is complicated by the fact that estimates of his significance for the history of science have varied between two extreme positions. Thus, for example, Erwin Schrödinger wrote of Hamilton:

> While these discoveries (Quaternions, etc.) would suffice to secure Hamilton in the annals of both mathematics and physics a highly honourable place, such pious memorials can in his case easily be dispensed with. For Hamilton is virtually not dead, he himself is alive, so to speak, not his memory. I daresay not a day passes—and seldom an hour—without somebody, somewhere on this globe, pronouncing or reading or writing or printing Hamilton's name. That is due to his fundamental discoveries in general dynamics. The Hamiltonian principle has become the cornerstone of modern physics, the thing with which a physicist expects *every* physical phenomenon to be in conformity. . . .
>
> The modern development of physics is continually enhancing Hamilton's name. His famous analogy between mechanics and optics virtually anticipated wave-mechanics, which did not have to add much to his ideas, only had to take them seriously—a little more seriously than he was able to take them, with the experimental knowledge of a century ago. The central conception of all modern theory in physics is "the Hamiltonian." If you wish to apply modern theory to any particular problem, you must start with putting the problem "in Hamiltonian form."
>
> Thus Hamilton is one of the greatest men of science the world has produced.[6]

In 1945 J. L. Synge lamented that Hamilton's fame was passing into eclipse.[7] Synge cited many aspects of this eclipse but stressed above all the neglect of Hamilton's contribution to the calculus of variations. He wrote: "Hamilton was, in fact, a great contributor—probably the greatest single contributor of all time—to the calculus of variations." (7; 15)

In 1940 E. T. Whittaker published a paper entitled "The Hamiltonian Revival,"[8] in which he maintained: "The nadir of Hamilton's reputation was touched about the beginning of the present century: since when, there has been a steady movement in the reverse direction: one after another, the significance of his great innovations has been appreciated. . . ."[9] Whittaker wrote of Hamilton in 1954: "After Isaac Newton, the greatest mathematician of the English-speaking peoples is William Rowan Hamilton. . . ."[10]

In 1937 E. T. Bell in his widely read *Men of Mathematics*[11] entitled the chapter on Hamilton "An Irish Tragedy." Herein Bell presented Hamilton's life as a tragedy, in a sense a magnificent failure.

This disparity of views concerning Hamilton, which in fact dates back to the nineteenth century, is central to Hamiltonian historiography. The main source of this disparity of views relates to Hamilton's work on quaternions. Hamilton believed that quaternions represented the mathematics of the future and consequently devoted more than twenty years of his life to them. The view of quaternions held by nearly all mathematicians of the present is however quite different; the consensus now is that the quaternion system is but one of many comparable mathematical systems, and though it is interesting as a rather special system, it offers little value for application. The historian of today must take the above evaluation of quaternions as most probably valid, though there remain sources of doubt. Statements qualifying or contradicting this evaluation—made by such important scientists as E. T. Whittaker,[12] George D. Birkhoff,[13] and P. A. M. Dirac[14]—instill some degree of caution in the historian, as do the two large volumes by Otto F. Fischer,[15] in which the author attempted to rewrite much of modern physics in terms of Hamilton's quaternions.

E. T. Bell's view of Hamilton as a tragic failure certainly stemmed from the fact that he felt that quaternions are of little interest to modern mathematics. Bell was convinced that Hamilton was the victim of a monomaniacal delusion; he stated "that Hamilton's deepest tragedy was neither alcohol nor marriage but his obstinate belief that quaternions held the key to the mathematics of the physical universe."[16] The problem of quaternions also stands behind much of what E. T. Whittaker wrote concerning Hamilton; Whittaker however passed over the problem by stressing Hamilton's contributions to mathematical physics and by arguing that quaternions "may even yet prove to be the most natural expression of the new physics."[17]

The present study must stand in the shadow of this dispute concerning Hamilton's greatness; nevertheless it is hoped that substantial progress toward a solution may be achieved in terms of the following analysis, which will be developed more fully later. It is not possible to argue that the quaternion system is the vectorial system of the present day; the so-called Gibbs-Heaviside system is the only system that merits this distinction. Nor is it legitimate to argue (as Whittaker has done) that the quaternion system will be the system of a future day. Both of these alternatives are unacceptable; nonetheless it can be argued (though in my opinion this has not previously been done) that Hamilton's quaternion system led by an historically determinable path to the Gibbs-Heaviside system and hence to the modern system. In what follows it will be shown that this was in fact the case, and thus it will become clear that Hamilton deserves immense credit for his work in quaternions, since this work led to the now widely used system of vector analysis. The reasons why this is so little known will also be discussed. If this analysis is found acceptable, it should clear up the major problem in Hamiltonian historiography.

II. *Hamilton's Life and Fame*

Though in general a detailed discussion of a scientist's life need not be included in a study such as this, it is of necessity otherwise in regard to Hamilton and quaternions. The reason for this is that the fame attained by Hamilton during his lifetime strongly influenced subsequent events. Some indication of the importance of Hamilton's fame in this history may be attained by a comparison of the title pages of Hamilton's and Grassmann's first major works. The title page of Grassmann's *Ausdehnungslehre* of 1844 contained the following:

Hermann Grassmann
Lehrer an der Friedrich-Wilhelms-Schule zu Stettin

By contrast, the title page of Hamilton's *Lectures on Quaternions* contained:

> SIR WILLIAM ROWAN HAMILTON, LL.D., M. R. I. A., FELLOW OF THE AMERICAN SOCIETY OF ARTS AND SCIENCES; OF THE SOCIETY OF ARTS FOR SCOTLAND; OF THE ROYAL ASTRONOMICAL SOCIETY OF LONDON; AND OF THE ROYAL NORTHERN SOCIETY OF ANTIQUARIES AT COPENHAGEN; CORRESPONDING MEMBER OF THE INSTITUTE OF FRANCE; HONORARY OR CORRESPONDING MEMBER OF THE IMPERIAL OR ROYAL ACADEMIES OF ST. PETERSBURGH, BERLIN, AND TURIN; OF THE ROYAL SOCIETIES OF EDINBURGH AND DUBLIN; OF THE CAMBRIDGE PHILOSOPHICAL SOCIETY; THE

NEW YORK HISTORICAL SOCIETY; THE SOCIETY OF NATURAL SCIENCES AT LAUSANNE; AND OF OTHER SCIENTIFIC SOCIETIES IN BRITISH AND FOREIGN COUNTRIES; ANDREWS' PROFESSOR OF ASTRONOMY IN THE UNIVERSITY OF DUBLIN; AND ROYAL ASTRONOMER OF IRELAND.

William Rowan Hamilton was born of undistinguished ancestry on the midnight between August 3 and 4, 1805, in Dublin, Ireland. He was orphaned at age fourteen, but had ceased to live with his parents from the age of three, at which time he had been sent to live with his uncle, James Hamilton, an Anglican clergyman serving Trim, Ireland. Hamilton's uncle, a man of education and intelligence, directed his nephew's preuniversity education. The success of the uncle as tutor and the brilliance of Hamilton as student were manifested in many ways, of which the best known is that at age thirteen, Hamilton "was in different degrees acquainted with thirteen languages. . . ." [18] These languages were Greek, Latin, Hebrew, Syriac, Persian, Arabic, Sanskrit, Hindoostanee, Malay, French, Italian, Spanish, and German. The study of languages was however only one of Hamilton's interests, for he also read in geography, religion, mathematics, astronomy, and the best of English and foreign literature. At age sixteen he began Laplace's *Mécanique Céleste* and detected an error therein. It is interesting but not significant for this study that the error found by Hamilton was in Laplace's demonstration of the law of the parallelogram of forces. (2,I; 661–662)

In 1823 Hamilton entered Trinity College of Dublin University. He had placed first in the entrance exam and had decided that his calling was to science. His record at the University bordered on the incredible. In his second year he was awarded an *optime* for his knowledge of Greek, and in his third year another *optime* for his knowledge of mathematical physics. The winning of even a single *optime* was very rare. Upon winning the second *optime*, Hamilton "became a celebrity in the intellectual circle of Dublin; and invitations, embarrassing from their number, poured in upon him. . . ." (2,I; 209) Hamilton resolved to attempt to win in his final year the University Gold Medals in both classics and in science. This wish was not fulfilled, for during the summer after his third year Hamilton was offered the honor of becoming Andrews' Professor of Astronomy at the University of Dublin and Royal Astronomer of Ireland. His student days were also distinguished by success in creative endeavors. He wrote numerous poems and received honors for some of them. Researches in science begun in his seventeenth year on certain questions in mathematical optics led to his now famous "Theory of Systems of Rays," which was read in 1824 and published with further developments four years later.[19] Other important papers in the same line of development came in 1830, 1831, and 1837. His aim in these papers (which extend to over three hundred pages) was to reduce optics to a mathematical science in terms of his "Characteristic Function." The success of his mathematical methods in optics led Hamilton to extend these methods for use in dynamics. The distinguished historian of mechanics René Dugas has summarized the nature and importance of Hamilton's work in optics and dynamics:

> In short, jealous of the formal perfection which Lagrange had been able to give to dynamics, and which optics lacked, Hamilton undertook the rationalisation of geometrical optics. He did this by developing a formal theory which was free of all metaphysics and which, moreover, succeeded in accounting for all the experimental facts. . . .
>
> Then, returning to dynamics, Hamilton presented the law of *varying action* in a form very like that which he had discovered in optics. Thus he reduced the general problem of dynamics (for conservative systems) to the solution of two simultaneous equations in partial derivatives, or to the determination of a single function satisfying these two equations.
>
> . . . .
>
> Hamilton's guiding idea is continuous from his optical work to his work in dynamics—in this fact lies his greatness and his power. Here was a synthesis that Louis de Broglie was to rediscover and turn to his own account; a synthesis that was, it appears, to be Schrödinger's direct inspiration.[20]

These works certainly contributed to Hamilton's fame; Jacobi was probably thinking of them when in 1842 he referred to Hamilton as "le Lagrange de votre pays." (2,III; 509) Often however such highly mathematical papers do not produce popular fame. It was otherwise for Hamilton; in 1832[21] he predicted on a theoretical basis two new phenomena in optics, internal and external conical refraction. At Hamilton's request, Humphrey Lloyd, friend and colleague of Hamilton, attempted to verify Hamilton's prediction. In this he was completely successful, finding both predicted phenomena and a third which had not been predicted. (2,I; 635) Graves wrote that this discovery "excited at the time a very considerable sensation among scientific men in England and on the Continent. . . ." (2,I; 636) Whewell's praise for Hamilton was vigorous and immediate, and Airy referred to the discovery as "perhaps the most remarkable prediction that has ever been made. . . ." (2,I; 637) De Morgan writing in 1866 stated: "Opticians had no more imagined the possibility of such a thing, than astronomers had imagined the planet Neptune, which Leverrier and

Adams calculated into existence. These two things deserve to rank together as, perhaps, the two most remarkable of verified scientific predictions."[22] Plücker of Bonn wrote:

> No experiment in physics has made such a strong impression on my mind as that of conical refraction. A single ray of light entering a crystal and leaving as a luminous cone: this is something unheard of and without analogy. Mr. Hamilton predicted it, starting from the form of the wave which had been deduced by a long calculation from an abstract theory. I confess I would have had little hope of seeing an experimental confirmation of such an extraordinary result, predicted by the mere theory which Fresnel's genius had recently created. But since Mr. Lloyd had demonstrated that the experimental results were in complete accordance with the predictions of Mr. Hamilton, all prejudice against a theory so marvelously lofty has been forced to disappear. (2,I; 637)

The fame that came to Hamilton because of this discovery was increased by the fact that it was made, like nearly all the discoveries discussed thus far, before Hamilton had reached his thirtieth year.

Also illustrative of, and contributory to, Hamilton's popular fame was his close friendship with William Wordsworth and other literary figures such as Maria Edgeworth and Samuel Taylor Coleridge. Numerous letters passed between Hamilton and Wordsworth, and each visited the other on many occasions. Wordsworth had said that Hamilton was one of two men to whom he could look up (the other was Coleridge). To this Hamilton replied: "If I am to look down on you, it is only as Lord Rosse looks down in his telescope to see the stars of heaven reflected." (2,III; 237)

By 1835 Hamilton's fame was established. In that year he was knighted and received a medal from the Royal Society; in addition he finished a paper on algebraic couples, which is the first of Hamilton's publications to be of direct importance for the present study. In 1837 he was elected president of the Royal Irish Academy and held this position until his resignation in 1845, soon after his discovery (1843) of quaternions. The last twenty-two years of his life, from 1843 to 1865, were for the most part devoted to the development of quaternions.

Honors of all sorts continued to be bestowed on him. One of these deserves final mention. In 1865, the year of his death, Hamilton received notice that the newly founded National Academy of Sciences of the United States had elected him a Foreign Associate, along with fourteen other men. The members had voted to place Hamilton's name at the head of the list of Foreign Associates, presumably signifying that in their opinion he was the greatest living scientist. In this they were probably overly enthusiastic, but their judgment does attest to the fact, which is very significant for this study, that Hamilton's fame among his contemporaries was very great. At this same time the majority of Grassmann's scientific work had been completed, but he was still nearly unknown. His subtitle in his 1862 *Ausdehnungslehre* had changed only slightly; it was now "Professor am Gymnasium zu Stettin."

III. *Hamilton and Complex Numbers*

It was stated previously that Hamilton's discovery of quaternions was in the tradition of the work done on complex numbers, and in this regard the history of the geometrical representation of complex numbers and associated ideas was given. But there was a second line of development in studies on complex numbers that also led to quaternions. This line of development was established by Hamilton himself in his long and important essay published in 1837 and entitled: "Theory of Conjugate Functions, or Algebraic Couples; with a Preliminary and Elementary Essay on Algebra as the Science of Pure Time." This paper is important in itself; indeed one mathematician referred to it as a greater contribution to algebra than his discovery of quaternions.[23] Hamilton's paper is divided into three sections: the first section, which consists of "General Introductory Remarks," was written last; the second section, an essay "On Algebra as the Science of Pure Time" was written in 1835; and the third section, containing his "Theory of Conjugate Functions, or Algebraic Couples," was for the most part written in 1833. (2,II; 144) Neglect of the historical sequence of the composition has led to a number of historical misconceptions.

Hamilton began the paper by writing:

> The study of Algebra may be pursued in three very different schools, the Practical, the Philological, or the Theoretical, according as Algebra itself is accounted an Instrument, or a Language, or a Contemplation; according as ease of operation, or symmetry of expression, or clearness of thought, (the *agere*, the *fari*, or the *sapere*,) is eminently prized and sought for. The Practical person seeks a Rule which he may apply, the Philological person seeks a Formula which he may write, the Theoretical person seeks a Theorem on which he may meditate. (3; 293)

He then proceeded to state that the aim of this paper was theoretical.

> The thing aimed at, is to improve the *Science*, not the Art nor the Language of Algebra. The imperfections sought to be removed, are confusions of thought, and obscurities or errors of reasoning; not difficulties of application of an instrument nor failures of symmetry in expression. . . .
>
> For it has not fared with the principles of Algebra as with the principles of Geometry. No candid and intelligent person can doubt the

truth of the chief properties of *Parallel Lines*, as set forth by EUCLID in his Elements, two thousand years ago; though he may well desire to see them treated in a clearer and better method. The doctrine involves no obscurity nor confusion of thought, and leaves in the mind no reasonable ground for doubt, although ingenuity may usefully be exercised in improving the plan of the argument. But it requires no peculiar scepticism to doubt, or even to disbelieve, the doctrine of Negatives and Imaginaries, when set forth (as it has commonly been) with principles like these: that a *greater magnitude may be subtracted from a less*, and that the remainder is less than nothing; that *two negative numbers*, or numbers denoting magnitudes each less than nothing, may be *multiplied* the one by the other, and that the product will be a *positive* number, or a number denoting a magnitude greater than nothing; and that although the *square* of a number, or the product obtained by multiplying that number by itself, is therefore *always positive*, whether the number be positive or negative, yet that numbers, called *imaginary*, can be found or conceived or determined, and operated on by all the rules of positive and negative numbers, as if they were subject to those rules, *although they have negative squares*, and must therefore be supposed to be themselves neither positive or negative, nor yet null numbers, so that the magnitudes which they are supposed to denote can neither be greater than nothing, nor less than nothing, nor even equal to nothing. It must be hard to found a SCIENCE on such grounds as these. . . . (3; 294)

Hamilton then asked

whether existing Algebra, in the state to which it has been already unfolded by the masters of its rules and of its language, offers indeed no rudiment which may encourage a hope of developing a SCIENCE of Algebra: a Science properly so called; strict, pure, and independent; deduced by valid reasonings from its own intuitive principles; and thus not less an object of priori contemplation than Geometry, nor less distinct, in its own essence, from the Rules which it may teach or use, and from the Signs by which it may express its meaning. (3; 295)

Hamilton concluded "that the Intuition of TIME is such a rudiment" (3; 295) and elaborated on this idea by writing:

The argument for the conclusion that *the notion of time may be unfolded into an independent Pure Science*, or that *a Science of Pure Time is possible*, rests chiefly on the existence of certain priori intuitions, connected with that notion of time, and fitted to become the sources of a pure Science; and on the actual deduction of such a Science from those principles, which the author conceives that he has begun. (3; 296–297)

In the second section of this paper Hamilton attempted to develop the real number system on the basis of the intuition of the concept of time. In this way he believed he could justify the use of negative numbers as corresponding to steps in time.

It is generally believed that Hamilton's stress on time was derived from Kant. Such may not be the case, for Kant's name is never mentioned in the paper. In Hamilton's later exposition of these ideas in the Preface to his *Lectures on Quaternions* (4; [2]–[3]) he did mention Kant and wrote that reading Kant's *Critique of Pure Reason* "encouraged [him] to entertain and publish this view. . . ." (4; [2]) As early as 1827 Hamilton wrote, immediately after mentioning geometry: "The sciences of Space and Time (to adopt here a view of Algebra which I have elsewhere ventured to propose) became intimately intertwined and indissolubly connected with each other." (2,I; 229) From a number of statements made by Hamilton in letters it seems quite clear that he began reading Kant *four years after* making the above statement.[24] In 1835 Hamilton wrote: "and my own convictions, mathematical and metaphysical, have been so long and so strongly converging to this point (confirmed no doubt of late by the study of Kant's *Pure Reason*), that I cannot easily yield to the authority of those other friends who stare at my strange theory." (2,II; 142) It thus seems that at most Kant served as a catalyst for the development of his ideas and as a confirmation of them.

In the third part of the essay Hamilton presented his "Theory of Conjugate Functions, or Algebraic Couples." While the second part of his essay is generally considered of minor importance, the third part is universally admitted to be of great importance, for herein Hamilton developed complex numbers in terms of ordered pairs of real numbers in almost exactly the same way as it is done in modern mathematics.[25] The stress in this section was not on time, although the interpretation of the couples in terms of time was given. Hamilton at no point in the paper mentioned Warren or the geometrical interpretation of complex numbers; from this it seems probable that Hamilton believed (like Gauss) that the geometrical representation was an aid to intuition, but not a satisfactory justification for complex numbers. Essentially what Hamilton did in this section was to set up ordered pairs of real numbers (a, b) and define operations on them. These operations were all done in terms of the rules for real numbers. He then showed that the couples thus considered were equivalent to complex numbers of the form $a + bi$. He wrote:

> In the THEORY OF SINGLE NUMBERS, the symbol $\sqrt{-1}$ is *absurd*, and denotes an IMPOSSIBLE EXTRACTION, or a merely IMAGINARY NUMBER; but in the THEORY OF COUPLES, the same symbol $\sqrt{-1}$ is *significant*, and denotes a POSSIBLE EXTRACTION, or a REAL COUPLE, namely (as we have just now seen) the *principal square-root of the couple* $(-1, 0)$. In the latter theory, therefore, though not in the former, this sign $\sqrt{-1}$ may properly be employed; and we may write, if we choose, for any couple (a_1, a_2) whatever, $(a_1, a_2) = a_1 + a_2\sqrt{-1}$. . . . (3; 417–418)

Hamilton concluded the essay by writing:

> the present *Theory of Couples* is published . . . to show . . . that expressions which seem according to common views to be merely symbolical, and quite incapable of being interpreted, may pass into the world of thoughts, and acquire reality and significance, if Algebra be viewed as not a mere Art or Language, but as the Science of Pure Time. The author hopes to publish hereafter many other applications of this view; especially to Equations and Integrals, and to a Theory of Triplets and Sets of Moments, Steps, and Numbers, which includes this Theory of Couples. (3; 422)

The "Theory of Triplets" that he sought was of course the extension of the complex number system to three dimensions.

It is clear from this paper that Hamilton understood the nature and importance of the associative, commutative, and distributive laws.[26] The majority of mathematicians appreciated the significance of these laws only after number systems (especially quaternions) had been developed which did not obey them.

Hamilton's "Theory" was poorly received. Most mathematicians did not agree with Hamilton's stress on time, and a few felt the need for the development of complex numbers on a basis other than a geometrical one. That Gauss and Bolyai rejected the geometrical justification of complex numbers is almost certainly due to the fact that they both had previously discovered non-Euclidean geometry. When non-Euclidean geometry became known (after 1860), mathematicians then became interested in the development of complex numbers in terms of ordered pairs of real numbers.

Hamilton's "Essay" was an important event in the history of the discovery of quaternions for a number of reasons. First, it set Hamilton on the quest for higher complex numbers from another direction, in addition to the quest in terms of a method of analysis for three-dimensional space. Second, through his method of couples at least Hamilton himself became convinced of the legitimacy of complex numbers, and more importantly he also obtained a method that could be extended in such a way as to assure the legitimacy of higher complex numbers, formed for example by triplets or quadruplets (as in the case of quaternions). To put it another way, by this method Hamilton was prepared, perhaps to discover, more significantly to accept as legitimate, "four-dimensional" complex numbers (as quaternions), even if no geometrical justification were to be available. Support for the above analysis is found in a letter of 1841 from Hamilton to De Morgan:

> As to Triplets, I must acknowledge, that though I fancied myself at one time to be in possession of something worth publishing about them,

I never could resolve the problem which you have justly signalised as the most important in this branch of (future) Algebra: to *assign* two symbols Ω and ω, such that the one symbolical equation

$$a + b\Omega + c\omega = a_1 + b_1\Omega + c_1\omega$$

shall give the three equations

$$a = a_1,\ b = b_1,\ c = c_1$$

But, if my view of Algebra be just, it *must* be possible, in *some* way or other, to introduce not only triplets but *polyplets*, so as in some sense to satisfy the symbolical equation

$$a = (a_1, a_2, \ldots a_n);$$

a being here one symbol, as indicative of one (complex) thought; and $a_1, a_2, \ldots a_n$ denoting n real numbers, positive or negative; that is, in other words, n dates, in the chronological sense of the word, only excluding outward marks and measures, and the notion of cause and effect. (2,II; 343)

Moreover, after 1843 Hamilton stressed the importance of this point of view for his discovery of quaternions. He said in fact that he made this discovery one day when "being then fresh from a reperusal of my old essay, I renewed my attempts to combine my general notion of sets of numbers, considered as suggested by sets of moments of time, with geometrical considerations of points and lines in tridimensional space. . . ." [27]

IV. *Hamilton's Discovery of Quaternions*

At the end of his "Essay" of 1837 Hamilton stated that he was seeking a triplet system. He had, in fact, made definite attempts to find triplets as early as 1830. (4; [39]) The conjecture of that year entailed abandonment of the distributive property. That Hamilton was not alone in his quest is shown by Hamilton's own statement that John T. Graves had tried to form a higher complex number system for space "as early, or perhaps earlier than myself." [28] Hamilton and Graves had been in correspondence on the subject from at least 1836, at which time Graves sent Hamilton a system which was similar to one Hamilton had constructed in 1835. (4; [36]–[37]) In 1841 Hamilton received a letter from Augustus De Morgan, in which De Morgan asked Hamilton about his triplets. With this letter was a copy of De Morgan's 1841 paper "On the Foundation of Algebra," [29] in which De Morgan had included a brief discussion of triplets. (4; [41]–[42])

When by 1843 Hamilton began another intense search for triplets, the framework within which the search had to be conducted was clear to him. The following may be taken as an outline of the properties that he consciously hoped the new numbers would have.

1. The associative property for addition and multiplication. Thus if N, N', and N'' are three such numbers, then $N + (N' + N'') = (N + N') + N''$ and $N(N'N'') = (NN')N''$.

2. The commutative property for addition and multiplication. $N + N' = N' + N$ and $NN' = N'N$.

3. The distributive property. $N(N' + N'') = NN' + NN''$.

4. The property that division is unambiguous. Thus if N and N' are any given complex numbers, it is always possible to find one and *only one* number X (in general, a number of the same form as N and N') such that $NX = N'$.

5. The property that the new numbers obey the law of the moduli. Thus if any three triplets combine so that

$$(a_1 + b_1 i + c_1 j)(a_2 + b_2 i + c_2 j) = a_3 + b_3 i + c_3 j,$$

then

$$(a_1^2 + b_1^2 + c_1^2)(a_2^2 + b_2^2 + c_2^2) = (a_3^2 + b_3^2 + c_3^2):$$

6. The property that the new numbers would have a significant interpretation in terms of three dimensional space.

It is well known that ordinary complex numbers have all these properties, with the exception that their geometrical interpretation is for two-dimensional space. In one sense, then, the above is simply a detailed statement that Hamilton sought for new numbers which would be directly analogous to ordinary complex numbers. Of the above properties only the commutative property for multiplication had to be abandoned for quaternions. With limits as restrictive as these Hamilton could only be satisfied with quaternions, for, as C. S. Peirce proved in 1881, "ordinary real algebra, ordinary algebra with imaginaries, and real quaternions are the only associative algebras in which division by finites always yields an unambiguous quotient." [30] It is significant to ask at this point which of these properties are retained for the scalar (dot) and vector (cross) product multiplications in modern vector analysis. For the dot product the associative law for multiplication is not relevant, and both the law of the moduli and the unambiguity of division must be abandoned.[31] For the cross product the associative and commutative properties must be abandoned.[32] Moreover division is not unambiguous, and the law of the moduli fails as well.[33]

From the above comparison of the properties of quaternions and vectors it is evident that at least in some ways quaternions not only are simpler than modern vectors but also entail fewer innovations.

In the period immediately after their discovery quaternions were criticized because of the abandonment of the commutative property for multiplication. It is an interesting historical speculation in this regard as to what would have been said of a system in which the commutative and associative properties failed, and in which division was in general impossible, and in which moreover two different types of multiplication were defined. In any case it was in the context of the above properties that Hamilton in 1843 renewed his attempts to find triplets.[34]

On October 16, 1843, Hamilton discovered quaternions. Perhaps the best description of the circumstances surrounding this event is contained in a letter Hamilton wrote in 1865 to his son Archibald H. Hamilton:

> If I may be allowed to speak of *myself* in connexion with the subject, I might do so in a way which would bring *you* in, by referring to an *ante-quaternionic* time, when you were a mere *child,* but had caught from me the conception of a Vector, as represented by a *Triplet:* and indeed I happen to be able to put the finger of memory upon the year and month – October, 1843 – when having recently returned from visits to Cork and Parsonstown, connected with a Meeting of the British Association, the desire to discover the laws of the multiplication referred to regained with me a certain strength and earnestness, which had for years been dormant, but was then on the point of being gratified, and was occasionally talked of with you. Every morning in the early part of the above-cited month, on my coming down to breakfast, your (then) little brother William Edwin, and yourself, used to ask me, "Well, Papa, can you *multiply* triplets"? Whereto I was always obliged to reply, with a sad shake of the head: "No, I can only *add* and subtract them."
>
> But on the 16th day of the same month – which happened to be a Monday, and a Council day of the Royal Irish Academy – I was walking in to attend and preside, and your mother was walking with me, along the Royal Canal, to which she had perhaps driven; and although she talked with me now and then, yet an *under-current* of thought was going on in my mind, which gave at last a *result,* whereof it is not too much to say that I felt *at once* the importance. An *electric* circuit seemed to *close;* and a spark flashed forth, the herald (as I *foresaw, immediately*) of many long years to come of definitely directed thought and work, by *myself* if spared, and at all events on the part of *others,* if I should even be allowed to live long enough distinctly to communicate the discovery. Nor could I resist the impulse – unphilosophical as it may have been – to cut with a knife on a stone of Brougham Bridge, as we passed it, the fundamental formula with the symbols, i, j, k; namely
>
> $$i^2 = j^2 = k^2 = ijk = -1,$$
>
> which contains the *Solution* of the *Problem,* but of course, as an inscription, has long since mouldered away. A more durable notice remains, however, on the Council Books of the Academy for that day (October 16th, 1843), which records the fact, that I then asked for and obtained

leave to read a Paper on *Quaternions*, at the *First General Meeting* of the Session: which reading took place accordingly, on Monday the 13th of the November following. (2,II; 434–435)

Thus in a very dramatic manner Hamilton discovered and announced the discovery of quaternions. These are hypercomplex numbers of the form $w + ix + jy + kz$, where w, x, y, and z are real numbers, and i, j, and k are unit vectors, directed along the x, y, and z axes respectively. The i, j, and k units obey the following laws:

$$ij = k \qquad jk = i \qquad ki = j$$

$$ji = -k \qquad kj = -i \qquad ik = -j$$

$$ii = jj = kk = -1$$

It is to be noted that for two quaternions q and q', qq' does not in general equal $q'q$. The loss of commutativity in quaternions, while it is very important historically, is also significant mathematically, because this complicates calculations in which quaternions are used. All the other properties discussed above are satisfied by quaternions. Thus it may be verified that quaternion multiplication is associative and quaternion division is unambiguous. These are two important properties which bear special mention, since they are not preserved in the algebra of modern vectors.

There have been a number of discussions published on the mathematical details of Hamilton's procedure in his discovery; for the present purposes all that need be said is that Hamilton worked within the context that has been discussed above.[35]

Almost immediately after his discovery Hamilton stated that he "felt that it might be worth my while to expend [on quaternions] the labour of at least ten (or it might be fifteen) years to come." (2,II; 436) Hamilton actually spent the last *twenty-two years* of his life working almost exclusively on quaternions. The letters of the first few days after the discovery show that Hamilton felt that his system had importance for heat theory, electricity,[36] and spherical trigonometry. (2,II; 442) In 1851 he wrote: "In general, although in one sense I hope that I am actually growing *modest* about the quaternions, from my seeing so many peeps and vistas into future expansions of their principles, I still must assert that this discovery appears to me to be as important for the middle of the nineteenth century as the discovery of fluxions was for the close of the seventeenth." (2,II; 445)

In one sense at least Hamilton's discovery was epoch making, for quaternions were the first well-known consistent and significant number system which did not obey the laws of ordinary arithmetic. His "curious, almost wild" (as he called it [2,II; 441]) discovery may be compared to the discovery of non-Euclidean geometry. Both discoveries broke bonds set by centuries of mathematical thought. Immediately after 1843 other new number systems were discovered by Augustus De Morgan (who published five new systems),[37] John T. Graves (1844) (2,II; 454–455), and Charles Graves (1846).[38]

This section will be concluded by a discussion of Hamilton's publications on quaternions through the year 1847. On November 13, 1843, Hamilton read a paper on quaternions before the Royal Irish Academy, of which at least part was published in 1844.[39] Either this paper or the very similar paper in the July, 1844, issue of the *Philosophical Magazine* was his first publication on quaternions. (5,*25;* 10–13) Among the most important papers are his "On Quaternions," delivered November 11, 1844, to the Royal Irish Academy and published in 1847,[40] and the similar paper published in the July, 1846, issue of the *Philosophical Magazine.* (5,*29;* 26–31) In these papers Hamilton dealt with the fact that quaternions are not analogous to ordinary complex numbers in that the scalar part (the w of $w + ix + jy + kz$) does not indicate distance on an axis unless, as he had suggested earlier, quaternions be considered as four dimensional. Thus Hamilton (writing in the third person) stated:

> And on account of the facility with which this so called *imaginary* expression, or square root of a negative quantity, is constructed by a *right line having direction in space,* and having x, y, z for its three rectangular components, or projections on three rectangular axes, he has been induced to call the trinomial expression itself, as well as the line which it represents, a VECTOR. A *quaternion* may thus be said to consist generally of a *real* part and a *vector.* The fixing a special attention on this last part, or element, of a quaternion, by giving it a special name, and denoting it in many calculations by a single and special sign, appears to the author to have been an improvement in his method of dealing with the subject: although the general notion of treating the constituents of the imaginary part as coordinates had occurred to him in his first researches.
>
> Regarded from a geometrical point of view, this algebraically imaginary part of a quaternion has thus so natural and simple a signification or representation in space, that the difficulty is transferred to the algebraically real part; and we are tempted to ask what this last can denote in geometry, or what in space might have suggested it.[41]

The origin of the word *vector* (and the word *scalar*) is clear from the following quotation in the similar paper in the *Philosophical Magazine.*

> The algebraically *real* part may receive . . . all values contained on the one *scale* of progression of number from negative to positive infinity; we shall call it therefore the *scalar part,* or simply the *scalar* of the quaternion, and shall form its symbol by prefixing, to the symbol of the quater-

> nion, the characteristic Scal., or simply S., where no confusion seems likely to arise from using this last abbreviation. On the other hand, the algebraically *imaginary* part, being geometrically constructed by a straight line or radius vector, which has, in general, for each determined quaternion, a determined length and determined direction in space, may be called the *vector part*, or simply the *vector* of the quaternion; and may be denoted by prefixing the characteristic Vect., or V. We may therefore say that *a quaternion is in general the sum of its own scalar and vector parts*, and may write Q = Scal. Q + Vect. Q = S.Q + V.Q or simply Q = SQ + VQ. (5,29; 26–27)

From the above quotations it may be inferred that it was Hamilton who introduced the term *scalar* and also the term *vector* in its precise mathematical sense, although the similar term *radius vector* had been used for many years before.

The above quotations however have a far greater significance than this. In a sense they mark the beginning of modern vector analysis. Hamilton had introduced his symbols S and V because "separation of the real and imaginary parts of a quaternion is an operation of such frequent occurrence, and may be regarded as so fundamental in this theory. . . ." (5,29; 26) Hamilton illustrated the use of his symbols as applied to the product of the multiplication of two quaternions α and α', in which the scalar parts were 0. Letting $\alpha = xi + yj + zk$ and $\alpha' = x'i + y'j + z'k$, Hamilton wrote: "$S.\alpha\alpha' = -(xx' + yy' + zz')$; $V.\ \alpha\alpha' = i(yz' - zy') + j(zx' - xz') + k(xy' - yx')$. . . ." (5,29; 30) It is obvious that these are equivalent to the modern vector (cross) product and to the negative of the modern scalar (dot) product.[42] Hamilton and Tait made very frequent use of these symbols, using them in cases where the dot and cross product would now be used. Hamilton then proceeded to prove such equations as $S.\alpha\alpha' = 0$ when α and α' are parallel.

Another pair of very important papers appeared in 1846 and 1847.[43] In these papers Hamilton introduced the new operation

> . . . $\lhd$, defined with relation to these three symbols *ijk*, and to the known operation of partial differentiation, performed with respect to three independent but real variables *xyz*, as follows:
>
> $$\lhd = \frac{id}{dx} + \frac{jd}{dy} + \frac{kd}{dz};$$
>
> *this new characterestic will have the negative of its symbolic square expressed by the following formula:*
>
> $$-\lhd^2 = \left(\frac{d}{dx}\right)^2 + \left(\frac{d}{dy}\right)^2 + \left(\frac{d}{dz}\right)^2;$$
>
> of which it is clear that the applications to analytical physics must be extensive in a high degree. (5,31; 291)

The final paper of special importance is that delivered November 13, 1843, but published only in full (and certainly with additions) in 1848.[44] This very long paper was mainly devoted to developing quaternions in analogy to his 1837 treatment of complex numbers as number couples, and to relating his quaternions to the operations, principles, and elements of traditional mathematics.

In summary, by the end of 1847 Hamilton had published at least thirty-four papers in five different journals. Many of these were expository; some used quaternions for the solution of problems in geometry, mechanics, and astronomy. On a number of occasions new results were obtained by means of quaternions, but none seem to have been of a striking nature.

V. *Quaternions Until Hamilton's Death* (*1865*)

The aim of this section is to consider Hamilton's work on quaternions from 1847 to 1865 and, more important to us, to discuss the reception of quaternions from the time of their discovery until 1865.

In considering the reception of quaternions it must not be forgotten that, unlike many new discoveries, the discovery of quaternions was made by a man having considerable fame before the discovery was made. It was in 1842 that Jacobi referred to Hamilton as "le Lagrange de votre pays," and the *Athenaeum*, in discussing the 1842 meeting of the British Association, mentioned that "peculiar interest was excited by the presence of the three great astronomers, Bessel, Herschel, and Hamilton, who were seen seated together on the platform." [45] Hamilton's fame acted as a powerful force in spreading knowledge of quaternions and in forestalling criticism of them. Mathematicians who did not use quaternions were nonetheless cautious about attacking them.

The question of the reception of quaternions should be seen within the perspective provided by the history of the reception of similar systems. The rapidity with which new systems of ideas (as opposed to new experimental or technical results) are received is often exaggerated. Non-Euclidean geometry, Boolean algebra, Maxwell's theory, and even the calculus were only gradually appreciated and assimilated into scientific thought. Moreover complex numbers, the Grassmannian system, and the Gibbs-Heaviside (or modern) system of vector analysis were all slowly received. When Hamilton and the young Tait express their surprise that the quaternion system was only slowly being appreciated, their statements are to be taken with caution, for they express a lament that is nearly universal among the proponents of new systems of ideas.

It is also important to realize that to a certain extent the groundwork had been laid for quaternions before 1843. The work of Argand, Mourey, Warren, Gauss, and others had made the geometrical representation of complex numbers an accepted idea. The work of Gauss, Peacock, De Morgan, and others had helped to produce a more sophisticated view of algebra. Finally, the work of Bellavitis, Möbius, and Grassmann, and the fact that Wessel, Gauss, Argand, Servois, Français, Mourey, John T. Graves, De Morgan, and Hamilton had all, and for the most part independently, searched for triplets, indicates that there was a felt need for such a system as quaternions.

Quaternions, with their abandonment of commutativity, were in any case somewhat revolutionary in conception. Thus, even John T. Graves, perhaps the mathematician best prepared for accepting such a new idea, wrote in the following manner in response to Hamilton's letter announcing the discovery:

> There is still something in the system which gravels me. I have not yet any clear views as to the extent to which we are at liberty arbitrarily to create imaginaries, and endow them with supernatural properties. You are certainly justified by the event. You have got an instrument that facilitates the working of trigonometrical theorems and suggests new ones, and it seems hard to ask more; but I am glad that you have glimpses on physical analogies. (2,II; 443)

It seems that De Morgan and MacCullagh were troubled by the abandonment of commutativity, and if such great mathematicians were troubled, what must have been the reaction among lesser lights? Probably the initial reaction was opposition, passing with time into curiosity and a greater openness of mind. By 1847 quaternions had to some degree become known, and at the British Association meeting of that year a debate took place concerning their value. Hamilton described the debate in a letter to R. P. Graves:

> Being ready, however, at the commencement of the Meeting, I was told that I need not straiten myself as to time, the pressure of other Papers having not yet been felt; and Dr. Peacock said afterwards to me that I had given "a capital exposition": while from many other quarters it has been told me that he has expressed himself everywhere as favourable to my whole system. Mr. Jarrett brought forward again his Cambridge objections, and dwelt particularly on the possibility of making mistakes in the use of my new calculus. In reply to which I disclaimed the power of setting any limit to the faculty of making blunders; but said that the practical question on that point was, whether with a reasonable degree of attention a reasonable security against error could be attained; which I thought that my experience of the working of the quaternions enabled me to answer in the affirmative. The Rev. Richard Greswell, an Oxford man for whom I have a great respect, but who is not particularly scien-

tific, made some critical remarks in a not unfriendly spirit; but obliged me to disclaim a triplet which he attributed to me, of Positive, Negative, and Imaginary: and to state that with me the distinction between Positive and Negative was exactly the same as that between Future and Past, or between Past and Future. After some gentle skirmishes of this sort, rose Herschel; who said that his admiration of the quaternions had increased with every resumption of his study of them: and that although it might be difficult at first to master the extremely abstract conceptions, and the new algorithm which they involve, yet he was well convinced that it was worth the trouble. They appeared to him a bag, into which one need only insert his hand to draw forth treasures; he might call them a cornucopia of scientific abundance: and in a word, his earnest advice to mathematicians would be, to "study the quaternions." Sir John Herschel had remained at Oxford on purpose to make this statement; and immediately afterwards he started off for Collingwood. Mr. Airy, seeing that the subject could not be cushioned, rose then to speak of his own acquaintance with it, which he avowed to be none at all; but gave us to understand that what he did not know could not be worth knowing. He warned all persons, if they should use the method, to do so with the extremest caution; professing to regard me as believing it to be a right one, solely on the ground of the agreement of its results, so far as they have been yet obtained, with those of the older method. What was obscure was to him as if it were erroneous, what was paradoxical was to him as if it were false; and he thought *that* system useless as an algebraical geometry, of which the expressions were so extremely difficult of geometrical interpretation. (2,II; 585–587)

Concerning this quotation it should be mentioned that Herschel's statement appeared in the *Athenaeum* in an even stronger form.[45] It also seems that Jarrett's, Greswell's, and Airy's objections were only partly valid, though forcefully delivered. MacCullagh's views seem to have changed around this time, for in late 1847, on the night before he committed suicide, he told a friend that in his opinion Herschel had not gone too far in his statement at the British Association meeting. (2,II; 595–596)

In 1846 Hamilton resigned the presidency of the Royal Irish Academy presumably to obtain more time for his quaternion researches, and in 1848 both the Royal Irish Academy and the Royal Society of Edinburgh awarded him medals for his discovery. It was also in 1848 that Hamilton gave a series of four lectures on quaternions at Dublin University. He had earlier, in 1845, given "a short sketch" of them. Among those who heard the 1848 lectures were George Salmon and Arthur Cayley, who was the first mathematician, besides Hamilton, to publish a paper on quaternions (Cayley's paper appeared in 1845).[46] George Salmon also became interested in quaternions and advanced far enough in their study that at one time Hamilton thought that Salmon might become his successor in quaternion studies. (2,III; 90) Salmon however went on to attain fame in other branches of mathematics, and Peter Guthrie Tait (who will be discussed in detail later) became Hamilton's successor in quaternion researches.

Hamilton's lectures of 1848 were eventually greatly expanded and published as his *Lectures on Quaternions* in 1853. Salmon aptly described the need for such a work when he wrote to Hamilton in 1852: "Your book will probably make the use of your method more general. At present it is a bow of Ulysses, which no one can bend but the owner." (2,III; 346)

Hamilton's *Lectures on Quaternions* consisted of a series of seven lectures, extending to 737 pages with an additional 64-page preface and a 72-page table of contents. In the preface Hamilton discussed the history of quaternions and showed in great detail how quaternions could be interpreted in terms of his concept of algebra as founded on the concept of time and in terms of quadruples of real numbers. These sections were neither easy nor probably interesting reading for most readers. Early in the work Hamilton wrote that he would use a "*metaphysical style* of expression." (4; 4) This was not a happy choice, and the work was further encumbered by his practice of introducing a multitude of new terms. Early in the work he set up an elaborate terminology by using such words and combinations as vector, vehend, vection, vectum, revector, revehend, revection, revectum, provector, . . . transvector, . . . factor, . . . profactor, . . . versor, . . . and quadrantal versor, which is of course a semi-inversor. Authors of later quaternion books expressed in three pages the mathematical ideas that Hamilton had presented in his first seventy-four pages. Hamilton dealt at some length with the addition, subtraction, multiplication, and division of quaternions. He showed how quaternions were to be applied to spherical geometry and trigonometry and showed the relations of quaternions to ". . . CO-ORDINATES, DETERMINANTS, TRIGONOMETRY, LOGARITHMS, SERIES, LINEAR AND QUADRATIC EQUATIONS, DIFFERENTIALS, AND CONTINUED FRACTIONS. . . ."[47] He concluded with some geometrical and astronomical applications and the introduction of the biquaternion.

It was without question a very long and a very difficult book to read. Herschel in 1853 wrote to Hamilton that the book would "take any man a twelvemonth to read, and near a lifetime to digest . . . ,"[48] and in 1859 Herschel wrote:

> Your deduction from Quaternions of Fresnel's Wave is one of those things which I have just knowledge enough to admire without enough to understand. But it set me again on reading you *Lectures on Quaternions*, and I got through the three first chapters of it with a much clearer

perception of meaning than when I attacked it some three or four years back, but I was again obliged to give it up in despair. Now I pray you to listen to this cry of distress. I feel *certain* that if you pleased you *could* put the whole matter in as clear a light as would make the Calculus itself accessible as an instrument to readers even of less "penetrating power" than myself, who, having once mastered the *algorithm* and the *conventions* so as to work with it, would then be better prepared to go along with you in your metaphysical explanations. (2,III; 121)

It may be recalled that Herschel had been a Senior Wrangler and was a man of great ability. If he could read no more than the first three chapters (129 pages), how far would the less able mathematicians advance?

Hamilton faced some difficulty in getting such a large, new work published. Thus in 1853 he wrote to a friend (Mortimer O'Sullivan):

You will I hope bear with me if I say, that it required a certain *capital* of scientific reputation, amassed in former years, to make it other than dangerously imprudent to hazard the publication of a work which has, although at bottom quite conservative, a highly revolutionary air. It was a part of the ordeal through which I had to pass, an episode in the battle of life, to know that even candid and friendly people secretly, or, as it might happen, openly, censured or ridiculed me, for what appeared to them my monstrous innovations. (2,II; 683)

The longest review of the work appeared in the *North American Review* for 1857, written probably by Thomas Hill.[49] Hill began by writing: "It is confidently predicted, by those best qualified to judge, that in the coming centuries Hamilton's Quaternions will stand out as the great discovery of our nineteenth century. Yet how silently has the book taken its place upon the shelves of the mathematician's library. Perhaps not fifty men on this side of the Atlantic have seen it, certainly not five have read it." [50] He then proceeded to compare Hamilton to Archimedes, Galileo, Descartes, Leibniz, and Newton in regard to the problem of who can judge their ideas.

Other sections praise the book and its author in such strong words as "The discoveries of Newton have done more for England and for the race, than has been done by whole dynasties of British monarchs; and we doubt not that in the great mathematical birth of 1853, the Quaternions of Hamilton, there is as much real promise of benefit to mankind as in any event of Victoria's reign." [51] After mentioning Pythagoras, Hill stated: "And if the world should stand for twenty-three hundred years longer, the name of Hamilton will be found, like that of Pythagoras, made immortal by its connection with the eternal truth first revealed through him." [52] The force of such statements was somewhat weakened when Hill admitted he had not read the work: "In looking over Hamilton's eight hundred pages on Quaternions (for we will not pretend to say we have read them. . . ."[53] The review was concluded with an extremely brief sketch of the mathematics of quaternions.

Hill's review of Hamilton's *Lectures on Quaternions* is certainly one of most favorable reviews ever published; any person interested in mathematics who read and accepted Hill's statements would feel inclined, if not obliged, to read Hamilton's book. In the review Hill showed more ability for rhetoric than for the analysis of mathematical ideas; he reviewed Hamilton, not his book, and thereby failed to heed his own advice, for he had argued that few if any could judge the significance of the book. The fate of Hamilton in this case is the not uncommon fate of men of fame; their books are reviewed favorably but uncritically.

At various times after 1843 Hamilton received letters or visits from mathematicians who had acquired an interest in quaternions. Among these were Thomas Penyngton Kirkman, who became a first-rate mathematician but lost interest in quaternions; Robert Carmichael, whose interest in quaternions was cut short by his death at age 31, but who had by that time published five papers on quaternions; and J. P. Nichol, Professor of Astronomy at Glasgow, who wrote to Hamilton that the discovery of quaternions seemed of no "less value than the momentous step shown us by Descartes. . . ." (2,II; 635)

In 1860 Nichol published a work called *Cyclopaedia of the Physical Sciences,* for which Hamilton wrote an exposition of quaternions.[54] The interesting aspect of this is that Nichol appended to the article a short note which stated that since these sections were written,

> the new branch of mathematics to which they relate appears to have attracted an increasing degree of attention, in our own and in foreign countries. The *Lectures* (Dublin, 1853) have for example, formed the subject of a favourable article in the *North American Review* for July, 1857: and, indeed, the Quaternions had been mentioned as among the sources of hope for the future progress of analytical mechanics, in the conclusion to a very beautiful volume (*A System of Analytic Mechanics,* &c., page 476. Boston, 1855) on that science; by Professor Benjamin Peirce of Harvard University, U.S., as follows: ". . . and much must soon become antiquated and obsolete as the science advances, and especially when we shall have received the full benefit of the remarkable machinery of Hamilton's Quaternions."[55]

Benjamin Peirce, who was the leading American mathematician of the time (and the teacher of Thomas Hill), will be discussed in detail later. At this point it need only be stated that in 1872 Peirce republished his *System of Analytical Mechanics* without altering the statement quoted above[56] *and* without introducing quaternions.

Such statements as those discussed above helped to lay the foundation for the spread of the quaternion system, but in a very indirect way. To advocate the use of a system is not of course equivalent to using the system, and the latter is the more important. In the late 1850's and 1860's the feeling gradually arose that quaternions were often praised but seldom used. In the last third of the century their use became more frequent, partly as a result of Hamilton's second book on quaternions and partly because elementary presentations of quaternions became available through the efforts of other mathematicians. De Volson Wood in an elementary exposition (written in 1880) perhaps went too far in his description of Hamilton's *Lectures* when he wrote of it: "but the style is so peculiar, being diffuse and hesitating as if he *labored* to make his readers understand, and the arrangement being such as to separate parts of essential principles, that, probably, comparatively few of its readers, without other aids, have mastered its principles."[57] Charles Graves, Hamilton's colleague at Dublin, wrote in his éloge for Hamilton: "Students of his lectures on Quaternions have sometimes complained that he has claimed from them too much attention to the metaphysics of the subject, and has stopped them in their career of building up, in order that they might contemplate afresh the plan of the structure."[58]

Some information is available concerning the early reception of quaternions in other countries. Although in 1861 Hamilton wrote that France had been very slow in accepting quaternions (2,III; 129), it was the Frenchman Alexandre Allégret who in 1862 published the first book on quaternions not written by Hamilton.[59] Hermann Hankel stated in 1867 that Hamilton wrote in a very unconventional way for continental readers and that the only source of information for them at that time was Allégret's work, which Hankel called an "indeed insufficient presentation."[60] In Italy Giusto Bellavitis became interested in quaternions and in 1858 and 1862 published papers relating to quaternions.[61] Finally, interest must have spread to Russia, for Tait wrote to Hamilton in 1862: "Professor Bolzani is here and we have had long discussions on the subject [quaternions]. He is immensely interested in it, and has given a course of lectures on it in Russia, of which he has promised to send copies when printed." (2,III; 149) Thus there was some interest in quaternions among the non-English speaking countries, although naturally it was less than that in Britian and in America.

By 1856 Hamilton had begun work on his *Elements of Quaternions,* which was published in a nearly finished form in 1866, the year after his death.[62] Hamilton was aware that his earlier *Lectures* had not provided a successful presentation of quaternions. In 1859 John Herschel encounaged Hamilton to make the *Elements* an introductory work with examples and problems (2,III; 121), and this approach seems to have been Hamilton's intention. But in 1862 he wrote to a friend (A. S. Hart): "I want to finish a *Book of Reference*—a *short* one, unluckily for sale, it cannot now be—but my intention for myself, and *hope* as regards other writers, is that the *Elements* may be *cited*, almost like the *στοιχεῖα* of Euclid, in future treatises or memoirs on the Quaternions." (2,III; 139) The wish to make it a "*Book of Reference*" was fulfilled, for the *Elements* (even in its unfinished state) was one and one-half times as long as the *Lectures*.[63] This is not altogether sad; a number of mathematicians could have composed an elementary work (and many later did), but probably only Hamilton could have written such a work as the *Elements*.

Hamilton paid much attention to the foundations for the quaternion system. This was valuable in that later workers in quaternions could in their researches turn to, or in their publications refer to, the *Elements* for the detailed and rigorous proof of any fundamental property. Hamilton did not stress his concept of algebra as the "Science of Pure Time," or his ideas on the foundation of complex and higher complex numbers in terms of ordered sets of numbers. In a letter (1861) to De Morgan, Hamilton described many of the important characteristics of the *Elements:*

> As a *book*, I am far better satisfied with the new volume than with the *Lectures*, which, however, is not saying much, for I am very much *dis*-satisfied with them, in point of composition and arrangement—the *συντάξις*, &c.
>
> It must, I own, require a somewhat resolute patience to read the *Lectures through;* but I trust that, although parts may conveniently be omitted at first reading, the new work is so arranged, and *subdivided*, as to be quite easily accessible to any student who has a competent (not a profound) knowledge of former mathematics, and wishes to understand the subject. No new *notation*, for instance, is introduced—and after all, I do not employ many such—without a series of examples following, in numbered sub-articles, to render its meaning and its use familiar. And generally I hope that the progress is well *graduated* throughout; while I have avoided what may be called *talk*, with some care; and have only *one* metaphysical remark, in four and a half lines of a note! (2,III; 568)

The work is certainly clearer than the *Lectures*, although belabored explanations occur frequently in both books. A primary aim of the *Elements* was to do for quaternions what others had done for ordinary numbers and to some extent for complex numbers—to develop them in terms of all the major aspects of mathematics. Thus the logarithms of quaternions were discussed, quaternions raised to quaternion powers, the solution of equations in quaternions, and so on. That the majority of the applications of quaternion methods were geometrical rather than physical was a natural result of the fact that Hamilton (as he himself admitted [2,III; 150]) had lost contact with many developments in physics. This partiality was in some ways unfortunate, for interest in quaternions and vector analysis was to be stronger among physicists than mathematicians.

It is noteworthy that many parts of the *Elements* could be taken over later by vector analysts. Thus for example the first section treated quaternions with zero scalar components (hence vectors) and included a treatment of vector differentiation. Hamilton had planned to devote a large section of the work to his operator $\triangleleft$; however by the time of his death this section had not been written. The development of this operator, so important for physical science, was mainly left for Tait. Hamilton noted in a footnote: ". . . Professor Tait, who has already published tracts on *other* applications of Quaternions, mathematical and physical, including some on Electro-Dynamics, appears to the writer eminently fitted to carry on, happily and usefully, this new branch of mathematical science: and likely to become in it, if the expression may be allowed, one of the chief successors to its inventor." [64] Five hundred copies of the *Elements* were printed in 1866 (2,III; 202), and in the next year Tait's *Elementary Treatise on Quaternions* appeared. Tait had begun this book by 1859, but at Hamilton's request he had withheld its publication until after the publication of Hamilton's *Elements*. (2,III; 133)

VI. *Summary and Conclusion*

The status of quaternion analysis at the time of Hamilton's death in 1865 may now be summarized. By the end of 1865 one hundred and fifty papers had been published on quaternions; one hundred and nine (or 73 percent) were by Hamilton. The remaining forty-one were by fifteen other authors, four of whom were not British. Two books had been published; the first, written by Hamilton, was long and difficult to read; the second, written by Allégret, was less than one-tenth as long and written in French. Two other books had been written; one of these (Hamilton's *Elements*) was long, thorough, and of great value as a reference work for mathematicians; the other (Tait's *Treatise*) was short, readable, and of great value as a text for those who wished to become acquainted with quaternion methods. Together the two works complemented each other. Quaternions were not in 1865 an established branch of mathematics, but by 1865 a foundation for this had certainly been laid.

Hamilton died on September 2, 1865. He continued to work on his *Elements of Quaternions* until a few days before his death. Charles Graves wrote in an éloge that Hamilton's "diligence of late was even excessive—interfering with his sleep, his meals, his exercise, his social enjoyments. It was, I believe, fatally injurious to his health." [65] There is certainly something tragic in the thought of the brilliant Hamilton devoting the last twenty-two years of his life to quaternions, which are *now* of little interest. But judgments on Hamilton's place in the history of science require a more sophisticated basis than that quaternions now seem to be of little importance. The judgment must be made on two bases. The first is whether or not Hamilton acted with insight in light of what was known and could be known in mathematical and physical science at that time. Enough has been said on this that the reader may judge for himself. The second point is whether it has historically turned out that Hamilton's ideas led in any way to fruitful developments. Enough will be said on this that the reader may judge for himself.

Notes

[1] *A Collection of Papers in Memory of Sir William Rowan Hamilton*, ed. David Eugene Smith (*Scripta Mathematica* Studies, No. II). (New York, 1945).

[2] Rev. Robert Perceval Graves, *The Life of Sir William Rowan Hamilton*, 3 vols. and an *Addendum* (Dublin, 1882–1891).

[3] Sir William Rowan Hamilton, "Theory of Conjugate Functions, or Algebraic Couples; with a Preliminary and Elementary Essay on Algebra as the Science of Pure Time" in *Transactions of the Royal Irish Academy, 17* (1837), 293–422.

[4] Sir William Rowan Hamilton, *Lectures on Quaternions* (Dublin, 1853). Three types of page reference are used: (10) means the tenth page of the preface, x means the tenth page of the table of contents, and 10 means the tenth page of the text.

[5] Sir William Rowan Hamilton, "On Quaternions, or on a New System of Imaginaries in Algebra" in *Philosophical Magazine*, 3rd. Ser., *25* (1844), 10–13, 241–246, 489–495; *26* (1845), 220–224; *29* (1846), 26–31, 113–122, 326–328; *30* (1847), 458–461; *31* (1847), 214–219, 278–293, 511–519; *32* (1848), 367–374; *33* (1848), 58–60; *34* (1849), 294–297, 340–343, 425–439; *35* (1849), 133–137, 200–204; *36* (1850), 305–306.

[6] Erwin Schrödinger as quoted in "The Hamilton Postage Stamp: An Announcement by the Irish Minister of Posts and Telegraphs." (1; 82)

[7] J. L. Synge, "The Life and Early Work of Sir William Rowan Hamilton." (1; 17)

[8] E. T. Whittaker, "The Hamiltonian Revival," *Mathematical Gazette, 24* (1940), 153–158.

[9] *Ibid.*, 154.

[10] E. T. Whittaker, "William Rowan Hamilton" in *Lives in Science* (New York, 1957), 61.

[11] E. T. Bell, *Men of Mathematics* (New York, 1937).

[12] See for example Whittaker's "Hamiltonian Revival."

[13] George D. Birkhoff, "Letter from George D. Birkhoff" included in the papers in the "Quaternion Centenary Celebration" in *Proceedings of the Royal Irish Academy, 50* (1944–1945), 72–75.

[14] P. A. M. Dirac, "Application of Quaternions to Lorentz Transformations" in *Proceedings of the Royal Irish Academy, 50* (1944–1945), 261–270.

[15] Otto F. Fischer, *Universal Mechanics and Hamilton's Quaternions, A Cavalcade* (Stockholm, 1951) and *Five Mathematical Structural Models in Natural Philosophy with Technical Physical Quaternions* (Stockholm, 1957).

[16] Bell, *Men of Mathematics*, 404. Tradition has it that Hamilton overindulged in alcohol; there is to my knowledge no unambiguous, direct evidence for this in the writings of Hamilton's contemporaries. The statements that seem relevant are very difficult to interpret; thus Hamilton's chief biographer, R. P. Graves, both was and wrote like a Victorian clergyman. It is clear that Hamilton's marriage was not happy or was his home life well ordered, but it is far from clear that these factors seriously hindered his creativity.

[17] Whittaker, "Hamiltonian Revival," 158.

[18] [Rev. Robert Perceval Graves], "Sir William Rowan Hamilton" in *Dublin Uni-*

versity Magazine, 19 (1842), 94. For proof that Graves wrote the above paper, see (2,II; 344).

[19] William Rowan Hamilton, *Transactions of the Royal Irish Academy, 15* (1828), 69–174.

[20] René Dugas, *A History of Mechanics*, trans. J. R. Maddox (Neuchatel, Switzerland, 1955), 400–401.

[21] Sir William Rowan Hamilton, "Third Supplement to an Essay on the Theory of Systems of Rays" in *Transactions of the Royal Irish Academy, 17* (1837), 1–144.

[22] Augustus De Morgan, "Sir W. R. Hamilton" in *Gentleman's Magazine, 220* (1866), 133. To verify that this unsigned obituary notice was written by De Morgan, see (2,III; 216–217).

[23] C. C. MacDuffee, "Algebra's Debt to Hamilton." (1; 25)

[24] (2,I; 478). For further information concerning Hamilton's reading of Kant's *Critique of Pure Reason* see (2,I; 545, 582, 585) and (2,II; 87–88, 96–97, 342).

[25] It should be mentioned that at nearly the same time Johann Bolyai developed a similar representation of complex numbers. Bolyai wrote down his ideas in 1837 but did not publish them. Hamilton presented his ideas in 1833 and published them in 1837. Paul Stäckel refers to Bolyai's development as inferior to Hamilton's. See Paul Stäckel, *Wolfgang und Johann Bolyai, Geometrische Untersuchungen*, 2 pts. (Leipzig and Berlin, 1913). See Part I, *Leben und Schriften der Beiden Bolyai*, pages 130–133, for Stäckel's commentary; and Part II, *Stucke aus den Schriften der Beiden Bolyai*, pages 223–233, for the document itself.

[26] See, for example, equations 65, 75, 112, 191, 195, and 196 in the second section of the work cited in note (3) above, and equations 6, 56, 57, and 58 in the third section.

[27] (2,II; 574). It seems doubtful that Hamilton's method of "polytets" led directly to the *discovery* of quaternions. However it is probable that this method played a major part in his *acceptance* of quaternions and moreover supported his search for them by indicating the possibility and legitimacy of higher complex numbers. That this is an important point is indicated from the history of ordinary complex numbers, where the difficult task was not *discovery* but *acceptance.*

[28] (4; [35]). John T. Graves was a mathematician and the brother of R. P. Graves, Hamilton's biographer.

[29] Augustus De Morgan, "On the Foundation of Algebra" in *Transactions of the Cambridge Philosophical Society, 7*, pt. 2.

[30] This is proved in Charles Saunders Peirce's addenda to Benjamin Peirce's "Linear Associative Algebra" in *American Journal of Mathematics, 4* (1881), 97–229. The above quotation from C. S. Peirce is from page 229.

[31] Thus $(\mathbf{i}+\mathbf{j})\cdot(2\mathbf{i}+2\mathbf{j})=4$, but so also does $(\mathbf{i}+\mathbf{j})\cdot(4\mathbf{i})$. The term "dot product" will be used throughout without qualification; though this is legitimate historically, it is questionable mathematically. The term "product" in current mathematics is generally not used when the entity resulting from the combination (by multiplication) of two factors is not a member of the same class as the factors, that is, when closure is not preserved. The dot product is now classified as a bilinear functional.

[32] To see that the associative property fails, consider $\mathbf{i}\times(\mathbf{i}\times\mathbf{j})=\mathbf{i}\times\mathbf{k}=-\mathbf{j}$, whereas $(\mathbf{i}\times\mathbf{i})\times\mathbf{j}=0$. To see that the commutative property fails, consider $\mathbf{i}\times\mathbf{j}=\mathbf{k}\neq\mathbf{j}\times\mathbf{i}=-\mathbf{k}$.

[33] Thus, to show that division is not unambiguous, consider $\mathbf{i}\times(\mathbf{i}+\mathbf{j})=\mathbf{i}\times\mathbf{j}=\mathbf{k}$.

[34] It is historically interesting, but probably not historically significant, that in 1841 Hamilton delivered a paper "On the Composition of Forces" in *Proceedings of the Royal Irish Academy, 2* (1844), 166–168. In this paper Hamilton proved (he claimed for the first time) that "the resultant force coincides *in direction* with the diagonal of

the rectangle constructed with lines representing x and y as sides." (*Ibid.*, 168.) He said that the previous (Laplace) proof of the parallogram of forces theorem had dealt only with the magnitude of the resultant force.

[35] The main primary documents dealing with this point are (1) the short summary of his discovery written on the day of the discovery. This is printed (only) in *Proceedings of the Royal Irish Academy, 50* (1944–1945), 89–92. (2) His letter of the next day (October 17, 1843) to John T. Graves printed (only) in the *Philosophical Magazine*, 3rd Ser., *25* (1844), 490–495. (3) Hamilton's mainly historical preface to his *Lectures on Quaternions*. The main historical discussions of this point are (1) C. C. MacDuffee, "Algebra's Debt to Hamilton" (1; 25–35); (2) Peter Guthrie Tait, "Quaternions" in *Encyclopaedia Britannica*, 9th ed., vol. XX (1890), 160–164; (3) Edmund Taylor Whittaker, "The Sequence of Ideas in the Discovery of Quaternions" in *Proceedings of the Royal Irish Academy, 50* (1944–1945), 93–98.

[36] (2,II; 439–440). These are vague glimmerings. Thus he wrote in regard to the quaternion $v + jx + jy + kz$, "xyz may determine *direction and intensity;* while v may determine the *quantity* of some agent such as electricity. x, y, z are *electrically polarized, v electrically unpolarized*. . . . The Calculus of Quaternions may turn out to be a CALCULUS OF POLARITIES." (2,II; 439–440)

[37] See the abstract of De Morgan's "Memoir on Triple Algebra" in (2,III; 251–253). De Morgan's paper was written in 1844 and published in *Transactions of the Cambridge Philosophical Society, 8*, pt. 3.

[38] See (4; [38]). Charles Graves' "On Algebraic Triplets" was published in the *Proceedings of the Royal Irish Academy, 3* (1847), 51–54, 57–64, 80–84, 105–108.

[39] Sir William Rowan Hamilton, "On a New Species of Imaginary Quantities Connected with the Theory of Quaternions" in *Proceedings of the Royal Irish Academy, 2* (1844), 424–434.

[40] Sir William Rowan Hamilton, "On Quaternions" in *Proceedings of the Royal Irish Academy, 3* (1845–1847), 1–16.

[41] Sir William Rowan Hamilton, "On Quaternions" in *Proceedings of the Royal Irish Academy, 3* (1847), 3.

[42] Thus if $\boldsymbol{\alpha} = (x\mathbf{i} + y\mathbf{j} + z\mathbf{k})$ and $\boldsymbol{\alpha}' = (x'\mathbf{i} + y'\mathbf{j} + z'\mathbf{k})$ are two vectors in the modern sense, then $\boldsymbol{\alpha} \cdot \boldsymbol{\alpha}' = xx' + yy' + zz'$ and $\boldsymbol{\alpha} \times \boldsymbol{\alpha}' = \mathbf{i}(yz' - zy') + \mathbf{j}(zx' - xz') + \mathbf{k}(xy' - yx')$.

[43] Sir William Rowan Hamilton, "On Quaternions" (communicated July 20, 1846) in *Proceedings of the Royal Irish Academy, 3* (1847), 273–292, and (5,*31;* 278–293).

[44] Sir William Rowan Hamilton, "Researches respecting Quaternions. First Series" in *Transactions of the Royal Irish Academy, 21* (1848), 199–296.

[45] In the *Athenaeum*, for July 7, 1847, as quoted in (2,II; 587).

[46] In a later part of this work a detailed statistical study is given on the number of papers published, per five-year period, on quaternions.

[47] This is extracted from the summary of the seventh lecture as given in the contents (4; xxxvi).

[48] (2,II; 683). De Morgan made a very similar statement. (2,II; 683).

[49] Unsigned article, "Review of Sir William Rowan Hamilton's Lectures on Quaternions" in *North American Review, 85* (1857), 223–237. My evidence for attributing the authorship to Hill is that this is done in Robert Edouard Moritz, *On Mathematics and Mathematicians* (New York, 1958), 279. David Eugene Smith and Jekuthiel Ginsburg in *A History of Mathematics in America before 1900* (Chicago, 1934) state that Thomas Hill (1818–1891) was a student of Benjamin Peirce (who was an early advocate of quaternions). Hill became president of Harvard in 1862.

[50] *Ibid.*, 223.

[51] *Ibid.*, 228.

[52] *Ibid.*, 226.

[53] *Ibid.*, 229.

[54] Sir William Rowan Hamilton, "Quaternions" in Nichol, *Cyclopaedia of the Physical Sciences*, 2nd ed. (London and Glasgow, 1860), 706–726.

[55] *Ibid.*, 726.

[56] See Benjamin Peirce, *A System of Analytical Mechanics* (New York, 1872), 476–477.

[57] De Volson Wood, "Quaternions" in *Analyst*, 7 (1880), 33.

[58] Charles Graves, "Éloge" on Sir William Hamilton in *The Mathematical Papers of Sir William Hamilton*, vol. I, ed. A. W. Conway and J. L. Synge (Cambridge, England, 1931), xiii. The eloge was given before the Royal Irish Academy in 1865.

[59] Alexandre Allégret, *Essai sur le calcul des quaternions* (Paris, 1862), 7 + 72 pp.

[60] Hermann Hankel, *Theorie der complexen Zahlensysteme* (Leipzig, 1867), 196.

[61] Giusto Bellavitis, "Del Calcolo dei quaternioni di W. R. Hamilton e delle sue relazioni col metodo delle equipollenze" in *Atti del Reale Istituto Veneto di Scienze, Lettere ed Arti* (1858), 334–342, and same author and title, *Memorie di Matematica e Fisica della Società Italiana delle Scienze* (Modena), 2nd Ser., *1* (1862), 126–186.

[62] Sir William Rowan Hamilton, *Elements of Quaternions* (London, 1866), 59 + 762 pp.

[63] Though in terms of pages the two books are comparable, the *Elements* contains by my rough estimate 453,740 words as compared to 304,000 for the *Lectures*. This figure includes the long preface in the *Lectures*, for which there is no counterpart in the *Elements*.

[64] Sir William Rowan Hamilton, *Elements of Quaternions*, vol. II, 2nd ed., ed. Charles Jasper Joly (London, 1901), 350.

[65] C. Graves, "Éloge," xv.

CHAPTER THREE

Other Early Vectorial Systems, Especially Grassmann's Theory of Extension

I. *Introduction*

Hamilton and those who worked with ordinary complex numbers were not the only mathematicians of the time who were searching for vectorial systems. Indeed, at least six other men from four countries were developing systems that were more or less vectorial in character. These men were August Ferdinand Möbius, Giusto Bellavitis, Hermann Günther Grassmann, Adhémar Barré (better known as Comte de Saint-Venant), Augustin Cauchy, and Reverend Matthew O'Brien. By far the most important of these men for the present study is Grassmann, whose first publication of his system, his *Ausdehnungslehre,* came in 1844, the year in which Hamilton published his first paper on quaternions.

It should be remarked that the significance of these men in the history of vector analysis is different from that of Hamilton, for it was only the ideas of Hamilton (as will be shown) that had a major influence on the development of the modern system of vector analysis. The fact that these men had little or no direct influence on subsequent developments raises the important methodological question, "Why should these men be included in the history of vector analysis?"

To this question at least two replies may be given. First, the ideas of these men are of such great originality and merit that they demand attention. It could in fact be argued that Grassmann's creation surpasses that of Hamilton in profundity and perfection. While this reason for their inclusion is certainly important, a second reason seems even more decisive. Chronology may be distinguished from history in that whereas the former presents events in isolation, the latter aims at delineating the "trends of the times" and the causes for events. Coincidence is certainly not the explanation for the fact that more than ten men in six countries, working in the period from the 1790's to the 1850's, sought to create vectorial systems. Though few of these men knew of the ideas of any of the others, nonetheless some factors in the mathematics and physics of this period must have motivated their search. Thus the discussion of these men will be aimed at describing the form of the trend of which they were part and at seeking out the causes that led to their investigations and results.

II. *August Ferdinand Möbius and His Barycentric Calculus*

Though he neither constructed an original vectorial system nor decisely influenced anyone who did, nevertheless August Ferdinand Möbius holds an important position in the history of vector analysis for a number of reasons. The chief of these is that Möbius constructed a mathematical system (his barycentric calculus) which is in many ways similar to vectorial systems. In his system of space analysis, geometrical entities (points rather than vectors) were dealt with directly and (as he hoped) advantageously. Möbius' positioned points, with which numerical magnitudes are usually associated, are added in such a way that both position and magnitude are included in the addition. Thus both in aim and in method his system has kinship with vectorial systems.[10] Furthermore his system as later independently discovered by Grassmann was put foward by Grassmann as an integral part of a system that included vectors. Möbius also carried on an important correspondence with Grassmann (which will be discussed later), and in his later years Möbius did some work directly in the vectorial tradition (although only a small part of this was published during his lifetime). Thus the work of Möbius admits to discussion from a number of points of view.

August Ferdinand Möbius (1790–1868) was born in Schulpforta, educated at Leipzig, Göttingen, and Halle, and began his teaching career in 1815 at Leipzig University, becoming ordinary professor of mechanics and astronomy in 1844, a position he held until his death. He made important contributions to mathematics, mechanics, and astronomy.[11]

The most significant work by Möbius for this study is his *Der barycentrische Calcul, ein neues Hülfsmittel zur analytischen Behandlung der Geometrie dargestellt und insbesondere auf die Bildung neuer Classen von Aufgaben und die Entwickelung mehrerer Eigenschaften der Kegelschnitte,* published at Leipzig in 1827. Herein Möbius gave a full exposition of his calculus of points. Möbius came upon the first notions of his system in 1818, and by 1821 he had decided to publish them in book form.[12] In 1823 he published his first discussion of his new method as a short Appendix to his *Beobachtungen auf der Königlichen Universitäts-Sternwarte zu Leipzig.*[13]

An idea of the nature of Möbius' system may be gained from the following brief exposition.[14] In the foreword to his *Der barycentrische Calcul* Möbius pointed out that the physical concept of center of gravity, or centroid (Schwerpunkt), had been useful ever since Archimedes' time for the discovery of geometric truths. (2; iii) He went on to state:

> The present researches also proceed from the same elementary and purely geometrical concept of the center of gravity. What first stimulated these researches was consideration of the fruitfulness of the law, that each system of weighted points has only one center of gravity, and that thus, in whatever sequence one brings the points into connection, the result is that one and the same point must always be found. The simple technique, by means of which I was able to prove more geometrical laws, stimulated me to find a suitable algorithm for still greater simplification of such investigations. (2; iv)

Early in his first chapter Möbius stated that in his system a line segment from a point A to a point B would be designated by AB, whereas a line from B to A would be designated by BA, or $-AB$. He then indicated how collinear segments were to be added. (2; 3–5) Möbius extended this sign principle to and constructed addition laws for figures determined by more than two points, for example, to triangles (ABC) and pyramids ($ABCD$). (2; 20–23) It is surprising that Möbius nowhere in this work presented addition laws for noncollinear segments and also that he proceeded from such equations as $BDE + BEC + BCD = 0$ to $ABDE + ABEC + ABCD = 0$ (2; 24) without viewing this as a multiplication by A, as Grassmann was to do later.

The central theorem in his book was the following:

> Given any number (ν) of points $A, B, C, \ldots N$ with coefficients $a, b, c, \ldots n$ where the sum of the coefficients does not equal zero, there can always be found one (and only one) point S—the centroid—which point has the property that if one draws parallel lines (pointing in any direction) through the given points and the point S, and if these lines intersect some plane in the points $A', B', C', \ldots N', S'$, then one always has:
>
> $$a.AA' + b.BB' + c.CC' + \ldots + n.NN' = (a + b + c + \ldots + n)SS',$$
>
> and consequently, if the plane goes through S itself, then
>
> $$a.AA' + b.BB' + c.CC' + \ldots + n.NN' = 0. \text{ (2; 9–10)}$$

He later stated that in place of AA', BB', . . . he would write A, B, . . . , and hence the above expression becomes $a.A + b.B + c.C + \ldots + n.N = (a + b + c + \ldots + n)S$. Thus Möbius dealt with weighted points or points with numerical coefficients (positive or negative) which were added as point masses are added in computing the center of gravity of a body. On such a basis Möbius proceeded to his discovery of homogeneous coordinates [15] and to a new treatment of many parts of geometry.

Möbius' highly original and well-presented work was well received, with Cauchy, Jacobi, Dirichlet, Steiner, Plücker, and Gauss all taking some interest in it.[16] However his methods never attained widespread use, and no second edition of the work appeared (except for the republication of the work in 1885 in the first volume of his collected works). Möbius did however continue to publish papers on this subject throughout his life.

In 1843 Möbius published the method for adding and subtracting noncollinear vectors in his *Die Elemente der Mechanik des Himmels*.[17] In a letter to Baltzer he had stated that he was introducing this concept in his lectures in the winter of 1841–1842. Reinhardt implied that Möbius discovered this method himself, but Reinhardt also showed that the method was presented in a June 23, 1835, letter to Möbius from Bellavitis (who will be discussed in the next section). (1,IV; 717–718) It was also in the early 1840's that Möbius came into contact with Grassmann; the discussion of this relationship will best be postponed to the Grassmann sections.

The final work by Möbius that demands attention is his "Ueber geometrische Addition und Multiplication," written in 1862 and revised in 1865, but published only in 1887 in the fourth volume of his collected works. (1,IV; 659–697) This work was mainly derived from Grassmann's work, with changes in notation. Though the work as published in 1887 could have had some influence on the history of vector analysis, such influence is unlikely. Nevertheless his treatment is of interest, since it represents Möbius' final judgment as to what form of vector analysis is most useful.

After the usual discussion of the addition of vectors, Möbius presented his "geometrische Multiplication," which for two vectors AB and CD was represented by $\overline{AB \cdot CD}$. This product was to be numerically equal to the area of the parallelogram, or twice the area of the triangle, determined by these two vectors. The product of the multiplication was the parallelogram or triangle itself and was considered to be an entity that could assume any position in space provided that it always remained parallel to the plane of AB

and CD. The usual sign conventions were used. (1,IV; 663–665) This definition is similar but not identical to the modern definition of the vector (cross) product, since in Möbius' geometrical multiplication the product is a plane figure, not another vector. The two products are of course equal numerically.

It may be noted that any vector uniquely determines a set of mutually parallel planes (the planes perpendicular to that vector) and similarly any plane figure determines a vector (or a set of mutually parallel vectors) perpendicular to that plane figure. Thus the two products are similar but not identical (the two products will be compared in more detail later). Möbius was well aware of these two points of view; indeed, in treating the addition of plane figures he showed that this addition could be represented by the addition of vectors perpendicular to those figures if the lengths of the vectors are equal to the areas of the plane figures. (1,IV; 671–673) Möbius then proceeded to the geometrical product of a vector and a parallelogram, and concluded that the figure produced was the parallelepiped determined by the vector and two adjacent sides of the parallelogram. (1,IV; 673 ff) This was followed by his definition of the projective product of two vectors, which was symbolized by $\underline{AB \cdot CD}$ and is equivalent to modern scalar (dot) product. (1,IV; 678 ff)

The next section treated the relation between the geometric and projective products of vectors in a plane. Herein Möbius began by considering four coplanar vectors, u, v, a, a', whose positions were only limited by the restriction that u and v were to be mutually perpendicular. He then gave the equation $\underline{au} \cdot \underline{a'v} + \overline{av} \cdot \overline{a'u} = 0$, which equation he attempted to prove. It is clear that the left-hand member of this equation represents a number, but it seems that Möbius never defined the product of two areas; hence the interpretation of the right-hand member of the equation is by no means clear. (1,IV; 682–683) In fact, Möbius at a number of points in this manuscript failed either to develop a suitable symbolism or to see that a given expression was ambiguous.[18] He concluded by developing the projective multiplication of solid figures. (1,IV; 685–697)

Though Grassmann was never mentioned, it is clear that Möbius' treatment was mainly derived from Grassmann's efforts. Hamilton is likewise not mentioned, although Möbius had known of his system since at least 1859.[19] It is noteworthy that Möbius only occasionally dealt with matters relevant to his system of point analysis. In general his treatment is almost completely lacking in comments that allow one to judge how Möbius himself viewed his own work; whether, for example, he viewed it as presenting a system superior to Hamilton's. It is clear however that Möbius eventually came to understand and appreciate the significance of the two Grassmann products, which when limited to three-dimensions were precursors of the modern dot and cross products.

III. *Giusto Bellavitis and His Calculus of Equipollences*

Giusto Bellavitis, an Italian mathematician born in 1803, was at the time of his death in 1880 a professor at the University of Padua and a senator in the Kingdom of Italy.[20] He was a man of broad interests but in some cases of narrow view, whose fame in general and relevance for the present study derives from the creation of his calculus of equipollences. What this is may be understood from the following section of a paper (1835) which contained his earliest full explanation of this method.

4. 1.° A straight line (retta) expressed as usual by two letters is understood as taken from the first letter to the second, so that *AB* and *BA* should not be regarded as the same entity, but as two equal quantities having opposite signs.

2.° Two straight lines are called *equipollent* if they are equal, parallel and directed in the same sense.

3.° If two or more straight lines are related in such a way that the second extremity of each line coincides with the first extremity of the following, then the line which together with these forms a polygon (regular or irregular), and which is drawn from the first extremity of the last line, is called their *equipollent-sum* (composta-equipollente). This is represented by the signs + interposed between the lines to be combined, and with the sign $\triangleq$ indicating the equipollence. Thus we have

$$AB + BC \triangleq AC,$$

$$AB + BC + CD \triangleq AD, \text{ etc.}$$

Such equipollences continue to hold when one substitutes for the lines in them, other lines which are respectively equipollent to them, however they may be situated in space. From this it can be understood how any number and any kind of lines may be *summed*, and that in whatever order these lines are taken, the same equipollent-sum will be obtained. . . .

5.° In equipollences, just as in equations, a line may be transferred from one side to the other, provided that the sign is changed. . . .

6.° The equipollence $AB \triangleq n.CD$, where n stands for a positive number, indicates that *AB* is both parallel to and has the same direction as *CD*, and that their lengths have the relation expressed by the equation $AB = n.CD$.

5. Let us restrict ourselves now to lines situated in the same plane. The *inclination* of the line *AB* is the angle *HAB*, which this line forms with

the horizontal AH drawn from left to right, with the qualification that positive angles are measured from the right upwards, and from 0° to 360°. . . .

2.° The angle or inclination of CD on AB is equal to the inclination of CD minus that of AB.

3.° The equipollence

$$AB \triangleq \frac{CD.EF}{GH}$$

requires not only that the lengths AB, CD, etc., should be such as to satisfy the equation into which the equipollence is changed by converting the equipollence sign into an equal sign, but also that

$$\text{inc. } AB = \text{inc. } CD + \text{inc. } EF - \text{inc. } GH \ldots.$$

The line equipollent to 1 is considered as horizontal, that is, as having no inclination. . . .

6. *Fundamental Theorem.* In equipollences, terms are transposed, substituted, added, subtracted, multiplied, divided, etc., in short, all the algebraic operations are performed which would be legitimate if one were dealing with equations, and the resulting equipollences are always exact. As was said in 5.°, non-linear equipollences can only be referred to figures in a single plane.[21]

The first point to be made concerning this passage is that Bellavitis herein described geometrical entities that are in all ways equivalent in behavior to complex numbers as geometrically represented. Numerous mathematicians, including Bellavitis, recognized this equivalence.[22] It should however be stressed that Bellavitis did not view his system as based on the complex number system; indeed, throughout his entire life he was opposed to imaginary numbers as algebraic entities.[23] Thus he viewed his lines as essentially geometrical *entities*, rather than as geometrical *representations*.

One major significance of Bellavitis is that in his papers and book on equipollences he gave numerous, and in many cases ingenious, applications of his method to mathematical and physical problems, so that in a limited sense he was a rival of Grassmann and Hamilton. Bellavitis' chief follower was Charles Ange Laisant, who during the last quarter of the nineteenth century devoted much energy to making the system known in France. It is also of significance that Bellavitis, as the founder of a vectorial system which was different at least in spirit from other systems of the times, may be viewed as another representative of the early nineteenth-century tendency to construct vectorial systems. It is interesting in this regard to note that Bellavitis made a long, unsuccessful attempt to extend his system to three-dimensional space. (3; 158–159)

Bellavitis came to his system in connection with attempts to give a geometrical justification for complex numbers. Thus he stated: "It was while considering a geometrical representation of imaginary numbers proposed by Buée that there came to me (in 1832) the first idea of the method of equipollences, but then I thought that geometrical truths could not rest on the theory of imaginary numbers, which entities I had for some years opposed as unworthy of belonging to a science based on reason alone."[24] Bellavitis had in fact published articles attacking the views of various mathematicians on imaginaries. (3; 159, 163) His first publication (1832) relating to his method, but not containing it in a fully developed form, contained a method of deriving properties of points in a plane from the properties of points in a straight line (for example, he deduced the Pythagorean theorem by means of this method).[25] In this paper he did not use the term "equipollence," but he seems to have used it in an orally presented communication of 1832 at the Ateneo Veneto.[26] A second publication appeared in 1833,[27] and a third in 1835, in which Bellavitis used the term "equipollence" and gave a full exposition of his system.[28] The long quotation given at the beginning of this section was taken from that paper.

IV. *Hermann Grassmann and His Calculus of Extension: Introduction*

Hermann Grassmann was a brilliant mathematician, whose creations in vectorial analysis can only be compared to those of Hamilton. It will be shown that his system could have led to modern vector analysis, but it did not. The reasons for and the nature of this failure will be traced.

Grassmann presented his system in a number of different forms; in fact he wrote four works in which his system was presented, and these four differ substantially among themselves, and moreover, two of Grassmann's most important followers (Schlegel and Hyde) chose a fifth form for their presentation of his system. The part of his system that is encountered in some mathematics books of today is a small and by no means characteristic part. Because of these circumstances a number of Grassmann's works will be discussed rather fully. Many details concerning Grassmann's life and the reception accorded his writings have been included in the hope that they may make clear both how and why his system failed to make as significant an impression on the times as it might have.

What Grassmann created was above all a mathematical system, not just a new mathematical idea or theorem. His creative act cannot be compared with such mathematical discoveries as the Pythagorean theorem or Newton's version of the calculus. Rather it is best thought of as comparable to such creations as non-Euclidean geometry or Boolean algebra. Thus it was natural that Grassmann chose to introduce his system, not by means of a paper, but rather by means of a long and complicated book. Grassmann's system is so broad and so general that there are serious difficulties in trying to summarize it; nevertheless a summary is necessary for the present purposes. The main stress however will be placed on the discussion of such ideas as Grassmann's form of the scalar (dot) and vector (cross) products, which have counterparts in modern vector analysis.

Hermann Günther Grassmann (1809–1877) was born and lived the majority of his life in Stettin (or Szczecin), a town in Pomerania on the Oder River a short distance inland from the Baltic. He was the third of the twelve children born to Justus Günther Grassmann (1779–1852), who, though trained mainly in theology, taught mathematics and physical science at the Stettin Gymnasium.[29] Grassmann, unlike Hamilton, was no prodigy; in fact, quite the opposite. Grassmann's father, it is said, often remarked that he would be happy if Hermann became a gardener or a craftsman. (5; 9)

In 1827 Grassmann entered the University of Berlin, where during his six semesters he mainly studied philology and theology, being especially influenced by the Church historian Neander and Schliermacher; he attended no mathematical lectures, though he did read some mathematical textbooks written by his father.[30] After leaving Berlin, Grassmann returned to Stettin and pursued studies in mathematics, physics, natural history, theology, and philology, preparing himself for the state examinations required for teachers of these subjects. In 1834 he accepted a position at a Berlin technical school, a position which the great geometer Jacob Steiner had just vacated to take a position at the University of Berlin. (5; 45) Grassmann stayed in Berlin for slightly more than a year, during which time he seems to have had little contact with Steiner. At the beginning of 1836 he returned to Stettin and spent the remainder of his life teaching there in various schools, in all cases below the university level despite his constant attempts to attain a university post.

In order to improve his standing as a teacher of science and mathematics, Grassmann in 1839 wrote the Berlin scientific examination committee that he would like to write a work to prove his competence. Thus Grassmann began work on a study of the tides entitled *Theorie der Ebbe und Flut.*[31] He completed his study in

1840, and the chief reader of the essay, Carl Ludwig Conrad, received it on April 26, 1840. The essay was of considerable length (over two hundred pages as printed in Grassmann's works) and was not carefully read by Conrad, for he had returned it by May 1, 1840, the day of Grassmann's examination. Thus it is not surprising that Conrad failed to see the importance of the work.[32] Yet Grassmann's *Theorie der Ebbe und Flut* is important, for it contained the first presentation of a system of spatial analysis based on vectors.

Information concerning the origin of this work may be gained from a letter written by Grassmann in 1847 to Saint-Venant concerning a paper published by the latter in 1845, in which Saint-Venant communicated results identical to results discovered earlier by Grassmann (this paper will be treated later). In this letter Grassmann wrote:

> As I was reading the extract from your paper on the geometric sum and difference, which was published in the *Comptes rendus*, I was struck by the marvelous similarity between your results and those discoveries which I made even as early as 1832. . . .
>
> I conceived the first idea of the geometric sum and difference of two or more lines and also of the geometric product of two or three lines in that year (1832). This idea is in all ways identical to that presented in the extract from your paper. But since I was for a long time occupied with entirely different pursuits, I could not develop this idea. It was only in 1839 that I was lead back to that idea and pursued this geometrical analysis up to the point where it ought to be applicable to all mechanics. It was possible for me to apply this method of analysis to the theory of tides, and in this I was astounded by the simplicity of the calculations resulting from this method. (5; 42–43)

Thus in 1832 Grassmann attained the first ideas for his system, ideas which were developed in the work (1839–1840) on tidal theory. Grassmann explained in greater detail what he had found in 1832 and indicated the source thereof in the foreword to his *Ausdehnungslehre* of 1844.[33] Grassmann wrote:

> The first impulse came from the consideration of negatives in geometry; I was accustomed to viewing the distances *AB* and *BA* as opposite magnitudes. Arising from this idea was the conclusion that if *A*, *B*, *C* are points of a straight line, then in all cases $AB + BC = AC$, this being true whether *AB* and *BC* are directed in the same direction or in opposite directions (where *C* lies between *A* and *B*). In the latter case *AB* and *BC* were not viewed as merely lengths, but simultaneously their directions were considered since they were oppositely directed. Thus dawned the distinction between the sum of lengths and the sum of distances which were fixed in direction. From this resulted the requirement for establishing this latter concept of sum, not simply for the case where the distances were directed in the same or opposite directions, but also for any other case. This could be done in the most simple manner, since the law that

$AB + BC = AC$ remains valid when A, B, C do not lie in a straight line.

This then was the first step to an analysis which subsequently led to the new branch of mathematics, which is presented here. I did not however then realize how fruitful and how rich was the field that I had opened up; rather that result seemed scarcely worthy of note until it was combined with a related idea.

While I was pursuing the concept of geometrical product, as this idea had been established by my father (in his *Raumlehre*, pt. I, 174 and his *Trigonometrie*, p. 10), I concluded that not only rectangles, but also parallelograms, may be viewed as products of two adjacent sides, provided that the sides are viewed not merely as lengths, but rather as directed magnitudes. When I joined this concept of geometrical product with the previously established idea of geometrical sum the most striking harmony resulted. Thus when I multiplied the sum of two vectors by a third coplanar vector, the result coincided (and must always coincide) with the result obtained by multiplying separately each of the two original vectors by the third vector and adding together (with due attention to positive and negative values) the two products. [Thus $A(B + C) = AB + AC$.]

From this harmony I came to see that a whole new area of analysis was opening up which could lead to important results. This idea remained dormant for some time since the demands of my occupation led me to other tasks; also I was initially perplexed by the strange result that though the other laws of ordinary multiplication (including the relation of multiplication to addition) were preserved in this new type of multiplication, yet one could only exchange factors if one simultaneously changed the sign (i.e. changed + to − and − to +).

A work on tidal theory, which I undertook at a later time, led me to Lagrange's *Mécanique analytique* and thereby I returned to those ideas of analysis. All the developments in that work were transformed through the principles of the new analysis in such a simple way that the calculations often came out more than ten times shorter than in Lagrange's work.

This encouraged me to apply the new analysis to the difficult theory of the tides; there were numerous new concepts to develop and to clothe in the new analysis; the concept of rotation led to geometrical exponential magnitudes, to the analysis of angles and of trigonometric functions, etc. I was delighted how thorough the analysis thus formed and extended, not only the often very complex and unsymmetric formulae which are fundamental in tidal theory, come out as the most simple and symmetric formulae, but also the technique of development parallels the concept.

. . . .

Thus I feel entitled to hope that I have found in this new analysis the only natural method according to which mathematics should be applied to nature, and according to which geometry may also be treated, whenever it leads to general and to fruitful results. Thus I decided to make the presentation, extension and application of this analysis a task of my life. When I came to devote all my free time to this task, many of the gaps, left in the more casual earlier development, were filled up. Thus the following ideas (which are presented in this book) resulted: the sum of several points is their centroid; the product of two points is the vector connecting them; the product of three points is the surface area determined by them; the product of four points is the spatial magnitude (a pyramid) determined by them.

The conception of centroid as sum led me to examine Möbius' *Barycentrische Calcul*, a work of which until then I knew only the title; and I was not a little pleased to find here the same concept of the summation of points to which I had been led in the course of the development. This was the first, and as I subsequently learned, the only point of contact which my new system of analysis had with the one that was already known. Since however the concept of a product of points does not occur in that work and since with this concept, when it is combined with that of the sum of points, the development of the new analysis begins, I found that I could not expect any advancement of my ideas from that source. (4,I,I; 7–10)

From this quotation we learn first of all that Grassmann's discussion of the addition of lines is only partly within the "parallelogram of forces" tradition. Thus what Grassmann was discussing was the sum of two directed lines and only that, whereas those who dealt with the parallelogram of forces were not conceptually adding lines but were taking the geometrically determined diagonal as the representation of the resultant of two forces. Similarly those who dealt with sums of complex numbers were not directly conceptually adding lines but representing a sum as a line. Thus to some extent Grassmann's view was new. In any case Grassmann explicitly stated that he was unaware of the geometrical representation of complex numbers until Gauss in a letter of December 14, 1844, referred him to Gauss' paper of 1831.[34]

It is obvious that Möbius' *Barycentrische Calcul* could have suggested some of Grassmann's ideas; thus for example Möbius had the concept that the line *AB* (the line from A to *B*) was equal to the negative of the line *BA* (from *B* to A). Möbius had moreover come to view (before Grassmann) not only lines but also triangles and tetrahedra as positive or negative. It is thus very interesting that Grassmann stated that he had known of Möbius' work only by name until he had laid the foundation for his system. It is above all important to note that Grassmann was not anticipated by Möbius in the key concept for his system, that of geometrical multiplication. Although Möbius wrote such things as $AB = -BA$ (where A and *B* represent points), he failed to conceive of this as a multiplication of points.

In the above quotation Grassmann stated concerning geometrical multiplication that the idea came from his father and in a footnote referred the reader to two of his father's books, "*Raumlehre* Theil II, p. 194" and "*Trigonometrie*, p. 10."[35]

Justus Günther Grassmann made the following statements in a footnote in the first of these books:

> The rectangle itself is the *true geometrical product*, and the construction of it, as it appears in §53, is really *geometrical multiplication.* If the concept of multiplication is taken in its purest and most general sense, then one comes to view a construction as something constructed from elements already constructed and in fact constructed in the same way. Thus multiplication is only a construction of a higher power. In geometry the point is the original "producing" element; from it through construction the line emerges; if we take as the basis of a new construction the finite line constructed from the point, and if we treat it in the same manner as we formerly treated the point, then the rectangle emerges. Just as the line came from the point, so the rectangle comes from the line.
> The situation is the same in arithmetic. In this case the unit is the original "producing" element. The unit must simply be viewed as given. From this, through counting (arithmetic construction), number appears. If this number is taken as the basis of a new counting (setting it in place of the unit), then arises the arithmetic connection to multiplication, which is nothing else but a number of a higher order, a number of which the unit is also a number. Thus it could perhaps be said that the rectangle is a (finite) line in which, in place of the "producing" points, a finite line has been substituted. Thus the following laws may be suggested: *A rectangle is the geometrical product of its base and height and this product behaves in the same way as the arithmetic product.* (4,I,II; 507)

In the second book (the *Trigonometrie*) the elder Grassmann wrote in a footnote:

> If the concept of product is taken in its most pure and general sense, then it is viewed as the result of a synthesis in which an element (produced from an earlier synthesis) is set in the place of the original element and treated in the same way. The product must arise likewise from what resulted from the first synthesis, just as this arose from the original elements. In arithmetic the unit is the element, counting is the synthesis, and the result is number. If this number, as the result of the first synthesis, is set in the place of the unit, and treated in the same way (i.e., counted), then the arithmetic product appears, and this may be viewed as a number of a higher order than a number of which unity is already a number. In geometry the point is the element, the synthesis is the motion of the point in some direction, and the result, the path of the point, is the line. If this line, produced by the first synthesis, is set in the place of the point and treated in the same way (i.e., moved in some other direction), then a surface is produced from the path of the line. This is a true geometrical product of two linear factors and appears in the first place as a rectangle, insofar as the first direction shares nothing with the second. If the surface is set in place of the point, then a geometrical solid is produced as the product of three factors. This is as far as one can go in geometry since space has only three dimensions; no such limitation appears in arithmetic. For further information, see my *Raumlehre,* pt. II (Berlin, 1824). (4,I,II; 507–508)

These quotations justify attributing to Justus Günther Grassmann the honor for being the first to publish the idea of a purely geometrical product. He of course failed to extend the concept into a full-blown system; this his son did, and this was clearly a far greater achievement. It is noteworthy that the first of these statements was published before 1832, that is, before Hermann Grassmann reached the first insights leading to his system, whereas the second and more sophisticated statement appeared after 1832. This fact may however be of small importance, for the quotations may do no more than suggest what might have been communicated between a father and a son who was living in his father's house in 1832. No information seems to be available concerning the interesting question as to the source of J. G. Grassmann's concept of a geometrical product. It would for example be very exciting to find a link between his statements and the geometrical representation of complex numbers tradition or the "parallelogram of forces" tradition.

As Grassmann's *Theorie der Ebbe und Flut* is considered, it is important to keep the following points in mind. First, it was not published until 1911; hence it is at a lower level of historical interest than the first publication of his system, his *Ausdehnungslehre* of 1844. Second, it is of great interest as a source of information concerning how Grassmann came to the much fuller system, which was published in 1844. Third, the presentation of his system was somewhat biased, as he repeatedly stated, by the fact that it was tied to the specific physical problem of tidal theory. In this regard it would be interesting to be able to determine how much of Grassmann's full system was developed with physical applications in mind. At least this much seems clear: his original ideas were engendered, like those of Hamilton, in the course of a mathematical quest, but, unlike Hamilton, Grassmann seems to have forged many of his mathematical tools with physical applications (to tidal theory) in mind.

V. *Grassmann's* Theorie der Ebbe und Flut

The study began with an introduction in which Grassmann surveyed the history of tidal theory and indicated particularly his relationship to Laplace's treatment of the subject in *La Mécanique celeste*. In the first chapter Grassmann stated that he had developed new mathematical methods which he would use in the work. He felt forced however to comment: "Since the scope of the present work does not allow me to present all the principles of the methods of *geometrical analysis* which I have employed, I will borrow the laws (*though they could be developed independently*) from the related laws of algebraic analysis." (4,III,I; 18) This suggests that by

1840 he had developed a substantial part of the system as fully presented in 1844.

Grassmann then considered the law of inertia, which he expressed by the equation "$\frac{dp}{dt} \doteq$ Const." (4,III,I; 18), and he noted that the equality sign with the dot above it signified geometrical equality, that is, that the velocity remained constant in both magnitude and direction. Proceeding to the parallelogram of forces and velocities, he arrived at the statement: "*The velocity* (*S*) *imparted by the combined force is the geometrical sum of the velocities* (*P*, *Q*) *imparted by the individual forces*, or expressed as a formula, $S = P \dot{+} Q$ where $\dot{+}$ is the sign of geometrical summation." (4,III,I; 19) After indicating that more forces (or velocities) could be added in the same way, Grassmann turned to "the laws of geometrical addition and subtraction." (4,III,I; 19) He proved that the commutative and associative laws hold for the addition and subtraction of vectors (Strecken) and argued that since all laws for addition in ordinary algebra can be derived from these two laws, all those laws could now be assumed for vectorial addition. (4,III,I; 20) It should be noted here and kept in mind for later discussions that Grassmann was among the first to realize the full significance of these laws as well as their correlates for multiplication. Grassmann then began the development of the differential calculus of vectors, concluding with the statement: "*All laws of algebraic differentiation and consequently also of integration are likewise valid in geometrical analysis, insofar as they are in fact obtained from no other operations except addition and subtraction.*" (4,III,I; 21) After this he sketched the method for taking partial derivatives for vectors and restated the law of inertia as "$\frac{d^2}{t^2} p \doteq 0$" (which is equivalent, in modern notation, to $\frac{d^2p}{dt^2} = 0$), and he pointed out that this represents three ordinary equations. (4,III,I; 21–22)

Grassmann applied his methods to the physics of the problem and gave for example the vectorial formulation of the center of gravity. From this he returned to define the geometrical product of two vectors.

> By the *geometrical product of two vectors*, we mean the surface content of the parallelogram determined by these vectors; we however fix the position of the plane in which the parallelogram lies. We refer to two surface areas as geometrically equal only when they are equal in content and lie in parallel planes. By the *geometrical product of three vectors* we mean the solid (a parallelepiped) formed from them. As the *sign of*

> *geometrical multiplication* we choose either the point or the sign $\dot{\times}$, while we indicate ordinary algebraic multiplication by writing the two quantities next to each other or by using the sign $\times$. (4,III,I; 30)

As was natural, he showed that the product of three vectors which lie in the same or in parallel planes is zero. He proved both forms of the distributive law and showed that commutativity was replaced by anticommutativity. (4,III,I; 30) Also the expression for the geometrical product was given as the product of the lengths of lines multiplied by the sine of the angle between them (with attention to the sign), and finally the addition of directed areas (Flächenräume) was defined.

Grassmann's geometrical product is quite similar to the modern vector (cross) product. It will be advantageous to specify the similarities and differences for these two products, since the Grassmann product will occur again and since this difference in products constitutes one of the most significant differences between Grassmann's system and the modern system. The two products are identical numerically and likewise correspond in regard to sign conventions. The most significant difference is that in the modern multiplication the product is of the same kind as the factors (all are vectors), whereas in the Grassmann multiplication the product is not of the same nature as the factors. Thus in the Grassmann multiplication of two vectors the product is not another vector, but rather a directed area. It is true that this area (or the set of geometrically equal areas) determines a vector (or set of vectors) perpendicular to that area, and that the vector (or vectors) so defined is precisely that vector which is the product in the modern form of the multiplication. Conversely, the vector (or set of parallel vectors) defines a plane (or set of parallel planes) which is precisely that plane in which the directed area of the Grassmann product lies. Josiah Willard Gibbs and Edwin Bidwell Wilson maintained that the vector product has greater advantages for physical applications, whereas they held that the Grassmann form of the product has a superiority for other uses, as in projective geometry and in multiple algebra.[36]

After reaping some of the harvest made possible by this new technique, Grassmann stated:

> *In fact, there is scarcely a concept in mechanics which without these geometrical concepts can be determined in a general and a simple way. Nevertheless it was not mechanical considerations which led me to these concepts; rather I came to them purely on a geometrical basis, so that the true foundation of these concepts (which I cannot now discuss) and the establishment of their natural development must be given only in a purely geometrical context.* (4,III,I; 33)

Soon after this Grassmann defined what he called the linear product.

> By the *linear product* of two vectors we mean the algebraic product of one vector multiplied by the perpendicular projection of the second onto it. We select the sign $\overset{\frown}{\cdot}$ as the sign of linear multiplication, so that by definition $a \overset{\frown}{\cdot} b = ab \cos (ab)$. From this definition and since $\cos (ab) = \cos (ba)$, we see that $a \overset{\frown}{\cdot} b = b \overset{\frown}{\cdot} a$. (4,III,I; 40, 212)

He then proved that the distributive law holds for this product, and explained that the linear product of $(a \dotplus b \dotplus c)$ with $(a_1 \dotplus b_1 \dotplus c_1)$ where a, b, c and a_1, b_1, c_1 are two sets of mutually perpendicular vectors and a is parallel to a_1, b to b_1, c to c_1, is equal to $aa_1 + bb_1 + cc_1$. (4,III,I; 40) Finally he stated that since the commutative and distributive laws were preserved in this multiplication, all laws of algebraic multiplication were applicable. It should be clear that this product is identical to the modern scalar product.

Throughout the remainder of his work Grassmann applied these methods to tidal theory with some success. In later portions he introduced other new, but less relevant, mathematical methods, including the linear vector function. (4,III,I; 79 ff.)

In summary, Grassmann presented in this work a strikingly large part of vector analysis. Vector addition and subtraction, the two major kinds of vectorial products, vector differentiation, and the elements of the linear vector function were all presented in forms either equivalent or nearly equivalent to their modern counterparts. Grassmann's work was more than the first important system of vector analysis; it was also the first major work in the new algebra that was dawning at that time. The revolutionary nature of this new algebra may be compared with that of non-Euclidean geometry. Paradoxically Grassmann's development of his system as applied to three dimensions led him to present his system in his 1844 *Ausdehnungslehre* in a form applicable to n-dimensions, which in one sense constituted a discovery of non-Euclidean geometry.[37] This was a magnificent achievement for a mathematician in his early thirties who had never attended a university mathematical lecture.

VI. *Grassmann's* Ausdehnungslehre *of 1844*

Between 1840 and 1844 Grassmann published some language textbooks and a long mathematical paper which did not pertain to his new method. Around Easter, 1842, Grassmann began to turn his full energies to the composition of his *Ausdehnungslehre* (5; 91–92), and by the fall of 1843 he had finished writing it. (5; 95) That Grassmann required such a short time for the compositon of the book is doubly striking since he wrote it while involved in numerous other activities and since, according to a remark in his foreword, the presentation was reworked in many different forms before he selected the form actually employed. (4,I,I; 16) Difficulty in choosing the best form of presentation was caused in large part by the fact that sometime after 1839, and probably in 1842,[38] Grassmann discovered new implications and extensions of his earlier discoveries. He wrote in the foreword:

> When I thus proceeded to work out, coherently and from the beginnings, the results that I had found, being careful to appeal to no principle proven in any branch of mathematics, it resulted that the analysis which I discovered did not touch, as it had seemed before, on the area of geometry. Rather I soon realized that I had come upon a region of a new science of which geometry itself is only a special application.
>
> It had for a long time been evident to me that geometry can in no way be viewed, like arithmetic or combination theory, as a branch of mathematics; instead geometry relates to something already given in nature, namely, space. I also had realized that there must be a branch of mathematics which yields in a purely abstract way laws similar to those in geometry, which appears bound to space. By means of the new analysis it appeared possible to form such a purely abstract branch of mathematics; indeed this new analysis, developed without the assumption of any principles established outside of its own domain and proceeding purely by abstraction, was itself this science.
>
> The essential advantages which were attained through this conception were 1) in relation to the form—now all principles which express views of space are entirely omitted, and consequently the beginnings of the science were as direct as those of arithmetic—and 2) in relation to content—the limitation to three dimensions is omitted. Only herein do the laws come to light in their full clarity and generality and reveal themselves in their essential mutual relationships, and many regularities, which for three dimensions present themselves either not at all or only obscurely, now present themselves with this generalization in full clarity. (4,I,I; 10–11)

Thus in 1844 Grassmann's book appeared under the full title: *Die lineale Ausdehnungslehre, ein neuer Zweig der Mathematik dargestellt und durch Anwendungen auf die übrigen Zweige der Mathematik, wie auch auf die Statik, Mechanik, die Lehre vom Magnetismus und die Krystallonomie erläutert.* As it was mentioned before, Grassmann had no array of titles to put after his name, as had Hamilton when nine years later in 1853 he published the first full treatment of his system. Under Grassmann's name on the title page appeared only "Lehrer an der Friedrich-Wilhelms-Schule zu Stettin." Some idea of the number of copies of Grassmann's book that were printed, as well as some idea of its reception, may be obtained from the following quotation taken from a letter written to Grassmann in 1876 by the publisher of his first *Ausdehnungslehre:* "Your book *Die Ausdehnungslehre* has been out of print for some time. Since your work hardly sold at all, roughly 600 copies were in 1864 used as waste paper and the remainder, a few odd copies, have now been sold with the exception of one copy which remains in our library." [39]

As Grassmann's book is discussed, it is important to keep in mind that not only its richness but also its readableness must be considered. To put it briefly, the analysis of the work will lead to the conclusion that the work is a classic, that it is very difficult to read, and that it contains a large portion of modern vector analysis, which was however deeply embedded within a far broader system, and embedded in such a way that it could be extracted only with difficulty.

Grassmann included at the beginning of his work a philosophical introduction which is both interesting and historically significant, for it is one of the major barriers beyond which many mathematicians of his time did not pass.

Grassmann began this introduction by stating:

> The primary division in all the sciences is into the real and the formal. The former represent in thought the existent as existing independently of thought, and their truth consists in their correspondence with the existent. The formal sciences on the other hand have as their object what has been produced by thought alone, and their truth consists in the correspondence between the thought processes themselves.[40]

Concerning proof in the formal sciences Grassmann stated: "Proof in the formal sciences thus does not proceed outside of the sphere of thought itself; rather it remains purely in the combination of different acts of thought. Thus it is that the formal sciences need not begin with axioms; instead definitions form their foundation." (4,I,I; 22) The special relevance of this view becomes clear with the following two statements:

> Pure mathematics is thus the science of the particular existent which has come to be through thought. The particular existent, viewed in this sense, we name a thought-form or simply a form. Thus pure mathematics is the theory of forms. (4,I,I; 23)
>
> Before we proceed to the division of the theory of forms, we have to separate out one branch which has hitherto incorrectly been included in it. This branch is geometry. From the concepts set out above, it is evident that geometry as well as mechanics refers to a real existent; for geometry this is space. This is clear since the concept of space can in no way be produced by thought, but rather comes forth as something given. (4,I,I; 23)

What Grassmann believed he had discovered was thus a system, a purely formal system, which was above and independent of geometry and indeed of all mathematics as known at that time. It was a sort of universal algebra[41] in which the elements had no necessarily real content but could take on geometrically significant content when such was desirable. His "Theory of Forms" was indeed marvelously general and abstract.

Grassmann continued in the introduction to lay the philosophical foundation for his system. He specified four concepts with which his theory of forms could develop all of mathematics; these were the discrete and the continuous, the equal and the different. (4,I,I; 24–25) According to Grassmann, from the combination of the last of each of these pairs of opposites—that is, the continuous and the different—comes forth his "Ausdehnungslehre," that doctrine which includes the notion of difference (for example, the different dimensions of space) and in which the elements (for example, points) vary *continuously* and hence produce different entities (as the line). (4,I,I; 26–27) Thus Grassmann stated:

> Space theory [die Raumlehre] may again serve as an example. Here in two different directions from an element all the elements of a plane are produced when the producing element progresses in any way according to two directions. The totality of points (elements) producible in this way is comprised in one element [the plane]. The plane is thus the system of the second order; in it there is an infinite collection of directions dependent on those two original directions. If a third independent direction is added, then by means of this direction infinite space (as the system of the third order) is produced. Up to now one could not go beyond three directions; however in the pure theory of extension the number of directions can increase to infinity. (4,I,I; 29)

The above gives some idea of the philosophical basis that Grassmann attempted to construct for his system. The introduction was concluded with a discussion concerning the best form of presentation for the work; it need only be noted that Grassmann was well aware that a reader's entry into his system would not be easy.

Whatever may be the absolute value of Grassmann's philosophical underpinning of his system, it is important to keep in mind that it seems to have allowed Grassmann himself to feel at ease with new mathematical ideas, for example, noncommutative multiplications and n-dimensional spaces. In this regard it may be compared with Hamilton's speculations on algebra as the science of pure time. Finally, Grassmann's penchant for the abstract and the general did certainly lead to many ideas of value, a fact evident in the next section of his book.

This section was entitled "Survey of the General Theory of Forms." He began with the statement:

> By the general theory of forms I mean that series of truths which relate to all branches of mathematics in the same way, and thus presume only the general concepts of equality and difference, connection and separation. The general theory of forms should thus precede all special branches of mathematics. Since however that general branch does not as such exist yet, and since we cannot omit it without entangling ourselves in useless ramblings, we have no choice but to develop this subject insofar as we need it for our science. (4,I,I; 33)

This statement introduced what was one of the most difficult sections of the book for the reader of the time; in this case it was mainly the richness of content and the newness of the concepts that made this very important section difficult. Herein Grassmann proceeded to construct the basis of his theory of forms by postulating certain initially contentless forms (or magnitudes) and joining these forms by certain connections (Verknüpfungen). Thus given two forms a and b, Grassmann first introduced a connection symbolized by $\frown$ to yield a new form $a \frown b$. Grassmann specified that this connection was of such a nature that the following two equations could be set up:

$$a \frown b = b \frown a$$

$$(a \frown b) \frown c = a \frown (b \frown c) = a \frown b \frown c$$

He showed that if equations were set up in which any number of forms were connected, then one could in all cases rearrange the members and omit the parentheses. Grassmann called such a connection a synthetic connection (4,I,I; 34–36) and from it proceeded to his analytical connections. For Grassmann an analytical connection connects two forms in such a way that when the resultant form is synthetically connected with one of the original forms, the other original form results. Thus when $\smile$ is taken as the sign for analytical connection and a and b as the two forms, then $(a \smile b)$ synthetically connected to b gives a, or, in short, $(a \smile b) \frown b = a$. Grassmann then showed that the following equations are true:

$$a \smile b \smile c = a \smile c \smile b = a \smile (b \frown c)$$

$$a \smile (b \smile c) = a \smile b \frown c$$

Grassmann further specified this form of analytic connection by the assumption that the result of an analytic connection was to be unique or unequivocal; thus, for example, if $a \smile b = c$, then if a or b remains constant and the other changes, the result (c) must change. Finally he showed that these results could be extended to connections of any number of forms. (4,I,I; 36–38)

These statements must have appeared very strange to the reader of the times. At the very end of this development Grassmann stated what should have considerably enlightened the matter: when these connections obeyed the above laws, they could be called addition and subtraction.

Grassmann then showed that two new forms, the indifferent and the analytic form, could be produced. When any form was analytically connected to itself—for example, $a \smile a$—then the indifferent form resulted (indifferent in the sense that a could have any value). He represented this new form by $\backsim$ and stated that when this form is connected to b, producing $(\backsim \smile b)$, then the new form $(\smile b)$, called the analytic form, emerges. Finally Grassmann admitted that the indifferent form could be called "Null" (i.e., zero) and the analytic form could be called the negative form. (4,I,I; 38–40)

Grassmann went on to develop two further connections, one of which was synthetic and the other analytic. Both of these connections were of such a sort that when a form was applied through such a connection to two forms connected in either of the previous manners, a transformation would take place without a change in the result. His new kind of synthetic connection he symbolized by $\overset{\frown}{\frown}$ and postulated the following: $(a \frown b) \overset{\frown}{\frown} c = a \overset{\frown}{\frown} c \frown b \overset{\frown}{\frown} c$. Such a connection, Grassmann stated, could be called multiplication. (4,I,I; 41–43) Note that this implies that Grassmann will call any operation that is distributive a multiplication, and what he sought in his work was, broadly speaking, distributive operations, which need not be either commutative or associative.

Grassmann then introduced the second kind of analytical connection (i.e., division). This was limited by the law expressed in the equation

$$\frac{a \mp b}{c} = \frac{a}{c} \mp \frac{b}{c}$$

He pointed out that unlike the former type of analytic connection, three forms connected by division are not in all cases uniquely determined (thus, for example, if $\frac{a}{b} = c$ and $a = 0$, then b is not uniquely determined). (4,I,I; 43–44)

To the above summary it need only be added that according to Grassmann similar forms (forms of the same order, thus two points, two vectors, and so forth) when connected by either of the first two kinds of connection give in general forms of the same order, whereas in multiplication forms of the same or different orders may be combined and in general produce forms of a higher order. Thus the product of two vectors (forms of the first order) is a directed area (a form of the second order).

The section just discussed combined with the introduction to present a nearly impassable barrier for most mathematicians of the time. Ideas such as these were so new, so abstract, and at first sight so useless that it is not difficult to understand why Möbius regarded the work as simply unreadable, Baltzer found it made him dizzy, and Hamilton was led to write to De Morgan that to be able to read Grassmann he would have to learn to smoke.[42] However these passages are not only characterized by obscurity and abstractness; they also reveal (to a modern mathematician) the brilliance of a powerful mind. Grassmann had come to understand the associative, commutative, and distributive laws more fully than any earlier mathematician, and this understanding is reflected in the construction of his system.

Grassmann allowed his initially contentless forms to assume many values—numbers, points, vectors, oriented areas, and so forth; moreover he eventually[43] developed sixteen kinds of multiplicative connections. All these forms and connections were understood in terms of the laws set out by Grassmann in this section, and thus this section was one of the most important. Having laid the foundation, we may turn to the main contents of the book.[44]

The first chapter of part one of Grassmann's book centered on the production of his various systems. New systems were created by taking an element from another system and allowing it to undergo a new variation; thus for example a point (the element) is moved in a single direction (the variation) to produce a straight line (a system of the first order). If this line now varies by moving in some (rectilinear) direction, a plane (a system of the second order) is produced. According to Grassmann the same process can be extended to form systems of any order (hence his "any" dimensional spaces). In this chapter and within this framework Grassmann introduced vectors, proved that they obey his previously developed addition and subtraction laws, and explained vectorial addition. He furthermore proved various algebraic properties for his systems. Typical of his statements is the following: "*Every vector of a system of the mth order can be expressed as the sum of m vectors, which belong to the m given independent manners of change of the system. This expression is unique.*"[45]

The first chapter was concluded, as were all others, with a section on application of the theoretical ideas presented. For this chapter the applications were mainly two: Grassmann showed how his ideas could be used to lay a secure foundation for geometry and he gave the vectorial method of representing the center of gravity of a body as well as the laws of force and velocity for the center of gravity.

The second chapter of this part was entitled "Outer Multiplication of Vectors." Grassmann's outer product is essentially equivalent to what had been called the geometrical product in his 1840 work, and hence it is similar to the modern vector (cross) product. Grassmann mentioned his linear product (renamed inner product) only in an offhand manner in the foreword to the book. Since this was the first published definition of the modern scalar (dot) product, we shall give it in full. After Grassmann had concluded a discussion of his outer product, he wrote:

> Besides this concept there is another that likewise relates distances to fixed directions.
>
> Namely, when I took the perpendicular projection of the one distance onto the other, the arithmetic product of this projection and the distance (onto which the projection was made) represented the product of those distances, provided that the multiplicative relation to addition was valid. But this product was of an entirely different kind than the first, in that the factors could be exchanged without changing the sign and the product of two mutually perpendicular vectors was zero. I called the first the outer product and the second the inner product, since the latter has non-zero values only when the directions approach one another, that is, when the distances lie partly in each other.[46]

Grassmann's outer product was presented in his *Ausdehnungslehre* of 1844 in a manner both different from and more general than the presentation in his study of the tides. Grassmann began in a simple way by stating:

> We shall begin with geometry in order to secure an analogy according to which the abstract science must proceed, and thus obtain a clear idea to escort us along the unknown and arduous path of abstract development. We go from the vector to a spatial form of higher order when we allow the entire vector, that is each point of the vector, to describe another vector which is heterogeneous to the first, so that all points construct an equal vector. The surface area produced in this way has the form of a parallelogram. Two such surface areas which belong to the same plane are designated as equal if the direction of the moved vector lies in both cases on the same side (for example, on the left side) of the vector produced through the motion. When in the two cases the corresponding vectors lie on opposite sides, then the surface areas are designated unequal. Thus we have a simple and general law:
>
> *If in the plane a vector moves successively along any series of vectors, then the total surface area produced thereby (provided that the signs of the individual surface elements are set in the given manner) is equal to*

that area which would be produced if the vector had moved along the sum of those vectors. (4,I,I; 77–78)

Statements as concrete as the above were not typical of the work nor always desirable if we consider the nature of the work, for if geometrical intuition had been allowed to enter too early, it could have hindered the understanding of later developments which went beyond the usual three-dimensional geometry. To illustrate the above statement, Grassmann considered three coplanar parallel lines (*cd*, *ef*, and *ab*) which were cut by three pairs of

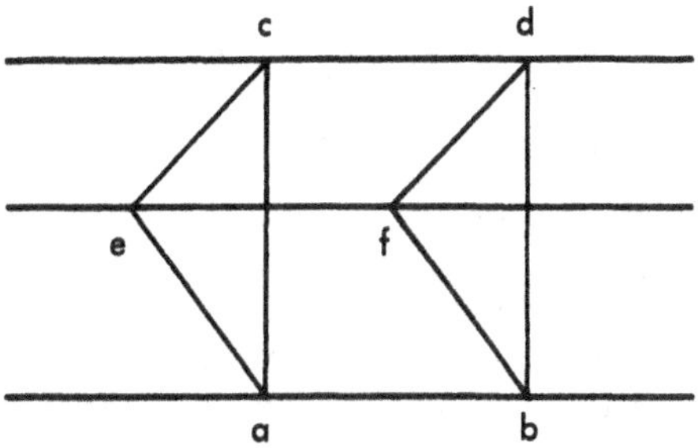

parallel lines: *ec* and *df*, *ea* and *fb*, *ca* and *db*. Letting *ab* represent a vector constrained to move parallel to itself, he pictured *ab* moving along *ac* until it reached line *cd*. Thus the area *acdb* was swept out. It will be seen by simple geometry that this area is equal to the sum of the areas of the parallelograms *aefb* and *ecdf*, that is, the areas swept out when *ab* moves along *ae* to *ef* and then along *ec* to *cd*. Moreover the resulting areas will be the same, he stated, if *ab* moves along *ae* until it coincides with *ef* or if it moves along *ac* and then *ce* as long as the reversal in direction *ac* is compensated for by taking the area resulting from *ab*'s motion along *ce* as negative. Grassmann then noted that since for example the motion of *ab* along *ac* gave the same area as when *ab* moved along *ae* and then *ec*, this was an operation that should be viewed as multiplication on the basis of his theory of forms.[47]

From the above one might easily infer that his outer product of 1844 was in all ways equivalent to his geometrical product of 1840. This is true only in part, since (as it will be shown presently) his geometrical product of 1840 was only one species of the more general outer product of 1844. Thus his geometrical product, similar to the modern vector (cross) product, was in the 1844 work embedded within a far broader system. This situation is typical of many of the elements of vector analysis developed in the *Ausdehnungslehre* of 1844 and is an important reason why it would have been difficult to extract the modern system of vector analysis from his system. At the end of his introductory section Grassmann stated: ". . . we can now return to our science in order to pursue it according to a purely abstract manner and independently of all spatial considerations." (4,I,I; 80)

At this point it will be helpful to consider in modern terms what Grassmann meant by his outer product. It was a product of base elements $e_1, e_2, e_3, \ldots, e_n$ which obey the following laws: letting x, y, z take on any values of n,

$$e_x e_y = -e_y e_x$$

$$e_x e_x = e_y e_y = 0$$

$$e_x(e_y + e_z) = e_x e_y + e_x e_z$$

It should be noted that the last equation need not have been given since the outer product must be distributive, for this property is contained for Grassmann in the definition of "product." The product of N such elements was considered to be an entity of the Nth order. Here again Grassmann's outer product differs from the modern cross product, since the result of the cross multiplication of two vectors is another vector (in Grassmann's terms, an entity of the first order), whereas the outer product of two vectors is an entity of the second order. This difference between the two products is especially important in that Grassmann's outer product allowed him to generate all sorts of new entities which he duly considered.

Grassmann proceeded to discuss the product of any number of vectors. Grassmann stated that a product such as $a.b.c\ldots$ was to mean that the vector a first moves along b (as before), then the resultant oriented area would move along c, and so on through orders higher than the third. Grassmann proved that such relations were distributive and that entities thus arising could be added under certain conditions. As each new entity arose, he checked its properties in relation to his general theory of forms.

Some of Grassmann's applications of his outer product merit discussion. Immediately after his full presentation of the outer product Grassmann showed how Varignon's principle could be represented in terms of the outer product in a very simple manner, and he in general discussed the representation of moments by this product.[48] He also showed how the outer product could be used in solving n first-degree equations in n unknowns. (4,I,I; 99–102) Then he considered what he called outer division, and through outer division ordinary numbers ("Zahlengrösse") entered his system for the first time. (4,I,I; 118–137)

In the last chapter of the first part he dealt with a process he called "Abschattung," or, loosely speaking, projection. Herein he was primarily concerned with the conditions under which the original equation could be definitely inferred from the resulting equation when an equation among extensive magnitudes was multiplied by an extensive magnitude. The results obtained in this chapter were applied to spatial projections and the solution of equations.

The majority of the ideas considered up to this point were given by Grassmann in the five chapters of the first section (Abschnitt) of his book, a section entitled "Extensive Magnitude" ("Die Ausdehnungsgrösse"). The second of the two sections of the book was entitled "Elementary Magnitude" ("Die Elementargrösse"). This section began with a lengthy discussion of elementary magnitudes, a term by which Grassmann referred mainly to points. Discussion of Grassmann's point analysis is relevant both directly, in that it was a prominent part of his system, and indirectly, in that it had numerous relationships to his system of vector analysis. His system of point analysis, having been discovered independently of Möbius, was presented differently and developed more fully.

Grassmann began this section by considering the addition and subtraction of elementary magnitudes, although he seems never to have stated precisely what he meant by elementary magnitudes except by implication. Early in the discussion he stated: "In order to bring this result into sharper focus, we will apply it to geometry and thus take the elements as point . . . ." (4,I,I; 158) Grassmann proceeded to write such equations as the following, in which Greek letters represent elementary magnitudes, and the *i*'s, *k*'s and *n*'s represent numbers:

$$i_1[\rho\alpha_1] + \ldots + i_n[\rho\alpha_n] = k_1[\rho\beta_1] + \ldots + k_n[\rho\beta_n] \qquad (4,I,I;\ 159)$$

This is essentially a vector equation (in which $\rho\alpha_1$, $\rho\alpha_2$ . . . are the vectors). Grassmann went from this to show that the equation remained true when any other element σ was substituted for ρ. He then established that the members of such an equation satisfied both the commutative and associative laws for addition and subtraction, that these operations gave unique results, and that the distributive law was satisfied for the multiplication of the element pairs such as $[\alpha\beta]$ with numbers. On the basis of these facts Grassmann showed that in the equation given above, the ρ's could be deleted to yield an equation such as $a\alpha + b\beta$ . . . where a, b, . . . represent numbers; α, β, . . . elements or points; and $a\alpha$ was considered a weighted point. (4,I,I; 160–163)

Grassmann, unlike Möbius, developed his system of point analysis in close relation to vectorial ideas and used in fact previously established vectorial laws in proving laws for points. He then showed that the sum of a number of weighted points could be represented by a single point whose weight was equal to the sum of the weights of the individual points. (4,I,I; 164) This led Grassmann to consider a point of 0 weight, which he showed was best interpreted as a vector. (4,I,I; 164–166) The final topic in the theoretical part of this chapter concerned the result obtained when a point or a multiple point was added to a vector. (4,I,I; 166–167)

The applied parts of the chapter dealt with such topics as geometrical methods for constructing sum points and physical applications of his point analysis. In this part of the chapter Grassmann made reference to Möbius:

> We are here at the first and only point at which our science touches on ideas already known. Namely in the barycentric calculus of Möbius an addition of simple and multiple points is presented. Möbius used it primarily only as an abbreviated method of expression but developed the same techniques of calculations which we presented (in greater generality) in the first paragraphs of this chapter. What is entirely lacking in Möbius presentation is the conception of sum as a magnitude in the case where the weights together total zero.
>
> What hindered the sagacious author of that work in viewing this sum as a vector of constant length and direction is certainly the strangeness of compounding length and direction into one concept. If that sum had been determined to be a vector, then the concept of the addition and subtraction of vectors for geometry (as we have presented it in Section I, Chapter I) would have arisen, and thus our work would have had a second point of contact with Möbius' work; also the barycentric calculus would have obtained a much freer and more general development. (4,I,I; 172)

In the second chapter of the second section of his book Grassmann introduced the outer multiplication of points. In the first chapter he had stated that $[\alpha\beta]$ should be interpreted as equal to $\beta-\alpha$, that is, as a vector from α to β. (4,I,I; 165–166) However $[\alpha\beta]$ should be viewed, not simply as the vector $\alpha\beta$, but as a vector limited to positions in the line through α and β, that is, (to use the modern term) as a line-bound vector. (4,I,I; 179) Since $[\rho\rho]=0$ (the line from a point to the same point is no line) and since $[\rho\alpha]=-[\alpha\rho]$ (because $[\rho\alpha]=\alpha-\rho=-(\rho-\alpha)=-[\alpha\rho]$), $[\alpha\beta]$ should be viewed as an outer multiplication. (4,I,I; 175)

It is not necessary for present purposes to follow in detail Grassmann's highly abstract and complicated development of the outer product for points. Rather it need only be noted that here as before Grassmann continually made use of vectors and vector relations in developing the outer product. Finally this section may be summarized through the use of Grassmann's own summary as presented in his paper of 1845 entitled "Kurze Uebersicht über das Wesen der Ausdehnungslehre." [49]

If A, B, C, D, are points, then we mean by

(1) $A.B$, the line, which has A and B as extremities, regarded as a definite part of the infinite right line determined by A and B;

(2) $A.B.C$, the triangle, whose vertices are A, B, C, regarded as a definite part of the infinite plane determined by A, B, C;

(3) $A.B.C.D$, the tetra[h]edron, whose vertices are A, B, C, D, regarded as a definite part of infinite space.

That is, we put $A.B = A_1.B_1$, when both products represent equal p'ts, with like signs, of the same right line; further

$$A.B.C = A_1.B_1.C_1,$$

when both triangles are equal parts, with like signs, of the same plane; and finally

$$A.B.C.D = A_1.B_1.C_1.D_1,$$

when both tetra[h]edrons have equal volumes, with like signs.[50]

The applied section of this chapter treated such topics as co-ordinate systems, transformations of co-ordinates, the equation of a plane in terms of points, and numerous applications to statics, including the representation of static moments and equilibrium conditions.

The materials discussed to this point constituted slightly more than two-thirds of Grassmann's book. The next to the last chapter dealt with what Grassmann called "Das eingewandte Produkt," or the regressive product. He pointed out that the outer product was characterized by the property that the product of two dependent factors is zero; since this property was not essential to the ideas of a product, a new product could be defined which was free of this restriction. (4,I,I; 206) Thus Grassmann introduced his regressive product, or, more accurately, he extended the concept of product in such a way that the regressive and outer products could be treated as two forms of one product. As Engel pointed out in his notes, this complicated the presentation. (4,I,I; 408)

An idea of Grassmann's regressive product may be gained from the following. It is natural to ask if any meaning may be assigned to the product of two factors, some of whose components are mutually dependent. Similarly any multiplication of factors may be viewed as taking place within a system of some order, for example, three- or four-dimensional space. When the sum of the orders of the factors exceeds the order of the space under consideration, the factors must be dependent. It was for this case that Grassmann defined the regressive product. Two quotations from Grassmann will explicate this further; the first quotation is from his "Kurze Uebersicht" and is more general than the second, which is from the latter part of the chapter under discussion and which gives a geometrical interpretation of the regressive product.

> Henceforth outer multiplication will be designated by simply writing the factors together; regressive multiplication, by a point placed between the factors. *We understand by the regressive product AB.AC, where A, B, C are any magnitudes, the product ABC.A, in which ABC is treated as a co-efficient belonging to A, provided that the product be referred to the field of lowest order in which A, B, and C lie at the same time.* (4,I,I; 310–311)
>
> The product of two line magnitudes in the plane is the point of intersection of the two lines joined with a part of that plane as factor; if for example *ab* and *ac* (where *a*, *b*, *c* represent points) are the two line magnitudes, then their product is *abc.a*. Furthermore the product of three line magnitudes in the plane is equal to the doubled surface content (set twice as factor) of the triangle enclosed by the lines, multiplied by the product of the three quotients, which express how many times each side is contained in its corresponding line magnitude; thus if *a*, *b*, *c* are those three points, and *mab*, *nac*, *pbc* (when *m*, *n*, *p* are numerical magnitudes) are the three line magnitudes, then their product is equal to
>
> $$mnp.abc.abc$$
>
> The product of two plane magnitudes in space is part of the intersecting edge multiplied by a part of space, for example, $abc.abd = abcd.ab$. Furthermore the product of three plane magnitudes is the intersection point of the three planes, multiplied by two parts of space, for example, $abc.abd.acd = abcd.abcd.a$. (4,I,I; 243–244)

The above description, which is sufficient for the present purpose, does scant justice to the complexities of this chapter. This is partly mitigated by the fact that some ideas in this chapter were deleted in the *Ausdehnungslehre* of 1862, and many of the developments were approached from a different point of view. The chapter concluded with geometrical applications.

The final chapter in the book was devoted mainly to applications of the previous ideas to geometry and to crystallography. The final section of the chapter was entitled "Remark on the Open Product." This section is of special importance since Gibbs argued that it contained "the key to the theory of matrices . . ." and the linear vector function.[51]

Before the reception accorded to Grassmann's work is discussed, it will be useful to state the major conclusions emerging from the above detailed discussion of his first *Ausdehnungslehre*. First, it should be evident that it was a work of great brilliance, a work that along with much else contained a large number of the ideas involved in modern vector analysis. It should also be clear that even if priority disputes are judged by date of publication (thus excluding Grassmann's *Theorie der Ebbe und Flut*), Grassmann's ideas as published in 1844 were far more extensive and richer than those of Hamilton at the same time. There is merit to Sarton's remark that Grassmann's *Ausdehnungslehre* of 1844 should be compared with Hamilton's *Lectures on Quaternions* of 1853, not with Hamilton's early researches.[52] Grassmann's book is a great classic in the history of mathematics, even though one of the most unreadable classics. The work was exceedingly abstract and very complicated and, perhaps most importantly, it departed from all the then current mathematical traditions. Even in the early sections of the book the reader encountered highly original ideas placed in a philosophic setting and at most tenuously attached to the mathematical ideas of the time.

In order for a mathematician of that time to derive the modern system of vector analysis from Grassmann's book, he would have to (1) read and understand the book (no small task), (2) delete major mathematical portions of the book (such as point analysis), (3) limit the presentation to three-dimensional space, (4) redefine some of the fundamental ideas (such as the outer product), (5) change the structure and emphasis of the work, (6) detach the presentation from the philosophical ideas contained in it, and (7) attach to it ideas already in the literature of the times, but unknown to Grassmann, such as the development of the geometrical representation of complex numbers and the theorems of Green and Gauss. No one at that time or at a later time accomplished these seven labors; that this was so is not surprising. It will presently be shown that more than twenty years passed before someone accomplished even the first of these, that is, before anyone read and understood and fully appreciated Grassmann's book.

VII. *The Period from 1844 to 1862*

In the period from the *Ausdehnungslehre* of 1844 to that of 1862 a number of events took place which shed light on the relation of Grassmann's work to the mathematical ideas of the times. These events, which will be discussed within the framework of the narrative of Grassmann's activities in this period, range in variety from the discovery and publication by others of ideas contained in Grassmann's system to the reading of Grassmann's book by Hamilton. The primary concern of this section is discussion of the reception accorded Grassmann's work.

Friedrich Engel accurately described the magnitude and form of this reception in the following statement:

> Thus Grassmann experienced what must be the most painful experience for the author of a new work: his book nowhere received attention; the public was completely silent about it; there was no one who discussed it or even publicly found fault with it. The mathematicians to whom he had sent the work expressed themselves as not unfriendly to it, to some extent even as benevolent to it, but no one really studied it. (5; 97)

Engel's statement is well illustrated by considering the reception given the book by the man generally esteemed as the most gifted mathematician of the period—Carl Friedrich Gauss. Grassmann apparently sent a copy of his book to Gauss, and the latter replied with a note of thanks dated December 14, 1844. Therein Gauss stated that he had worked on similar ideas nearly a half century before and had published some of his results in 1831; Gauss was probably referring to his work on the geometrical representation of complex numbers.[53] Gauss also stated that he was very busy and that he had concluded that to get at the kernel of the work it would be necessary to familiarize himself with the peculiar (eigenthümlichen) terminology used in the book. (4,I,II; 398) This it seems Gauss never did.

The reaction of Möbius to Grassmann's work should prove illuminating, for of all mathematicians of the period Möbius was in the best position to judge Grassmann's work. Thus in 1844 Grassmann visited Möbius at Leipzig, and in a letter of October 10, 1844, he asked Möbius to write a review of his book. He wrote: "Finally I take the liberty of asking you to write a review of the work for some critical journal, for I am convinced that just as now no one stands nearer the ideas expressed in the work than you, no one will be in a better position than you to judge the work so fundamentally and to bring to light both the weaknesses and whatever merits the book may contain." (5; 99) Möbius' reply was dated February 2, 1845, nearly four months later:

> I reply that I was sincerely pleased to have come to meet in you a kindred spirit, but our kinship relates only to mathematics, not to philosophy. As I remember telling you in person, I am a stranger to the area of philosophic speculation. The philosophic element in your excellent work, which lies at the basis of the mathematical element, I am not prepared to appreciate in the correct manner or even to understand properly. Of this I have become sufficiently aware in the course of

> numerous attempts to study your work without interruption; in each case however I have been stopped by the great philosophical generality. (5; 100)

However Möbius wrote that he had contacted a mathematician named Drobisch who was also a philosopher and hoped that he would write a review (he did not). Möbius went on to recommend that the best procedure might be for Grassmann himself to publish a review (he did!). (5; 100–101)

Engel recorded an illuminating exchange of letters between Ernst Friedrich Apelt (1812–1859), professor of philosophy at Jena, and Möbius. On September 3, 1845, Apelt wrote the following to Möbius:

> Have you read Grassmann's strange *Ausdehnungslehre*? I know it only from Grunert's *Archiv;* it seems to me that a false philosophy of mathematics lies at its foundation. The essential character of mathematical knowledge, its intuitiveness (Anschaulichkeit), seems to have been expelled from the work. Such an abstract theory of extension as he seeks could only be developed from concepts. But the source of mathematical knowledge lies not in concepts but in intuition. (5; 101)

Möbius in a letter dated January 5, 1846, responded:

> You ask me whether I have read Grassmann's *Ausdehnungslehre*. To this I answered that Grassmann himself presented me with a copy and that I have on numerous occasions attempted to study it but have never gone beyond the first sheets, since, as you mentioned, intuitiveness (Anschaulichkeit), the essential character of mathematical thought, is not to be found in the work. However I have been forcing myself through the work by skimming it on numerous occasions in regard to extension or generality as you would prefer to call it. From this I have come to feel that it can be influential in a good way for mathematics, especially in regard to the systematic presentation of its elements. To this belongs the addition and subtraction of lines when these are considered not only in relation to their length but also their direction. (5; 101)

Partly on the basis of the above letters and on the translation of "Anschaulichkeit" as intuitiveness Ernest Nagel suggested that a major reason for the poor reception of Grassmann's work was that "the contemporary scene was dominated by Kantian views on the indispensability of intuition for mathematics. . . ." [54]

To Nagel's views may be added the opinions of A. E. Heath. Writing in 1917, Heath argued that a major reason for the poor reception was the philosophical nature of the book, a condition compounded by a reaction against a philosophical form of presentation that had set in among mathematicians as a result of "the exaggerations of the metaphysical unification of knowledge in the schools of Schelling and Hegel." [55] In any case Grassmann's book provoked the following comment from Heinrich Richard Baltzer, to whom Möbius had recommended the book: "it is not now possible for me to enter into those thoughts; I become dizzy and see sky-blue before my eyes when I read them." (5; 102) Möbius wrote back to Baltzer: "If, as you write me, you have not relished Grassmann's *Ausdehnungslehre,* I reply that I have the same experience. I likewise have managed to get through no more than the first two sheets (Bogen) of his book." (5; 102)

Grassmann had sent a copy of the work to Johann August Grunert, founder and editor of the *Archiv der Mathematik und Physik.* Grunert encountered serious difficulties in reading the work and requested Grassmann to write a review of his book for publication in the *Archiv* and to give some simple examples. (5; 102–103) This Grassmann did, and in 1845 his "Kurze Übersicht über das Wesen der Ausdehnungslehre" appeared.[56] Grassmann's summary of his book was neither completely representative nor a major aid to understanding the work. It was however, as Grassmann himself stated, the only review of his book that was published. (4,I,II; 3)

Grassmann's other publication of 1845, his "Neue Theorie der Elektrodynamik,"[57] contained an important electrical discovery found through his new system of mathematics. This paper shared the fate of his *Ausdehnungslehre;* the significance of Grassmann's discovery was only realized after Clausius in the 1870's published the same result, without of course knowing of Grassmann's earlier discovery. (5; 104–105)

The only recognition that Grassmann received at this time for his mathematical investigations came in regard to a paper entitled "Die Geometrische Analyse." Möbius had written Grassmann on February 2, 1845, to notify him of an 1844 announcement of the Jablonowskischen Gesellschaft der Wissenschaft offering a prize for the creation of a system similar to (or for the extension of) the system sketched by Leibniz in his letter to Huygens of 1679, published for the first time in 1833. (5; 109) Möbius' letter put Grassmann in the rather delightful position of being aware of a prize offered for the completion of a task judged significant by a scientific group, which task Grassmann had already accomplished.

Grassmann thus came to write his *Die Geometrische Analyse geknüpft und die von Leibnitz erfundene geometrische Charakteristik.* The Jablonowskische Gesellschaft in 1846 awarded the prize to Grassmann's work[58] and published it in 1847 with an appendix by Möbius.[59] The work does not demand detailed analysis; it may be noted that Grassmann included a thorough treatment of his inner product (the modern dot product), which he had neglected to do in his *Ausdehnungslehre* of 1844. Though this paper had major defects in presentation, it was more readable than his *Ausdehnungslehre;* its fate however was the same: colossal neglect. (5; 111–118)

In 1847 Grassmann attempted to secure a university position. He wrote to Eichhorn, the Prussian minister for culture, and included with his letter copies of his principal mathematical writings. Eichhorn asked the noted mathematician Ernst Eduard Kummer to report on the significance of Grassmann's mathematical works. In Kummer's report scant praise was mixed with repeated statements about Grassmann's lack of clarity. Probably a major cause of Kummer's negative judgment was that Kummer had the works for less than a month. In any case Kummer joined the ranks of Gauss and others as great mathematicians who failed to appreciate the greatness of Grassmann's achievement. Thus no professorship was offered, nor in fact was one ever offered to Grassmann. (5; 123–130)

Three mathematicians of this period were forced, as it were, to take notice of Grassmann's work because of priority questions; they were Adhémar Barré, Comte de Saint-Venant (1797–1886), Augustin Cauchy (1789–1857), and Sir William Rowan Hamilton.

Saint-Venant was a French engineer, noted for his researches in elasticity, who in 1845 published a paper entitled "Mémoire sur les sommes et les différences géométriques, et sur leur usage pour simplifier la Mécanique."[7] This paper contained mathematical ideas similar or identical to some ideas fundamental to the Grassmannian system and to modern vector analysis. Saint-Venant began his paper by stating: "I call the geometrical sum of any number whatever of lines a, b, c, . . . given in magnitude, direction, and sense a line which is equal and parallel to the last side of a polygon of which the other sides are a, b, c, . . . placed end to end, each in its proper sense. If l is the last side, then I write $\bar{l} = \bar{a} + \bar{b} + \bar{c}$ . . . ." (7; 620) Saint-Venant then proceeded to define the "différence géométrique," or vector subtraction; the "différentielle géométrique"; and the "coefficients différentiels géométriques." (7; 620) This was followed by a definition of the sum of plane areas and of the "produit géométrique."

> I call the *geometrical product* (of a line b multiplied by a line a, designated $\bar{a}\,\bar{b}$) the area obtained, both in magnitude and direction, in forming a parallelogram from those two lines drawn from the same point. The positive face is that on which a is on the left and b is on the right. Thus $\bar{a}\,\bar{a} = 0$ and $\bar{b}\,\bar{a} = -\bar{a}\,\bar{b}$.
>
> I call the *geometrical product* of an area multiplied by a line the volume of the parallelepiped (or the oblique prism) having the area for a base and the sides equal and parallel to the given line. The volume is

considered negative when the sides are on the negative side of the base. $\bar{a}\,\bar{b}\,\bar{c}$ will designate the product of the area $\bar{b}\,\bar{c}$ multiplied by the line $\bar{a}$. (7; 621)

Thus Saint-Venant had discovered vectorial addition, subtraction, differentiation, and also a multiplication similar to the modern cross product, the major difference between them being that Saint-Venant's product was, like Grassmann's, not another vector, but a spatially oriented area. Saint-Venant then stated that geometric equations can be added, subtracted, and multiplied and made very brief mention of the possibility of integration and division for vectors. (7; 621) The ideas mentioned to this point were treated in two pages by Saint-Venant; thus his paper is best seen as a sketch of the fundamentals of a vector analysis. The final three and one-half pages of the paper suggested how his ideas could be applied to mechanics.

Though neither Saint-Venant's paper nor his letters to Grassmann (those published by Engel) allow us to answer the interesting question as to what motivated Saint-Venant to his creation, one letter does contain information concerning when Saint-Venant began to work out his ideas. In a letter of July 17, 1847, written in response to a letter (quoted earlier) in which Grassmann stated that his initial ideas came in 1832, Saint-Venant commented: "It was thus around 1832 that I first came to the idea of extending the use of algebraic signs to those geometrical operations which are performed on lines and areas in mechanics, but I published nothing until 1845." (5; 122) Although the question of the sources of ideas and motivation for Saint-Venant may not be answered definitively, some indication as to the possible sources may be given. It is very improbable that he was stimulated by the work of Möbius or of Grassmann's father, since Saint-Venant had great difficulty in reading German.[60] Since the writings of Argand, Servois, Buée, Mourey, Warren, and Gauss on the geometrical representation of complex numbers were all published before 1832 and were written in languages to which presumably Saint-Venant had access, it is possible, but direct evidence is lacking, that Saint-Venant was stimulated by one or more of these men.[61] Saint-Venant's influence on later writers was at most very small.[62]

Grassmann had heard of Saint-Venant's paper by 1847, and on April 18, 1847, Grassmann wrote to Augustin Cauchy and enclosed a letter for Saint-Venant whose address he did not know. Simultaneously Grassmann mailed to Cauchy two copies of his *Ausdehnungslehre* and requested Cauchy to give the letter and one copy of the book to Saint-Venant. (5; 120–121) The part of Grassmann's letter which contained statements concerning the history of his discoveries has already been quoted; the remainder included a statement of priority and a discussion of some materials not fully developed in his *Ausdehnungslehre* (in particular, the linear or inner product).

On July 17, 1847, Saint-Venant wrote to Grassmann that he had received the letter but not the book; he requested Grassmann to send him the book. (5; 122) Grassmann assumed that Cauchy had been delayed in giving the book to Saint-Venant, and on January 27, 1848, he wrote to Saint-Venant and sent him a copy of his "Die Geometrische Analyse," along with another paper. Grassmann did not send his *Ausdehnungslehre* nor did Cauchy give Saint-Venant the copy sent to him for Saint-Venant. Saint-Venant read Grassmann's "Die Geometrische Analyse" but did not correspond further with Grassmann at this time. (5; 122)

In 1853 another priority question was raised by a publication of that year in the *Comptes rendus*, authored by Augustin Cauchy, and entitled "Sur les clefs algébriques."[63] Cauchy's paper on his "clefs algébriques" or algebraic keys was primarily directed toward providing methods for dealing with algebraic problems, in general, the finding of unknowns in equations. A very simple example will illustrate both his method and its relevance. Given the equations $x + 3y = 11$ and $4x + 2y = 14$, we may introduce two algebraic keys, i and j, which behave so that $i \cdot i = j \cdot j = 0$ and $i \cdot j = -j \cdot i$. We then multiply each equation by one key and obtain $xi + 3yi = 11i$ and $4xj + 2yj = 14j$. If we now add these equations, we obtain an equation of the form $Ax + By = K$, where $A = i + 4j$, $B = 3i + 2j$, and $K = 11i + 14j$. From this we obtain the equation $(Ax + By)B = K(B)$. Since A, B, and K obey the same multiplication laws as i and j, the above equation becomes $ABx = KB$, which can be transformed into $x = \frac{KB}{AB}$. Thus we obtain

$$x = \frac{(11i + 14j)(3i + 2j)}{(i + 4j)(3i + 2j)} = \frac{(22ij + 42ji)}{2ij + 12ji} = \frac{-20ij}{-10ij} = 2$$ [64]

The significance of this is that Cauchy's keys are algebraically equivalent to Grassmann's extensive magnitudes in regard to the latter's outer multiplication. Moreover Grassmann developed in his *Ausdehnungslehre* nearly identical algebraic methods.[65]

Grassmann probably first learned of Cauchy's publications in the following way. Baltzer on June 14, 1853, wrote to Möbius concerning the identity of Cauchy's methods with those of Grassmann. Möbius then wrote to Grassmann on September 2, 1853, to inform him of Cauchy's publication and to suggest that Grassmann make a claim for priority. (5; 172–175) At the same time Möbius informed Grassmann (5; 172) of a paper of 1853 by Saint-Venant entitled "De l'interprétation géométrique des clefs algébriques et des déterminants."[66] The aim of Saint-Venant's paper was to show how Cauchy's keys could be interpreted geometrically and, in doing this, to show the mathematical relationship of Cauchy's ideas to those Saint-Venant had presented in his paper of 1845. Saint-Venant made no priority claim for himself or for Grassmann, though he mentioned Grassmann's linear or inner product in a footnote.

On February 19, 1854, Grassmann wrote to Möbius that illness had delayed him in seeing Cauchy's papers, but that he had finally managed to travel from Stettin to Berlin to read them. Grassmann had decided to claim priority through a letter to the French Academy and through a publication in Crelle's *Journal.* (5; 176–182) His letter to the French Academy was read on April 17, 1854, and stimulated the Academy to form a committee to investigate the priority question. The committee was composed of Lamé, Binet, and Cauchy (!); no decision was handed down by the committee, perhaps because of Cauchy's death in May, 1857. (5; 198) Grassmann's paper for Crelle's *Journal* appeared in 1855 under the title "Sur les différents genres de multiplication."[67] Herein Grassmann claimed priority over Cauchy and Saint-Venant and published some new results, in particular the definition of sixteen different kinds of multiplication.

On December 16, 1856, Saint-Venant (probably at Cauchy's request) wrote to Grassmann that he had never received the copy of the *Ausdehnungslehre* and had been unable to find a copy. He explained that this fact was the cause of his slighting Grassmann in his paper of 1853, and he requested Grassmann to inform him as to how a copy of the *Ausdehnungslehre* might be obtained. (5; 199–200) Grassmann replied in a letter of March 28, 1857, a letter that is typical of Grassmann's patience and good nature. Grassmann stated that his book was in the library of the French Institute, but wrongly classified; that he had believed Cauchy had passed on the copy of the *Ausdehnungslehre* to Saint-Venant, so that another copy had not been sent directly to him; and that he was sending to Saint-Venant a copy of his book and of a paper and was translating parts of both so that Saint-Venant could read them more easily. He also asked Saint-Venant to pass on the letter to Cauchy and to tell Cauchy of his high respect for him. There is no record of a reply from Saint-Venant. (5; 200–201)

In Grassmann's *Ausdehnungslehre* of 1862 he stated in regard to Cauchy: "I have no intention of accusing the famous mathematician of plagiarism. . . ." (4,I,II; 9–10) Victor Schlegel however in 1878 explicitly accused Cauchy of plagiarism (6; 38–39), though in a publication of 1896 he reversed his judgment.[68] Friedrich Engel believed that Schlegel went too far in his charge of plagiarism. Engel argued that Cauchy probably either did not read the *Ausdehnungslehre* or did not remember its contents. Engel however strongly criticized Cauchy for not answering Grassmann's priority claim. (5; 202)

Another alternative may however be suggested. Cauchy was well acquainted with Saint-Venant's paper of 1845 and used results from it in a paper of 1849 entitled "Sur les Quantités géométriques, et sur une méthode nouvelle pour la résolution des équations algébriques de degré quelconque."[69] Moreover this paper shows that Cauchy had a good knowledge of the work of Argand, Servois, and Buée on complex numbers; and papers by Cauchy of 1853 make clear that Cauchy knew of Hamilton's and Möbius' works.[70] Thus it can be suggested that Cauchy was knowledgeable in relevant works done independently of Grassmann and that the priority charge brought by Grassmann was appropriate only in regard to *part* of the methods published by Cauchy in his papers on his "clefs algébriques." It was perfectly possible for Cauchy to draw on the sources mentioned above for help in creating his methods. A residuum of originality of course would remain, and it is in regard to that residuum that a priority dispute would be in order.

Saint-Venant and Cauchy were not the only mathematicians who encountered Grassmann firmly lodged in a domain that they had previously viewed as their own discovery; great must have been Sir William Rowan Hamilton's surprise when in some unrecorded way he heard of the Stettin schoolmaster and his *Ausdehnungslehre* of 1844. Procuring a copy of the book, Hamilton set to reading it; his interesting comments are preserved in letters, in the margins of his copy of Grassmann's book, and in the historical preface to his *Lectures on Quaternions*. The latter comments, as fate would have it, comprised (with the exception of Möbius' comments) the only published discussion of Grassmann that appeared before the 1860's.

It was in late 1852 that Hamilton, who was then preparing the historical preface to his *Lectures*, read the *Ausdehnungslehre*. In a letter of October 26, 1852, Hamilton wrote to De Morgan: "a *very* original work. . . . which work, if any, the Germans, if they think *me* worth noticing, will perhaps set up in rivalship with mine, but which I did not see till long after my own views were formed and published." (8; 424) Hamilton continued to read Grassmann's book and on January 31, 1853, wrote the following to De Morgan:

> I have recently been *reading* (and it is curious that sometimes, when otherwise in mental activity, I seem to myself unable to read a page, or almost a sentence of German) more than a hundred pages of Grassmann's *Ausdehnungslehre,* with great admiration and interest. Previously I had only the most slight and general knowledge of the book, and thought that it would require me to learn to *smoke* in order to read it. If I could hope to be put in rivalship with Des Cartes on the one hand, and with Grassmann on the other, my scientific ambition would be fulfilled! But it is curious to see how narrowly, yet how completely, Grassman failed to hit off the Quaternions. He published in 1844, a little later than myself, but with the most obvious and perfect independence. (8; 441)

Hamilton's comments as given in a letter of February 2, 1853, are

> I am not quite so enthusiastic to-day about Grassmann as I was when I last wrote. But I have read through nearly all of what I could procure of his writings, including a subsequent commentary (in German) by Möbius. Grassmann is a great and most German genius; his view of *space* is at least as new and comprehensive as mine of *time;* but he has not anticipated, nor attained the conception of, the *quaternions,* even so nearly as I guessed he might have done, from a notion hastily taken up, of what might have been his meaning (and what it *was,* I *very* dimly know even *now*), in his doctrine of "eingewandte multiplikation." I quote from memory. His *outer* products (aüssere) I think that I *do* understand; and that is saying something for a person who has not learned to smoke. And even his *inner* products, published subsequently to the *outer* ones (in 1847), I can swallow pretty well. In fact, the "inner products" of Grassmann have much analogy to my *"scalar parts"* of a quaternion, and his "outer products" to my *"vector parts."* If the notion of *combining* them had occurred to him, *he might* have been led to the quaternions; but those he seems to me to have altogether failed to perceive. Yet I think that my own researches, or speculations, would have a better chance of being *appreciated* in these countries, if readers had first been put through a sufficient course (or dose) of Grassmann. I must say that I should not fear the comparison. You tolerate egotism in correspondence . . . . (8; 442)

De Morgan had asked Hamilton at one point if Grassmann's Christian name was Nebuchadnezzar. (8; 425) Hamilton picked up the joke and embellished it in his letter of February 9, 1853:

> if you have any curiosity to know anything of the result of my recent Nebuchadnezzarological reading (my daughter looking over my shoulder is amused at the folly of philosophers), it will be quite consistent with my humour to inform you. To the public I am likely to say but *little* at present about Grassmann; for I find that beyond the rule for *adding* lines, which he seems to have independently worked out, whereas I took it from Warren, we have scarcely a result in common, except one thing which *is* (in my view) important, namely, the inter-

pretation of $B - A$, where A and B denote *points*, as the *directed line AB*. He comes to this, in his page 139 of the *Ausdehnungslehre*, after *long* preparations, and ostrich-stomach-needing iron previous doses. I, knowing nothing of this result, as in any way arrived at by him, STARTED with the same interpretation in my Lectures, in 1848, having printed the same conception some years earlier, and having been familiar with it (*see* Pure Time) for a *long* time before. (8; 444)

Hamilton's published statement concerning Grassmann given in his *Lectures on Quaternions* of 1853 is as follows:

It is proper to state here, that a species of *non-commutative multiplication* for inclined lines (aüssere Multiplikation) occurs in a very original and remarkable work by Prof. H. Grassmann (Ausdehnungslehre, Leipzig, 1844), which I did not meet with till after years had elapsed from the invention and communication of the quaternions: in which work I have also noticed (when too late to acknowledge it elsewhere) an employment of the symbol $\beta - \alpha$, to denote the *directed line* (Strecke), drawn from the point α to the point β. Not withstanding these, and perhaps some other coincidences of view, Prof. Grassmann's system and mine appear to be perfectly distinct and independent of each other, in their conceptions, methods, and results. At least, that the profound and philosophical author of the Ausdehnungslehre was not, at the time of its publication, in possession of the theory of the *quaternions*, which had in the preceding year (1843) been applied by me as a sort of organ or *calculus for spherical trigonometry*, seems clear from a passage of his Preface (Vorrede, p. xiv.), in which he states (under date of June 28th, 1844), that he had not then succeeded in *extending the use of imaginairies from the plane to space;* and generally unsurmounted difficulties had opposed themselves to his attempts to construct, on his principles, a theory of *angles in space* (hingegen ist es nicht mehr möglich, vermittelst des Imaginären auch die Gesetze für den Raum abzuleiten. Auch stellen sich überhaupt der Betrachtung der Winkel im Raume Schwierigkeiten entgegen, zu deren allseitiger Lösung mir noch nicht hinreichende musse geworden ist).[71]

Finally, a remark made by Hamilton in a letter to J. T. Graves of September 30, 1856, may be cited. After mentioning that Grassmann's 1855 article in Crelle's journal had stimulated him in some recent researches, Hamilton stated that Grassmann "was well worthy to have anticipated me in the discovery of the quaternions; and it appears to me a very remarkable circumstance that he did *not.*" (8; 70)

By 1860 one mathematician of the British Isles (Hamilton), one mathematician of the German-speaking countries (Möbius), and one French mathematician (Saint-Venant) had come to appreciate to some extent Grassmann's work. Two other mathematicians, both of Italy, complete this small group. These two men were Luigi Cremona and Giusto Bellavitis.

In 1860 Luigi Cremona published in *Nouvelle Annales de mathématiques* a note concerning the solution of two problems that had been proposed in that journal.[72] As part of this note Cremona gave a brief exposition of Grassmann's ideas, which was prefaced with the following statements:

> ". . . I cannot refrain from mentioning a very expeditious and very curious method, of which the first idea seems to belong to Leibniz, but which has been truly established by Grassmann. Except for Möbius . . . and Bellavitis . . . I do not know of any geometers who have given Grassmann's researches the attention which they deserve.
>
> I will here reproduce the principle definitions and conventions of this ingenious theory which the author names *geometrical analysis*." [73]

At a later time Cremona used even stronger words in praise of Grassmann (5; 335) and included some of Grassmann's ideas in at least one of his books.[74]

Giusto Bellavitis came into contact with Grassmann during the 1850's; the occasion was a geometrical paper by the latter with which Bellavitis took issue. Bellavitis went on to study the *Ausdehnungslehre*, and through letters from Bellavitis in 1860 and 1862 Grassmann learned that the Italian mathematician was impressed by the *Ausdehnungslehre* and intended to acquaint his countrymen with it. (5; 106–107)

When one looks back on the decades before 1860, a number of generalizations emerge. It is striking first of all that Möbius, Bellavitis, Hamilton, Grassmann, Saint-Venant, and Cauchy should independently arrive at ideas that were in many cases similar. The motivation behind their investigations was in nearly every instance *geometric*, but their brilliance was primarily *algebraic*. Though Hamilton's ideas had already obtained much attention, Grassmann remained nearly unknown. Indeed Möbius mentioned in a letter of June 9, 1853, that he knew of only one mathematician, Bretschneider, who had read Grassmann's book completely through (5; 163n); and as Grassmann himself mentioned, the remaining copies of the first *Ausdehnungslehre* were used as wastepaper. (4,I,I; 18) The factors that caused this neglect have been amply discussed.

Finally a summary of Grassmann's other activities in this period will fill out the picture of Grassmann as a very active person. From 1844 to 1861 Grassmann published seventeen scientific papers, including important papers in physics, and a number of language and mathematics textbooks. He edited a political paper for a time and also published materials on the evangelization of China.[75] All this was done while he taught a heavy load and raised a family, for

Grassmann had married in 1849, and from this union eleven children came.

This period of his life was concluded with the publication of his second *Ausdehnungslehre.*

VIII. *Grassmann's* Ausdehnungslehre *of 1862 and the Gradual, Limited Acceptance of His Work.*

On October 31, 1861, Grassmann, no doubt with hopes running high, sent to Möbius a copy of the second *Ausdehnungslehre.* Eight years earlier in a letter to the same correspondent Grassmann had stated his intention of preparing a new work; there is evidence that by 1854 or 1855 he had begun writing. (5; 223) The 300 copies of the book bore the date 1862, were printed in the shop of Grassmann's brother, and were paid for by the author. (5; 223) Its title was *Die Ausdehnungslehre: Vollständing und in strenger Form bearbeitet.* Under Grassmann's name appeared the title "Professor am Gymnasium zu Stettin." [76]

In the foreword Grassmann discussed the poor reception accorded his earlier work and stated that the content of the new book was presented in "the strongest mathematical form that is actually known to us; this is the Euclidean. . . ." (4,I,II; 4) He elaborated on the content of the book and discussed the relation of his work to that of Gauss, Möbius, Bellavitis, Saint-Venant, and Cauchy. The final paragraph in the foreword was the following:

> For I remain completely confident that the labor which I have expended on the science presented here and which has demanded a significant part of my life as well as the most strenuous application of my powers, will not be lost. It is true that I am aware that the form which I have given the science is imperfect and must be imperfect. But I know and feel obliged to state (though I run the risk of seeming arrogant) that even if this work should again remain unused for another seventeen years or even longer, without entering into the actual development of science, still that time will come when it will be brought forth from the dust of oblivion and when ideas now dormant will bring forth fruit. I know that if I also fail to gather around me in a position (which I have up to now desired in vain) a circle of scholars, whom I could fructify with these ideas, and whom I could stimulate to develop and enrich further these ideas, nevertheless there will come a time when these ideas, perhaps in a new form, will arise anew and will enter into living communication with contemporary developments. For truth is eternal and divine, and no phase in the development of truth, however small may be the region encompassed, can pass on without leaving a trace; truth remains, even though the garment in which poor mortals clothe it may fall to dust. (4,I,II; 10)

As Grassmann himself pointed out, the form of presentation was the Euclidean. The book consisted of sets of theorems presented with a minimum of comment. Philosophical commentary had been banished (which was helpful), as were physical applications (which was not helpful). Concerning this form of presentation which Grassmann had also used in an elementary textbook of 1861, *Lehrbuch der Arithmetik,* Engel commented:

> Without a doubt this was a disastrous mistake. What was perfectly in place in the treatment of a subject so commonplace for all mathematicians as arithmetic, at least for readers who pursue mathematics as a science, was the most unsuitable form of presentation for a subject to which the reader was for the first being introduced. Though this form of presentation led to an admirable codification of the new concepts and laws of his theory of extension, still it was not a presentation likely to win followers for his ideas, let alone convert those who had not wished to read his first *Ausdehnungslehre.* Grassmann had actually only undertaken the immense task involved in the composition of the new work because he hoped that now at last he would find readers. It is actually puzzling that he could so deceive himself and be so mistaken in his choice of the means for attaining his aim. (5; 231)

The second *Ausdehnungslehre* was more than a reworking of the first: it was one-third longer and contained many new results. The most important of these was Grassmann's solution of the so-called "Pfaffian Problem." According to Engel this solution was only appreciated after 1877, a neglect which Engel ascribed to the fact that few mathematicians would seek such a solution in Grassmann's book and fewer would be able to disentangle it from the other materials surrounding it. (5; 232–233) The latter comment could with equal justice be applied in relation to vector analysis.

The reception accorded the second *Ausdehnungslehre* must have been a great disappointment to Grassmann, for at first his efforts seemed to have been in vain; he wrote in 1877, "this new work met with even less attention than the first." (4,I,I; 18) No reviews of it appeared (5; 231), and it seems probable that Grassmann received only one note of thanks from those to whom he sent copies.[77] The following statements of Engel may serve as a conclusion to the discussion of the early reception of the second *Ausdehnungslehre:* "Thus the second *Ausdehnungslehre* also remained for mathematicians a book with seven seals, and the abundance of entirely new developments which it contained was like something buried in the ground" and "As in the first *Ausdehnungslehre* so in the second: matters which Grassmann had published in it were later independently rediscovered by others, and only much later was it realized that Grassmann had discovered them earlier." (5; 232–233)

Thus Grassmann's works were almost totally neglected during

the first fifty-five years of his life; the first real ray of hope appeared in the second half of the 1860's. Grassmann in the foreword to the second edition (published in 1878) of his *Ausdehnungslehre* of 1844 described the events that led up to the recognition that he received in the last years of his life.

> Hermann Hankel was the first who, in his *Theorie der complexen Zahlensysteme* (Leipzig, 1867), stressed the fundamental significance of my *Ausdehnungslehre* (see pp. 16, 112, 119–140, and 140). Even more decisive was Clebsch's recognition. Shortly before his death Clebsch, in his "zum Gedächtniss an Julius Plücker, Göttingen, 1872" (see the notes set at the bottom of pages 8 and 28), emphasized the significance of my *Ausdehnungslehre* of 1844 in words of strong praise. In fact in his second note Clebsch stated: "In a certain sense the coordinates of the straight line, as well as a large part of the basis of the newer algebra, are already contained in Grassmann's *Ausdehnungslehre* of 1844. The more exact statement of these relations would however lead too far at present." In view of his loving and so constantly fruitful participation in the works of others, which trait distinguished this most eminent of the more recent mathematicians, it is certain that Clebsch would later have found space to present these relations and, as was his way, to fructify the *Ausdehnungslehre* with new and far-reaching ideas. But, alas, he was snatched away so suddenly in the midst of his powerful efforts on behalf of science.
>
> However three years earlier (in 1869) Victor Schlegel had begun to execute the ideas suggested by Clebsch. In his "System der Raumlehre nach den Prinzipien der Grassmann' schen *Ausdehnungslehre* und als Einleitung in dieselbe dargestellt von Victor Schlegel, Leipzig bei Teubner," of which the first part appeared in 1872 and the second in 1875, Schlegel has presented with great clarity, and in large measure independently, the meaning of the *Ausdehnungslehre*. This he did in a completely appropriate manner. It is especially to be emphasized that this book by Schlegel is the first which has viewed the essential ideas of the *Ausdehnungslehre* in their inner connections and has given them a presentation. (4,I,I; 18–19)

It was in November 1866 that Grassmann received a long letter from Hermann Hankel (1839–1873), who had been a student of Riemann and who was at that time privatdozent in mathematics at Leipzig. Hankel wrote that he was writing a treatise on complex numbers in which he planned to include a treatment of Hamilton's quaternion methods and also Grassmann's ideas, which he strongly praised. A series of letters passed between Hankel and Grassmann, and in 1867 the former's *Theorie der complexen Zahlensysteme* appeared. Roughly 10 percent of this book was devoted to Grassmann's system, on which Hankel bestowed much praise. Hankel's book was influential, and if he had not died in 1873, he might have done even more to make Grassmann's system known. Clebsch came to know Grassmann's work in the following way: Grassmann's oldest son, Justus, began studies in mathematics at Göttingen in 1869 and brought with him copies of the second *Ausdehnungslehre* to be given to Stern and Clebsch. (5; 311) Stern became enthusiastic about Grassmann's book and passed on his enthusiasm to Felix Klein, who said that it influenced his Erlanger Program of 1872. (5; 312) Clebsch's enthusiasm was unfortunately stilled by his death in 1872.

As Grassmann mentioned, Victor Schlegel (1843–1905) was one of the earliest proponents of his system; indeed Schlegel spent most of his life expounding and developing Grassmann's system. Because of Grassmann's isolation at Stettin and his teaching only elementary mathematical courses, it was unlikely that he would produce any students who would develop his system. However a number of his followers came from his "associates"; these included his sons Justus and Hermann, his brother Robert, and Schlegel himself, who for two years (1866–1868) taught with Grassmann in Stettin. Schlegel discussed this period in his *Hermann Grassmann: Sein Leben und seine Werke;* during his time in Stettin he had had many conversations with Grassmann, but Grassmann had only rarely mentioned his *Ausdehungslehre.* Schlegel's interest had however been stimulated to the point that he in 1869 after leaving Stettin decided to study Grassmann's works. Schlegel was delighted with Grassmann's methods and decided to publish a work explaining them. In 1871 he received encouragement in this project from Clebsch, and in 1872 appeared the first part of his *System der Raumlehre nach dem Prinzipien den Grassmann' schen Ausdehnungslehre und als Einleitung in Dieselbe;* the second part was published in 1875. (6; 61–62)

Schlegel attempted to explain Grassmann's results through their relations to elementary geometry (part I) and to the newer methods of higher geometry and algebra (part II). Engel commented on the content and significance of Schlegel's book in the following way:

> Actually Schlegel's work also was not a success. The author had definitely gone too one-sidedly for Grassmannian methods without practicing the necessary criticism. He thought that the people who argued for a progress of modern algebra as opposed to the theory of extension had done it only because they knew the theory of extension mainly by hearsay, whereas he had claim to what concerns knowing. On the other hand, the superiority of modern algebra over the theory of extension which others asserted was no fiction and consequently a large part of Grassmann's methods had become dispensible. Schlegel was not the man to put the old Grassmannian wine in new vessels; he was not able to present the ideas contained in the theory of extension from the point of view of the theory of invariants and to bring to light what was still new. (5; 324)

Engel also commented that Schlegel's biography of Grassmann, published in 1878, was more influential than his *System der Raumlehre*, despite (or perhaps because of) the fact that the biography went to extremes in praising Grassmann. (5; 324) Schlegel's partly successful efforts to make Grassmann's ideas known continued until Schlegel's death in 1905, by which time he had published over twenty-five papers in the Grassmannian tradition.

Three other men who developed an interest in Grassmann's work during the mid-1870's and published related works were Hermann Noth (1840–1882), William Kingdon Clifford (1845–1879), and W. Preyer (1841–1897); the latter wrote to Grassmann that during a trip to England (made in 1875 or 1876) he had discussed Grassmann's work with Sylvester, who was quite interested in it and had planned a publication in regard to it. (5; 330)

As more mathematicians became interested in Grassmann's ideas, the demand for copies of the first *Ausdehnungslehre* increased; none however were available, since the publisher had in 1864 used the remaining copies for wastepaper. Although the 300 copies of the second *Ausdehnungslehre* had not yet been sold, a second edition of the first *Ausdehnungslehre* was published in 1878 with three appendices and a new forward (left incomplete by Grassmann and completed by Schlegel).

Grassmann's activities after 1862 were many and diversified. He, like the young Hamilton, had a strong interest in and great talent for languages. It was only Grassmann however who made an important contribution to philology, a contribution which in fact rivals his mathematical work. A study of Sanskrit begun in 1849 (5; 155) culminated in the 1870's with the publication of his *Wörterbuch zum Rig-Veda* (1784 pages)[78] and his translation of the *Rig-Veda* (1123 pages).[79] These achievements of Grassmann did not go unrecognized, for in 1876 he was made a member of the American Oriental Society and received an honorary doctorate from the University of Tübingen. (5; 309)

In the period after 1862 Grassmann published textbooks on the German and Latin languages and on mathematics, as well as numerous religious and musical writings and a book on German botanical terminology. He also invented the Grassmann Heliostat at this time.[75] These activities, combined with his increasing interest in philology and increasing disappointment at the neglect of his mathematical creations, explain his diminished mathematical productivity during the late 1860's.

In the 1870's however Grassmann published a number of mathematical papers which Engel described as of inferior quality. Engel ascribed this to the fact that Grassmann had lost contact with the current mathematical literature and even to some extent with his own earlier ideas. (5; 315–317) In his "Die neuere Algebra und die Ausdehnungslehre" of 1874 [80] Grassmann attempted somewhat unsuccessfully to relate his ideas to new developments in algebra, particularly invariant theory; and in his "Zur Elektrodynamik" of 1877 [81] he claimed (justly) priority over Clausius in regard to the previously mentioned electrodynamical law published by Grassmann in 1845 and by Clausius in 1876.[82] "Die Mechanik nach den Prinzipien der Ausdehnungslehre" of 1872 [83] contained some of the results of his at that time still unpublished dissertation on tidal theory. One of the more controversial papers was his "Der Ort der Hamiltonschen Quaternionen in der Ausdehnungslehre" of 1877,[84] in which Grassmann attempted to show that quaternions could be derived from the units and multiplications discussed in his 1855 paper "Sur les différents genres de multiplication"; unfortunately his presentation was weakened by the fact that he knew Hamilton's ideas only from second-hand sources such as Hankel.

Grassmann's richly productive, but in ways tragic, life came to an end on September 26, 1877. It is regrettable both for mathematics and for Grassmann that the ideas of this brilliant, but isolated, schoolmaster were appreciated at such a late hour. Although the 1870's brought him a measure of fame, nevertheless by that time many of his ideas and methods had been (as Engel commented [5; 315]) rediscovered by others and integrated into different formulations and systems. Much of modern vector analysis did however lie embedded within and as yet unextracted from his system. But the fates decreed here as elsewhere: Grassmann's ideas exerted little or no influence on the later history of vector analysis. The irony is that they could have; the fact is (as will be shown) that they did not.

Much has been said in explanation of the neglect of Grassmann's works. Some of the factors that contributed to this neglect border on the incredible. Especially striking is the fate of Grassmann's ideas in relation to the small group who before 1870 recognized their significance. Two (Hankel and Clebsch) who "discovered" Grassmann died almost immediately thereafter; three (Möbius, Hamilton, and Bellavitis) had strong allegiances elsewhere (their own systems); and Schlegel was a man whose enthusiasm exceeded his critical facilities.

Though these and other factors help to explain the neglect of Grassmann's work, in a broad sense this neglect needs little explanation, for their discoveries were revolutionary and the historical pattern exhibited here is not uncommon.

Contemporary analyses of the history of scientific revolutions have uncovered many important patterns. Concerning discovery, it has been shown that there are no direct paths to discovery; frequently essentially irrelevant, often philosophic, ideas may play a decisive role. Thus it is probably not accidental that both Hamilton and Grassmann were supported (but not directed) in their discoveries by certain philosophic assumptions. While Hamilton meditated on the metaphysics of time, Grassmann developed ideas concerning space. The truth and direct relevancy of their speculations in this regard are beside the point; what is relevant is that these speculations supported them in their ventures into unexplored domains. The discovery of a new idea does not insure its acceptance, and philosophic notions that aid in the discovery of a new idea frequently hinder its acceptance. Eventually Hamilton and Grassmann realized this; the philosophic discussions that bulked large in their first books were banished from their later presentations.

Historical analyses of the acceptance of revolutionary ideas show that acceptance requires much time; this seems to be so almost by definition. But the statement that time is needed is only descriptive; it explains nothing. The real question is Why is time needed? What factors operate in time to bring about the acceptance of revolutionary ideas? Listed below are three of the numerous factors that frequently play significant roles in the acceptance of revolutionary scientific ideas.

(1) Acceptance is promoted when the discovery is taken up (or made) by someone who has already attained great fame in traditional pursuits or when it is associated with some event or nonscientific question of great importance.

(2) Acceptance is promoted when less startling discoveries are made which are more easily interpreted in terms of the revolutionary ideas than in terms of traditional doctrines.

(3) Acceptance is promoted when a need for the methods which the revolutionary ideas provide either arises or is recognized for the first time.

Examples of the action of these factors in such revolutions as those associated with Copernicus, Galileo, Lavoisier, Hutton, Young, Lobachevski, Darwin, and Einstein could be supplied in abundance. In the present case we shall examine these three factors only in relation to the acceptance of the ideas of Hamilton and Grassmann. Hardly a more striking example could be found to illustrate the first factor: in 1844 Hamilton had attained widespread fame on the basis of important, but not revolutionary, discoveries, and his ideas received immediate though limited attention; Grassmann was unknown however and suffered the fate of a man whose first great discovery is revolutionary. Concerning the second factor it is not accidental that many parts of Grassmann's system were rediscovered before the whole was appreciated. Columbus did not discover the North American *continent;* he discovered certain coastal areas and subsequent generations recognized a continent. Grassmann over a period of twelve years did discover a continent, but his readers neither expected to find nor wanted to find this continent. Concerning the third factor it is important to realize that the need for vectorial methods in 1800 was less than the need in 1900 and that although vectorial methods aid investigations, they are never indispensible. What can be done by vectorial methods can also be done by traditional methods, and thus time was needed during which scientists could see that the labor of learning vectorial methods was amply compensated.

We may now briefly trace the reception and development of Grassmann's ideas during the last two decades of the century. Roughly 150 papers and 9 books (excluding the publication of Grassmann's collected works) appeared from 1881 to 1900.[85] Although the majority of these publications appeared in German journals, a substantial number of papers may be found in the American, British, French, and Italian journals of the period.[86] Engel commented that there was a strong tendency among the authors of these papers to become unrealistically enthusiastic and narrow concerning the Grassmann methods. (5; 343) Felix Klein, who was well acquainted with Grassmann's methods, expressed the same view.[87]

The task of collecting and publishing Grassmann's complete works was begun in the 1890's. The stimulus for this came from Gibbs[88] and from Klein.[89] In 1892 Klein contacted Frederick Engel (1861–1941) and requested that he edit Grassmann's works and prepare a biography of Grassmann. (4,I,I; vi) The works began to appear in 1894, with the final section containing Engel's biography appearing in 1911. Though Engel was by no means an ardent follower of Grassmann's ideas, as he himself stated, (4,I,I; vi) he did nevertheless write a detailed, critical, and sympathetic biography of that great mathematician.

IX. *Matthew O'Brien*

Writing in 1892, the quaternionist C. G. Knott claimed "the antiquaternionic vector analysts of today [Gibbs, Heaviside, and Macfarlane] have barely advanced beyond the stage reached [in 1852] by O'Brien. . . ."[90] Knott's statement (which he frequently repeated) is obviously of great importance if it is true.

The Reverend Matthew O'Brien (1814–1855) was from 1844 to 1854 Professor of Natural Philosophy and Astronomy in King's College, London. In 1830 he had been admitted to Trinity College, Dublin, whence he proceeded to Cambridge to graduate as Third Wrangler in 1838.[91] During his short life O'Brien published roughly twenty papers and a few elementary books. The papers relevant to vector analysis were published during the last ten years of his life.[92]

Judgment on the significance of O'Brien's ideas is complicated, in part because the presentation of his vectorial ideas given in his early papers differs from that of his later papers. The last and longest of his papers was that published in the London *Philosophical Transactions* for 1852.[9] It is this paper that will be discussed with passing mention going to the important papers published earlier.

O'Brien began by stating that his aim was to provide a new notation for an operation of constant occurrence in mathematics, that of "*the translation of a directed magnitude.*" (9; 161) O'Brien then considered three instances of such translations: "*The generation of surface by the parallel motion of a right line . . . The effect produced on a rigid body by the translation of a force acting upon it . . . The effect produced by the translation of a force resulting from the actual motion of its point of application.*" (9; 162) From an analysis of these cases he concluded that the essential conception in each case "is that of *the effect produced by the translation of a directed magnitude* . . . ," and he noted that each of these effects may be represented as a product of the translated directed magnitude and the translation undergone. (9; 162) He stated that these translations could be classified as "lateral" (cases 1 and 2) or "longitudinal" (case 3) and that his new notation would be aimed at representing these two types of translations. (9; 162–163)

He then defined directed magnitude and the two types of translation: the translation of v along u is longitudinal when angle A is

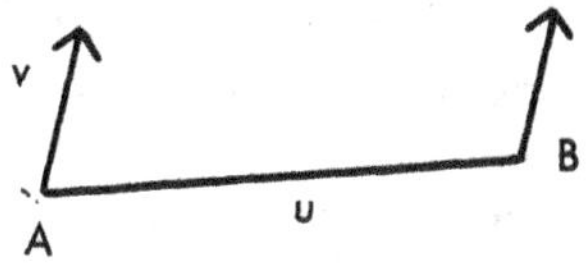

0° and lateral when angle A is 90°. (9; 163–164) O'Brien then introduced the "symbolical forms" $x.y$ and $x \times y$[93]; he assumed only that they were distributive functions of x and y. His choice of the

symbols . and × was explained as one stemming from the fact that they were obsolete symbols for multiplication. (9; 164–165) O'Brien defined "addition" of directed magnitudes as "putting together," while he differentiated between "simultaneous addition," used for directed magnitudes with coincident origins, and "successive addition," used for two directed magnitudes, one of which has its origin at the end point of the other. (9; 165–166) After a discussion of equality and of the suitability of directed magnitudes for the representation of physical entities, O'Brien introduced three "directed units" α, β, γ which go out from the origin along the x, y, z axes respectively; any directed magnitude u was thus to be represented in the form "$u = x\alpha + y\beta + z\gamma$." (9; 166–167)

O'Brien went on to discuss the significance of such an expression as $d\alpha$. He noted that for two directed magnitudes α and α' separated by an infinitesimal angle, "$d\alpha$ is the expression for an indefinitely small line *at right angles to* α." (9; 168) This property bears mention, since it served as the starting point for his vectorial system as presented in his earlier papers published during the 1840's. The results obtained in these papers were very similar to the results obtained in the paper presently under discussion, but the approach was quite different.

O'Brien showed that both lateral and longitudinal effects of translations are distributive and identified $u.v$ as the notation to be used to represent the lateral effect of the translation of v along u, and $u \times v$ to represent the longitudinal effect of the translation of v along u. (9; 169–170) He established the rules to be used for ordinary numbers in regard to these two multiplications and went from this to show that $u.v = v.u$ and $u \times v = v \times u$. (9; 170–172) Thus O'Brien had established that $u.v$ and $u \times v$ were both distributive and that the latter was a commutative multiplication; he never however investigated associativity. He also failed, up to this point in the paper, to state whether $u.v$ and $u \times v$ were numerical magnitudes or directed magnitudes (for example, vectors or vectorial planes). This ambiguity was only partly clarified by his later explanations.

O'Brien then introduced his term "directrix" explaining that in a lateral translation a plane was determined and that the directrix for a lateral translation was to be a line of a certain length perpendicular to that plane; thus the directrix for $\alpha.\beta$ was to be γ; for $\beta.\gamma$, α; for $\beta.\alpha$, $-\gamma$; and so on. It then becomes clear (though only in part from O'Brien's direct statements) that the magnitude of the product of two directed magnitudes linked by "." is $mn \sin \theta$ where m and n are the lengths of the two magnitudes and θ is the angle between the two directed magnitudes. O'Brien proceeded to introduce the symbol "D," meaning "the directrix of," and wrote, for example, $D(\alpha.\beta) = \gamma$. (9; 173–176)

Proceeding to longitudinal translations, he wrote, "$\alpha \times \alpha = 1$, $\beta \times \beta = 1$, $\gamma \times \gamma = 1$"; $\alpha \times \beta$, $\alpha \times \gamma$, $\beta \times \gamma$, . . . are naturally taken as equal to zero. Hence "$u \times v = mn \cos \theta$" where m and n are the magnitudes of u and v, and θ is the angle between them. (9; 176) O'Brien then considered the results of the repetition of the operation $D\alpha$, for example, in the expression $(D\alpha.)^2\alpha'$ where α and α' are any two unit directed magnitudes. In this regard he concluded that if and only if α and α' are mutually perpendicular, then $(D\alpha.)^2\alpha' = -\alpha'$, and thus that $(D\alpha.)$ has one of the properties of $\sqrt{-1}$.

The second part of the paper dealt with applications of the system; in his applications to physical optics O'Brien introduced the symbol Ω "to denote the operation $\alpha \frac{d}{dx} + \beta \frac{d}{dy} + \gamma \frac{d}{dz}$. . . ." He used this in the expressions Ωu, $\Omega \times v$, and $\Omega.v$. (9; 204–205) The above exposition of O'Brien's thought should give some ideas of his system; the question that must be dealt with is whether Knott was correct in his belief that O'Brien's system is essentially that of Gibbs and Heaviside, and hence is the now current system. If Knott was correct, then O'Brien deserves great credit and must be called the father of modern vector analysis.

One aspect of this question is whether O'Brien attained conceptions of the scalar (dot) product and the vector (cross) product. He must be given credit for having the scalar product, although his justification of it is hardly traditional. Concerning the vector product the question is far from simple. His product $u.v$ agrees in numerical magnitude with the modern vector product; however O'Brien never clearly stated what it represented. As we mentioned previously, it seems that he viewed it as a numerical quantity; at another point in the paper he seems however to have viewed it as a directed and oriented parallelogram. (9; 180) It is clear that he viewed his product $Du.v$ as a directed line segment, equal in magnitude, direction, and sense to the modern product.

But still this product in his system does not in all cases correspond with the modern product. The reason for this is O'Brien's failure to understand associativity and the related fact that his α, β, γ do not correspond with the modern **i**, **j**, **k**. This becomes evident from statements made by O'Brien in a footnote to his paper in which he compared his system to that of Hamilton. Herein he compared Hamilton's i, j, k with his α, β, γ and noted that Hamilton's i, j, k are both units of direction and have the property of corresponding to square roots of minus one, whereas his α, β, γ do not have the latter property. (9; 178) The modern **i**, **j**, **k** do have this property in some cases, specifically for example in $\mathbf{i} \times (\mathbf{i} \times \mathbf{v})$ where **v** has no **i** component. But O'Brien realized that he could express this as $(D\alpha.)^2 v = D\alpha.(D\alpha.v)$ where α is a unit vector and v contains no α component.

O'Brien's failure to understand associativity was manifested in this footnote when he stated that Hamilton's i^2 always equals -1, whereas his $(D\alpha.)^2$ equals -1 only when it is applied to a directed magnitude perpendicular to α. (9; 178) What is striking is that O'Brien should have expected this, for the two forms are not comparable, since $(i^2) = (ii)$, whereas $(D\alpha.)^2 = D\alpha.(D\alpha. \ldots)$. The first means i times i; the latter means apply $D\alpha$ to the quantity attained after $D\alpha$ has been applied to some other quantity. Thus O'Brien failed to see that his system was not associative for this product; indeed it seems not to have been a question for him. He never in the course of the paper wrote a triple product except in the form $(D\alpha.)^2\ v$. He had in fact no symbolism for expressing the equivalent of $(\mathbf{i} \times \mathbf{j}) \times \mathbf{k}$ (written in modern terms); this could only be written $(D\alpha.D\beta.)\gamma$, which would have been meaningless in his system. Thus one can at most attribute to O'Brien a very near approach to the modern vector (cross) product.

In summary, O'Brien should be viewed a forerunner of Gibbs and Heaviside, but he did not anticipate them in the construction of the modern system. His system was rather primitive as compared to theirs, was developed on a different and less satisfactory foundation, and differed from theirs in at least one major point. His system failed to include a treatment of associativity. If O'Brien's life had not ended in 1855, this last defect might have been remedied, but in neither his last paper nor in any of the earlier ones was it in fact remedied.

O'Brien seems to have had no followers; neither Heaviside nor Gibbs ever mentioned him, even after Knott had called attention to his work. It would be interesting to know what view, if any, Hamilton had of O'Brien's ideas and likewise to know whether O'Brien created his system from Hamilton's system. The first question seems unanswerable since Hamilton referred to O'Brien only once, and then only to include his name in a list of mathematicians who had "at moments turned aside from their own original researches, to notice, and in some instances to extend, results or speculations of mine. . . ."[94]

The answer to the second question seems almost certainly to be that O'Brien created his system directly from Hamilton's. He never stated this, nor did he deny it. O'Brien certainly knew of Hamilton; indeed Hamilton probably taught O'Brien during the early 1830's when O'Brien studied at Trinity College, Dublin. Moreover, as we indicated before, Hamilton's work immediately attracted attention. He had published from 1844 to 1846 eight papers (or eight sections of one paper) in the *Philosophical Magazine* before O'Brien's first paper[95] appeared in the same journal. But this external evidence does no more than add further support to what is obvious from the internal similarities between their systems.

If we put priority disputes aside, O'Brien is found to make a fitting conclusion for this chapter, since he was indeed on the road that lead to Heaviside and Gibbs, and his work further illustrates the breadth and depth of the search for new vectorial systems.

Notes

[1] August Ferdinand Möbius, *Gesammelte Werke,* 4 vols. (Leipzig, 1885–1887).

[2] August Ferdinand Möbius, *Der barycentrische Calcul* (Leipzig, 1827).

[3] Anton Favaro, "Justus Bellavitis, Eine Skizze seines Lebens und wissenschaftlichen Werkens" in Schlömilch's *Zeitschrift für Mathematik und Physik, 26* (1881), 153–169.

[4] Hermann Günther Grassmann, *Hermann Grassmanns Gesammelte mathematische und physikalische Werke,* 3 vols. in 6 pts. (Leipzig, 1894–1911).

[5] Friedrich Engel, *Grassmanns Leben* comprises vol. III, pt. II of the work cited in note (4) above.

[6] Victor Schlegel, *Hermann Grassmann: Sein Leben und Seine Werke* (Leipzig, 1878).

[7] Adhémar Barré, Comte de Saint-Venant, "Mémoire sur les sommes et les différences géométriques, et sur leur usage pour simplifier la mécanique" in *Comptes rendus de l'Académie des Sciences, 21* (1845), 620–625.

[8] Rev. Robert Perceval Graves, *Life of Sir William Rowan Hamilton,* vol. III (Dublin, 1889).

[9] Rev. Matthew O'Brien, "On Symbolic Forms derived from the Conception of the Translation of a Directed Magnitude" in *Philosophical Transactions of the Royal Society of London, 142* (1852), 161–206.

[10] Rudolf Mehmke in his valuable book *Vorlesungen über Punkt und Vektorenrechnung,* 1 Band, Punktrechung, 1. Teilband (Leipzig and Berlin, 1913), 346, argued that Giovanni Ceva could be viewed as a "Vorläufer," a precursor of Möbius in regard to a system of analysis based on points.

[11] For information on Möbius' life and scientific activity the following were used: Richard Baltzer, "August Ferdinand Möbius" (1,I; v–xx) and Curt Reinhardt, "Ueber die Enstehungszeit und den Zusammenhang der wichtigsten Schriften und Abhandlungen von Möbius" (1,IV; 699–728).

[12] These facts were established by Reinhardt through his study of Möbius' unpublished writings. See (1,IV; 707–710).

[13] Published at Leipzig. For this work see (1,IV; 441–476), but for the important appendix see (1,I; 389–398).

[14] A much fuller exposition is available in R. E. Allardice, "The Barycentric Calculus of Möbius" in *Proceedings of the Edinburgh Mathematical Society, 10* (1892), 2–21. Brief expositions with historical comments are given in Josiah Willard Gibbs, "Multiple Algebra" in *Scientific Papers of J. Willard Gibbs,* vol. II (New York, 1961), 91–117; Julian Lowell Coolidge, *A History of Geometrical Methods* (New York, 1963), 148–150; Hermann Rothe, "Systeme geometrischer Analyse" in *Encyklopädie der mathematischen Wissenschaften,* vol. III, pt. 1, 2nd half, (Leipzig, 1931[?]), 1289–1293; Carl B. Boyer, *History of Analytical Geometry* (New York, 1956), 242 ff. A section of Möbius' book was translated by J. P. Kormes and published in *A Source Book in Mathematics,* vol. II, ed. David Eugene Smith (New York, 1959), 670–676.

[15] This discovery was made nearly simultaneously and independently by Etienne Bobillier, K. W. Feuerbach, and Julius Plücker. See Carl B. Boyer, *History of Analytical Geometry* (New York, 1956), 240–244.

[16] See Richard Baltzer's remarks in (1,I; xi–xiii).

[17] Published at Leipzig; later published in (1,IV; 1–318). For Möbius' treatment of vectorial addition, see (1,IV; 41).

[18] Thus, for example, Möbius wrote $A'B'C' = ABC \cdot \cos ABC\hat{\ }A'B'C'$ where $A'B'C'$ is the projection of the triangle ABC on some plane. The right-hand member of this equation is a triangle (fixed as parallel to some plane) multiplied by the cosine of the angle between the two planes, which must be a number, whereas the left-hand member would in general not be parallel to the right-hand member. It is clear that he was simply stating that the projection of a triangle on some plane gives an area which is equal to the original area multiplied by the cosine of the angle between the two planes. However his symbolism did not express this. (1,IV; 685)

[19] In 1859 Möbius published a paper entitled "Neuer Beweis des in Hamilton's *Lectures on Quaternions* aufgestellten associativen Princips bei der Zusammensetzung von Bögen grösster Kreise auf der Kugelfläche." (1,II; 55–70)

[20] Biographical information has been drawn mainly from the work cited in note (3) above and Charles Ange Laisant, "Giusto Bellavitis" in Darboux's *Bulletin des sciences mathématiques*, 2nd Ser., *4* (1880), 343–348.

[21] Giusto Bellavitis, "Saggio di applicazioni di un nuovo metodo di Geometria analitica (Calcolo delle equipollenze)" in *Annali delle Scienze del Regno Lombardo-Veneto. Padova, 5* (1835), 246–247.

[22] See for example (3; 155) or M. Laquière, "Observations sur l'origine naturelle et géométrique du Calcul des Equipollences" in *Association Française pour l'avancement des sciences. Comptes rendus,* (1881), 77. For evidence that Bellavitis himself recognized the equivalence see Giusto Bellavitis, "Sulle Origini del Metodo delle Equipollenze" in *Memorie del Reale Istituto Veneto, di Scienze, Lettere, ed Arti, 19* (1876), 449 ff.

[23] (3; 155). For his views in 1876 see Bellavitis, "Origini," 476–477. For his views in the 1830's see below. In regard to his narrowness of view it is interesting to note that Bellavitis exhibited the same tendency in relation to non-Euclidean geometry and, according to Favaro, crusaded against its acceptance. On this see (3; 159).

[24] Bellavitis, "Origini," 449.

[25] Giusto Bellavitis, "Sulla Geometria Derivata" in *Annali delle Scienze del Regno Lombardo-Veneto. Padova, 2* (1832), 250–253.

[26] Bellavitis, "Origini," 451.

[27] Giusto Bellavitis, "Sopra alcune applicasioni di un nuovo metodo di Geometria Analitica" in *Il Poligrafo. Giornale di Scienze, Lettere ed Arti. Verona, 13* (1833), 53–61. I have not seen this paper, for the above journal does not seem to be available in this country.

[28] Giusto Bellavitis, "Saggio di applicasioni di un nuovo metodo di Geometria analitica (Calcolo delle equipollenze)" in *Annali delle Scienze del Regno Lombardo-Veneto. Padova, 5* (1835), 244–259. I have inferred from some not completely clear comments made by Bellavitis ("Origini," 451) that the 1833 paper contained neither the term "equipollence" nor the presentation of his fully developed system.

[29] There are a number of good biographical studies of Grassmann. The definitive work is cited in note (5) above. A shorter, valuable, but somewhat overly enthusiastic book is cited in note (6) above. Also used was Friedrich Engel, "Hermann Grassmann" in *Jahresbericht der Deutschen Mathematiker-Vereinigung*, 19 (1910), 1–13; A. E. Heath, "Hermann Grassmann" in *Monist, 27* (1917), 1–21; A. E. Heath, "The

Neglect of the Work of H. Grassmann" in *Monist, 27* (1917), 22–35; A. E. Heath, "The Geometrical Analysis of Grassmann and Its Connection with Leibniz's Characteristic" in *Monist, 27* (1917), 36–56; George Sarton, "Grassmann" in *Isis, 35* (1944), 326–330; R. Sturm, E. Schröder, and L. Sohncke, "Hermann Grassmann: sein Leben und seine mathematisch-physikalischen Arbeiten" in *Mathematische Annalen, 14* (1879), 1–45; and Victor Schlegel's valuable article, mainly on the reception of Grassmann's work, "Die Grassmann'sche Ausdehnungslehre" in *Zeitschrift für Mathematik und Physik, 41* (1896), 1–21 and 41–59.

[30] See (5; 3–7) and (6; 16–30).

[31] This was published posthumously in 1911 as vol. III, pt. I, of the work cited in note (4) above.

[32] Concerning the details of this period in Grassmann's life see (5; 72–85) and for Conrad's statement see (4,III,I; 209).

[33] Though in this foreword Grassmann did not give the date 1832, there is every reason to believe that Engel was right in taking the statements from the foreword as referring to the 1832 period.

[34] See (4,I,II; 8–9) for Grassmann's statement and (4,I,II; 397–398) for the relevant part of Gauss' letter.

[35] (4,I,I; 8). These books, though intended for elementary instruction, were written from a rather philosophical point of view. On this see (6; 48) and (5; 2–7). The full references to Grassmann's father's two books are Justus Günther Grassmann, *Raumlehre, Ebene raumliche Grössenlehre* (Berlin, 1824) and *Lehrbuch der ebenen und sphärischen Trigonometrie* (Berlin, 1835). The first book is the second part of his *Raumlehre,* the first part having been published in 1817 under the title *Geometrische Combinationslehre.* These books seem to be quite rare; neither the Library of Congress, British Museum, nor Biblioteque Nationale list copies in their catalogues.

[36] For Wilson's discussion of these two products and for Wilson's statements concerning Gibbs' views see Edwin Bidwell Wilson, "On Products in Additive Fields" in *Verhandlungen des III. Internationalen Mathematiker Kongresses im Heidelberg, 1904.* (Leipzig, 1905), 202–215.

[37] Concerning this see Grassmann's statement of 1877: "Uber das Verhältniss der nichteuklidischen Geometrie zur Ausdehnungslehre" (4,I,I; 293–294), and for Engel's note thereon see (4,I,I; 413).

[38] This is the year suggested by Engel. See (5; 92).

[39] (5; 331). This is based on Engel's restoration of one unclearly abbreviated word in the original letter.

[40] (4,I,I; 22). Ernest Nagel gave an interesting, but brief, discussion of the degree of philosophical sophistication manifested by Grassmann in this work in his "The Formation of Modern Conceptions of Formal Logic in the Development of Geometry" in *Osiris, 7* (1939), 142–224. On Grassmann see especially pages 168–174.

[41] This is not an inappropriate term; indeed Whitehead's *Universal Algebra* included many of Grassmann's ideas.

[42] For their statements see (5; 102, 102, 207).

[43] In a paper published in 1855, which will be discussed later.

[44] Some discussion of the writings in English that attempt to explain Grassmann's system may be given at this point. There are five such writings: (1) Hermann Grassmann, "A Brief Account of the Essential Features of Grassmann's Extensive Algebra," trans. Wooster Woodruff Beman, in *Analyst, 8* (1881), 96–97 and 114–124; (2) Alexander Ziwet, "A Brief Account of H. Grassmann's Geometrical Theories" in *Annals of Math, 2* (1885–1886), 1–11 and 25–34; (3) Edward Wyllys Hyde, *The*

Directional Calculus (Boston, 1890); (4) Edward Wyllys Hyde, *Grassmann's Space Analysis* (London, 1906); (5) Joseph V. Collins, "An Elementary Exposition of Grassmann's 'Ausdehnungslehre'" in *American Mathematical Monthly*, 6 (1899), 193–198, 261–266, 297–301, and 7 (1900), 31–35, 163–166, 181–187, 207–214, 253–258. I have used all of these, but their usefulness is definitely limited. They did however set the standard in giving English equivalents for Grassmann's new terms, and I have thus used their terms. Item 1 is the only mathematical writing by Grassmann ever translated into English. It is useful in that in it he attempted to summarize his *Ausdehnungslehre* of 1844, but an adequate summary was beyond its scope; in fact only 32 of the 188 sections of the *Ausdehnungslehre* were summarized. Items 2, 3, 4, and 5 all stem from the *Ausdehnungslehre* of 1862, which embodied a presentation substantially different from that of the earlier book. Item 2 is short, general, and hence limited. Items 3 and 4 contain Grassmann's ideas in a simplified form, i.e., limited to three dimensions and with the stress on applications. Item 5 stays fairly close to the original text of 1862, as 3 and 4 do not; it however is marred by misprints and a few mathematical errors and, though surpassing the above works, still falls far short of being an adequate summary of the *Ausdehnungslehre* of 1862. Finally Alfred North Whitehead's *Universal Algebra* (Cambridge, England, 1898) and Henry George Forder's *Calculus of Extension* (Cambridge, England, 1941) should be mentioned, though both depart considerably from Grassmann's presentation and content, and hence were used sparingly.

[45] (4,I,I; 62). This may be compared with the statement found in current vector analysis books that any vector may be expressed in the form $a\mathbf{i} + b\mathbf{j} + c\mathbf{k}$ where a, b, c are unique.

[46] (4,I,I; 11–12). In a footnote Grassmann stated that this product would be developed in the second volume of the *Ausdehnungslehre*, but no second volume appeared. He did however develop this product in other writings.

[47] This was so because for Grassmann any distributive operation was by definition a multiplication. For Grassmann's presentation of the above section see (4,I,I; 77–79).

[48] Concerning this see (4,I,I; 93–99, 114–118).

[49] This was published in Grunert's *Archiv der Mathematik und Physik*, 6 (1845), 337–350, and republished in (4,I,I; 297–312). It will be discussed later. Here as elsewhere in the summary Grassmann focused on summarizing the results more than the methods.

[50] As translated by W. W. Beman and published as "A Brief Account of the Essential Features of Grassmann's Extensive Algebra" in *Analyst*, 8 (1881), 117.

[51] Josiah Willard Gibbs, *The Scientific Papers of J. Willard Gibbs*, vol. II (New York, 1961), 167. For a fuller discussion see pages 94–113, 162–167.

[52] George Sarton, "Grassmann" in *Isis*, 35 (1944), 327.

[53] The letter is given in part in (4,I,II; 397–398). It may be noted that, as we pointed out previously, Grassmann first became acquainted with the geometrical representation of complex numbers through this letter. It may also be noted that Gauss wrote, probably in July, 1831, a very short essay entitled "Geometrische Seite der Ternaren Formen." This was however only published posthumously in Carl Freidrich Gauss, *Werke*, vol. II (Göttingen, 1863), 305–310; therein Gauss developed an expression roughly equivalent to the scalar product.

[54] Ernest Nagel, "The Formation of Modern Conceptions of Formal Logic in the Development of Geometry" in *Osiris*, 7 (1939), 173–174.

[55] A. E. Heath, "The Neglect of the Work of H. Grassmann" in *Monist*, 27 (1917), 24.

[56] Hermann Günther Grassmann, "Kurze Übersicht über das Wesen der Ausdehnungslehre" in Grunert's *Archiv der Mathematik und Physik, 6* (1845), 337–350. Republished in (4,I,I; 297–312).

[57] Hermann Günther Grassmann, "Neve Theorie der Elektrodynamik" in Poggendorff's *Annalen der Physik und Chemie, 64* (1845), 1–18. Republished in (4,II,II; 147–160).

[58] Strictly speaking it was not a competition, since Grassmann's work was the only essay submitted! On this see (5; 119).

[59] Hermann Günther Grassmann, *Die Geometrische Analyse geknüpft und die von Leibnitz erfundene geometrische charakteristik* (Leipzig, 1847). For this work see (4,I,I; 321–398). For Möbius' essay, which was entitled "Die Grassmannische Lehre von Punktgrössen und den davon abhängenden Grössenformen," see (1,I; 613–633).

[60] See for example Saint-Venant's explicit statement to this effect. (5; 200).

[61] In this regard it should be noted that (1) one of Gauss' two publications of 1831 was in Latin (the other in German), and hence Saint-Venant might have had access to this work; (2) Saint-Venant designated vectors by a straight line above the letter symbolizing the vector (thus $\bar{a}$), and Argand was the first and only one of the men mentioned who used the same symbol; and (3) Saint-Venant knew of Warren's work by at least 1853, when he mentioned it in an article, "De l'interprétation géométrique des clefs algébriques et des déterminants" in *Comptes rendus de l'Académie des Sciences, 36* (1853), 583.

[62] Although Saint-Venant is rarely mentioned by later authors, he did influence his countryman Henri Resal, whose *Traité de Cinématique purè* (Paris, 1862) included a presentation of elementary vector addition, subtraction, and differentiation as well as one vectorial product. Resal's product was however not Saint-Venant's; it was instead the modern scalar (dot) product which he presumably created by going from Saint-Venant's $a\, b \sin \theta$ to $a\, b \cos \theta$. Neither Grassmann nor Hamilton is mentioned, and although references to Saint-Venant are few (see pages 18 and 64), they are sufficient to show that the latter was his source. Resal was followed in his limited use of vectorial methods in mechanics by Joseph Somoff. See Somoff, *Theoretische Mechanik . Kinematik,* trans. from Russian by Alexander Ziwet (Leipzig, 1878), x–xi and 46–51.

[63] Augustin Cauchy, "Sur les Clefs algébriques" in *Comptes rendus de l'Academie des Sciences, 36* (1853), 70–75 and 129–136. See also *Œuvres completes d'Augustin Cauchy,* 1st Ser., vol. XI (Paris, 1899), 439–445, and vol. XII (Paris, 1900), 12–21. Cauchy also published "Sur les avantages que présente, dans un grand nombre de questions l'emploi des clefs algébriques" in *Comptes rendus, 36* (1853), 161–169, or in *Œuvres,* 1st Ser., vol. XII (Paris, 1900), 21–30; and "Mémoire sur les différentielles et les variations employées comme clefs algébriques" in *Comptes rendus, 36* (1853), 38–45 and 57–68, or in *Œuvres,* 1st Ser., vol. XII (Paris, 1900), 46–63; and a memoir published as a separate publication under the title *Mémoire sur les clefs algébriques,* its title as republished in *Œuvres,* 2nd Ser., vol. XII (Paris, 1938), 417–466.

[64] Because of the simplicity of the above illustration, the usefulness of the method for the solution of a large number of simultaneous equations is not readily apparent. Cauchy's method is directly analogous and involves the introduction of n distinct keys for the solution of n equations. For Cauchy's more elaborate example see *Œuvres,* 1st Ser., vol. XII (Paris, 1900), 13 ff.

[65] Concerning this see (4,I,I; 99–102; 156–157).

[66] Comte de Saint-Venant, "De l'interprétation géométrique des clefs algébriques

et des déterminants" in *Comptes rendus de l'Académie des Sciences, 36* (1853), 582–585.

[67] Hermann Günther Grassmann, "Sur les différents genres de multiplication" in Crelle's *Journal für die reine und angewandte Mathematik, 44* (1855), 123–141. Republished in (4,II,I; 199–217).

[68] Victor Schlegel, "Die Grassmann'sche Ausdehnungslehre. Ein Beitrag zur Geschichte der Mathematik in den letzten fünfzig Jahren" in Schlömilch's *Zeitschrift für Mathematik und Physik, 41* (1896), 6.

[69] Augustin Cauchy, "Sur les Quantités géométriques" in *Comptes rendus de l'Académie des Sciences, 29* (1849), 250–257, or in *Œuvres d'Augustin Cauchy*, 1st Ser., vol. XI (Paris, 1899), 152–160. The method of solving equations given in this paper has no connection with his later method; the paper deals mainly with the interpretation of complex numbers according to the geometrical methods developed in 1845 by Saint-Venant for geometric lines.

[70] Cauchy, "Sur les Clefs algébriques" in *Œuvres*, 1st Ser., vol. XI (Paris, 1899), 443–445, for Hamilton; and Cauchy, "Sur la Théorie des moments linéaires et sur les moments linéaires des divers ordres" in *Œuvres*, 1st Ser., vol. XII (Paris, 1900), 6, for Möbius.

[71] Sir William Rowan Hamilton, *Lectures on Quaternions* (Dublin, 1853), (62) of preface.

[72] Luigi Cremona, "Solution des questions 494 et 499, méthode Grassmann et propriété de la cubique gauche" in *Nouvelle Annales de mathématiques, 19* (1860), 356–361.

[73] *Ibid.*, 357

[74] Concerning this see Victor Schlegel, "Die Grassmann' sche Ausdehnungslehre" in *Zeitschrift für Mathematik und Physik, 41* (1896), 48.

[75] See the list of his publications as given in (5; 356–367).

[76] Hermann Günther Grassmann, *Die Ausdehnungslehre: Vollstandig und in strenger Form bearbeitet* (Berlin, 1862). Republished as vol. I, pt. II, of the work cited in note (4) above.

[77] Grunert wrote this note and Möbius may have written a note. See (5; 231–232, 223–224).

[78] Hermann Günther Grassmann, *Wörterbuch zum Rig-Veda*, 6 parts (Leipzig, 1872–1875).

[79] *Rig-Veda*, trans. Hermann Günther Grassmann, 2 parts (Leipzig, 1876–1877). George Sarton discussed the significance of Grassmann's philological achievement in his "Grassmann" in *Isis, 35* (1944), 326–330.

[80] Hermann Günther Grassmann, "Die neuere Algebra und die Ausdehnungslehre" in *Mathematische Annalen, 7* (1874), 538–548, and in (4,II,I; 256–267).

[81] Hermann Günther Grassmann, "Zur Elektrodynamik" in Crelle's *Journal für die reine und angewandte Mathematik, 83* (1877), 57–64, and in (4,II,II; 203–210).

[82] This paper is also significant in that through it Gibbs first came to know of Grassmann's work. Concerning this see Gibbs' letter to Schlegel given in Lynde Phelps Wheeler, *Josiah Willard Gibbs* (New Haven, 1962), 108.

[83] Hermann Günther Grassmann, "Die Mechanik nach den Prinzipien der Ausdehnungslehre" in *Mathematische Annalen, 12* (1877), 222–240, and in (4,II,II; 46–72).

[84] Hermann Günther Grassmann, "Der Ort der Hamiltonschen Quaternionen in der Ausdehnungslehre" in *Mathematische Annalen, 12* (1877), 375–386, and in (4,II,I; 268–282).

[85] For further details see the next chapter.

[86] For a fuller discussion of period see Victor Schlegel, "Die Grassmann'sche Ausdehnungslehre" in *Zeitschrift für Mathematik und Physik, 41* (1896), 1–21, 41–59.

[87] Felix Klein, *Vorlesungen über die Entwicklung der Mathematik im 19. Jahrhundert* (in 2 parts, published as one volume), pt. 1, ed. R. Courant and O. Neugebauer; pt. 2, ed. R. Courant and St. Cohn-Vossen (New York, 1956), pt. 1, 181–182.

[88] See the letter to Gibbs from Hermann Grassmann, Jr., as given by Wheeler, *Gibbs*, 116–117.

[89] See Engel's remarks in (4,I,I; vi).

[90] Cargill Gilston Knott, "Recent Innovations in Vector Theory" in *Proceedings of the Royal Society of Edinburgh, 19* (1892), 212.

[91] The above information has been taken from Charles Parish, "Matthew O'Brien" in *Dictionary of National Biography*, vol. XIV (Oxford, 1917), 794, which seems to be the best of the extremely few sources on O'Brien.

[92] The following constitutes a complete bibliography of O'Brien's papers relevant to vector analysis. (1) O'Brien, Matthew. "Contributions towards a System of Symbolical Geometry and Mechanics" (Abstract of paper IV), *Philosophical Magazine*, 3rd Ser., *31* (1847), 139–143; (2) ______. "On the Symbolical Vibratory Motion of an Elastic Medium, whether Crystallized or Uncrystallized" (Abstract of paper V), *Philosophical Magazine*, 3rd Ser., *31* (1847), 376–380; (3) ______. "On a New Notation for expressing various Conditions and Equations in Geometry, Mechanics, and Astronomy," *Transactions of the Cambridge Philosophical Society*, 8 (1849), 415–428; (4) ______. "Contributions towards a System of Symbolical Geometry and Mechanics," *Transactions of the Cambridge Philosophical Society*, 8 (1849), 497–507; (5) ______. "On the Symbolic Equation of Vibratory Motion of an Elastic Medium, whether Crystallized or Uncrystallized," *Transactions of the Cambridge Philosophical Society*, 8 (1849), 508–523; (6) ______. "On the Interpretation of the Product of a Line and a Force," *Philosophical Magazine*, 4th Ser., *1* (1851), 394–397; (7) ______. "On Symbolical Statics," *Philosophical Magazine*, 4th Ser., 2 (1851), 491–495; (8) ______. "On Symbolical Mechanics," *Philosophical Magazine*, 4th Ser., 2 (1851), 121–125; (9) ______. "On Symbolic Forms derived from the Conception of the Translation of a Directed Magnitude," *Philosophical Transactions of the Royal Society of London, 142* (1852), 161–206. This is the work referred to in note (9) above.

[93] O'Brien worked within a tradition common among British algebraists of the day who were interested in treating algebra as a science of symbols. A number of important mathematicians worked within this tradition, including Peacock, Hamilton, D. F. Gregory, and De Morgan. The points of view of these men of course differed, and O'Brien shared most sympathies with Gregory, as he explicitly stated. (9; 163)

[94] Sir William Rowan Hamilton, *Lectures on Quaternions* (Dublin, 1853), p. (64) of the preface. See the footnote on that page for O'Brien's name. C. G. Knott stated: "In one of his early papers Hamilton refers to O'Brien's work in terms of high praise." Knott, "Recent Innovations," p. 213. Knott however gave no reference to this paper, and a search of all Hamilton's papers from 1844 to 1858 – including a check of his biography (Robert Perceval Graves, *The Life of Sir William Rowan Hamilton*, 3 vols. [Dublin, 1882–1889]) – produced no information whatsoever concerning such a paper and indeed made it doubtful that Hamilton was even familiar with O'Brien's work, since in his papers and letters he often did discuss papers by his contemporaries relating to quaternions.

[95] This paper was a synopsis of a paper delivered in 1846 to the Cambridge Philosophical Society and published in the *Transactions* of that Society in 1849.

CHAPTER FOUR

Traditions in Vectorial Analysis from the Middle Period of Its History

I. *Introduction*

As a first approximation the history of vector analysis may be divided into three periods. The first period may be characterized as the time when mathematicians searched for, discovered, and developed systems of hypercomplex numbers which could be used for space analysis. This period began in the late eighteenth century with Wessel; its conclusion is best set as 1865, the year in which Hamilton died. The second period may be described as the time when some of the vectorial systems of the first period were discussed, tested, and in some cases extended. This period was more a time of "recognitions" than of discoveries. Thus for example during this time a number of scientists recognized the need for vectorial methods and in some cases specified the characteristics to be looked for in a vectorial system. The year 1880 may be taken as the terminus for this period as well as the beginning of the third period, which extended to 1910. During the third period the modern system of vector analysis was created, discussed, and accepted.

The above periodization is of course not entirely satisfactory; its value and accuracy as well as its shortcomings will become apparent from later sections. Its immediate relevance is that in terms of it the function of the present chapter may be defined: this chapter attempts an analysis of the second, or middle, period of the history of vector analysis.

The central figures in this chapter are Tait, Peirce, Maxwell, and Clifford. The selection of the year 1880 as the terminus of this period should be viewed as somewhat arbitrary, even though it is true that Maxwell and Clifford died in 1879 and Peirce in 1880, and that Gibbs' first presentation of modern vector analysis appeared in 1881. Nevertheless the transitions that occurred in these men in the period 1865 to 1880 took place in the minds of lesser men at a later date.

More specifically, this chapter is aimed at presenting a discussion of the degree and kind of reception accorded to the two major traditions that emerged from the first period; these were the Hamiltonian and the Grassmannian traditions. The second of these has briefly been discussed for this period of time in the last chapter, which was natural since Grassmann died in 1878. However there is more to be said in this regard.

Evidence will be supplied to show that the Hamiltonian tradition became the more vigorous, and activities and transformations within this tradition will be especially analyzed, since it was from this tradition that the modern system of vector analysis arose.

The four men emphasized in this chapter may at first sight seem strange bedfellows. Tait was the great champion of quaternions after Hamilton's death; Maxwell, on the other hand, was opposed to quaternions. Similarly Peirce shared Tait's enthusiasm, while Clifford could not. Be that as it may, in this chapter it will be shown that these men represent various traditions within the vectorial tradition; in later chapters it will be shown that some of these traditions played a decisive role in the history of vector analysis.

II. *Interest in Vectorial Analysis in Various Countries from 1841 to 1900*

There are numerous ways in which an historical study of the acceptance of a mathematical system may be conducted. Among these various methods is the quantitative study of the number of works published in the system as related to certain variables, for instance, time and country of publication. In the present section such methods have been used in an attempt to analyze the level of interest in the Grassmannian and the Hamiltonian traditions in the period from 1841 to 1900. Works from these two traditions have been classified in terms of (1) nature of publication, (2) subject of publication, (3) time of publication as broken down into five-year intervals from 1841 to 1900, (4) country of publication, and (5) author in some cases. It is the author's belief that such a study does not replace historical interpretation; it is simply another form of analysis which can complement the more traditional techniques.[9]

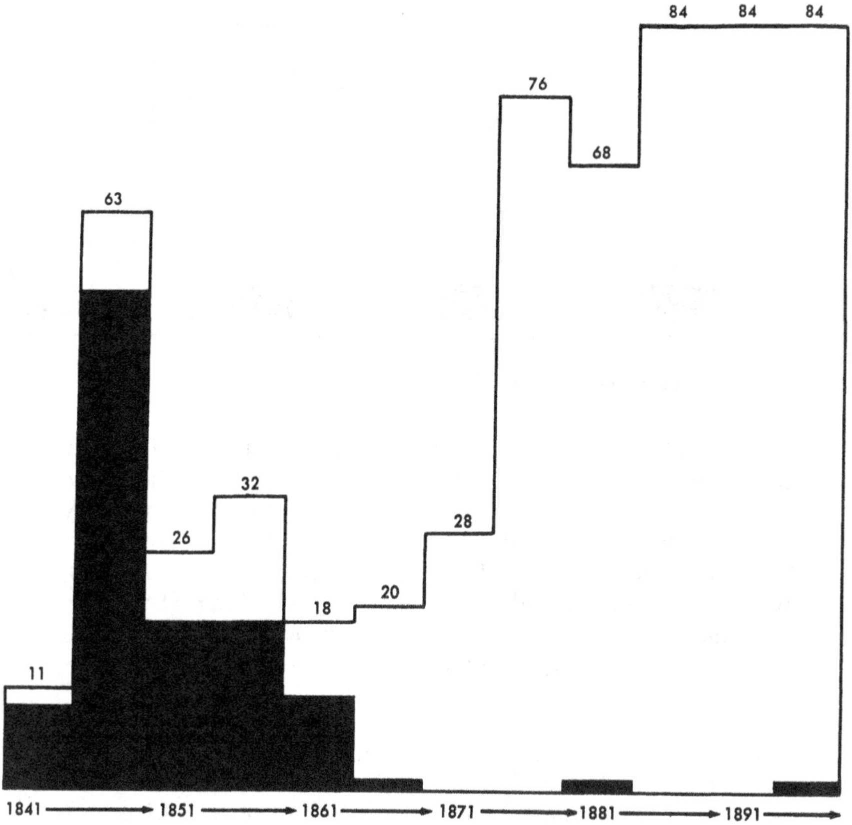

GRAPH I. Quaternion Publications from 1841 to 1900.

The first major finding derived from this study is that during the period from 1841 to 1900 there were 594 quaternion publications as compared to 217 Grassmannian analysis publications.[10] Hence 73.2 percent of the publications were in the quaternion tradition, or there were 2.73 quaternion publications for each Grassmannian publication.

The results obtained when only books were considered was strikingly similar. By actual count there were 38 quaternion books published from 1841 to 1900, whereas there were 16 books published during this period in the Grassmannian tradition. Thus 70.4 percent of the books dealt with quaternions, or there were 2.37 quaternion books for each book in the Grassmannian tradition. The quaternion books averaged 281 pages in length; the books of the Grassmannian tradition, 249 pages.[11] From these numbers it may

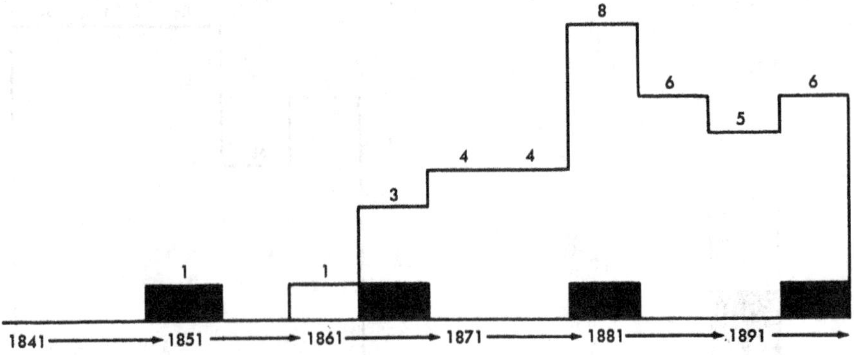

GRAPH II. Quaternion Books from 1841 to 1900.

be inferred that interest in the tradition begun with Hamilton was far greater than that begun with Grassmann. These numbers have been broken down into five-year intervals in Graphs I, II, IV, and V.

Graph I shows the number of quaternion publications in terms of five-year intervals from 1841 to 1900. The publications written by Hamilton himself are indicated by solid areas. Graph II presents the number of quaternion books for the same time intervals. The books by Hamilton (including a translation and a second edition) are indicated by solid areas.

From Graphs I and II the following conclusions may be drawn. Interest in quaternion analysis was at its highest level during the 1876–1900 period. The decrease in interest for the period 1881–1885 indicated by Graph I is balanced by the peaking of Graph II for the same interval. It is important to note that Hamilton wrote 73 percent of the pre-1866 quaternion publications and 19 percent of all quaternion publications.

It would of course be significant to compare the form of Graph I with a graph showing the rate of increase of mathematical publications during this time. Some idea of how the rate of increase of quaternion publications after 1870 compares with the rate of increase in mathematical publications in general may be obtained by means of the study made by H. S. White in 1915 based on an analysis of works listed in the journal *Fortschritte der Mathematik.*[12] Graph III shows the number of titles of mathematical articles and books published in the period 1868 to 1909.[13]

When Graphs I and III are compared, it seems at first sight that interest in quaternions was declining from 1876 to 1900, for it is evident that the percentage of mathematical literature that was devoted to quaternions decreases slightly. But this seems to be an

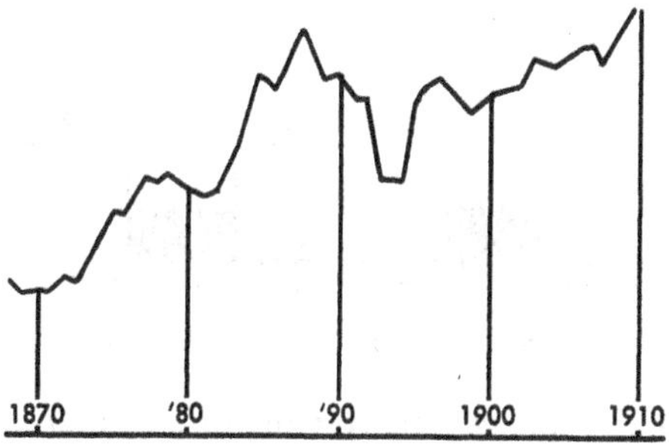

GRAPH III. Annual Number of Titles of Mathematical Articles and Books, 1868–1909.

erroneous conclusion, for even more striking than the increase in the number of mathematical publications during this period is the increase in the number of fields of mathematical research. Numerous fields—such as non-Euclidean geometry, mathematical logic, group theory, as well as many branches of applied mathematics—came into prominence in this period.

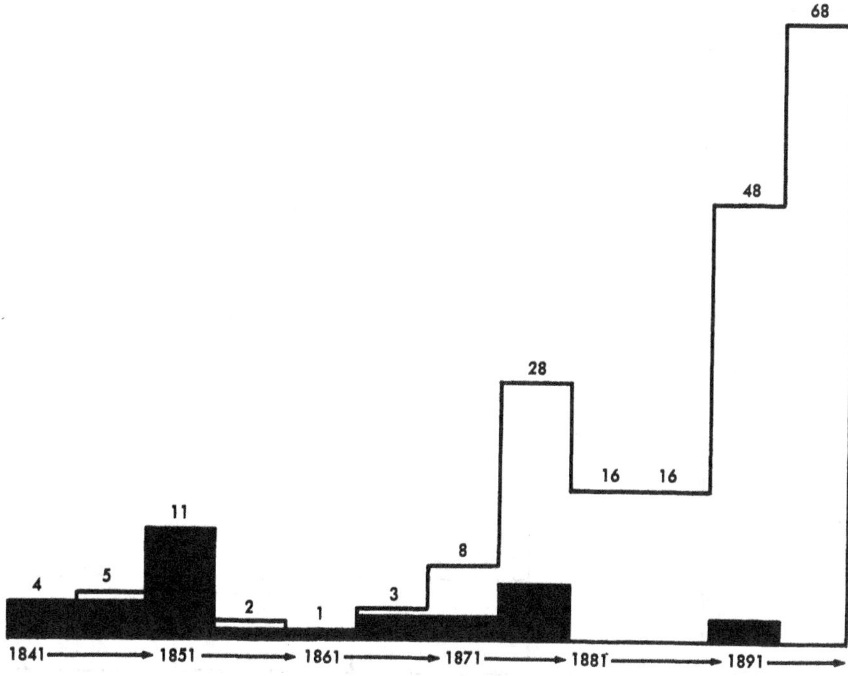

GRAPH IV. Grassmannian Analysis Publications from 1841 to 1900.

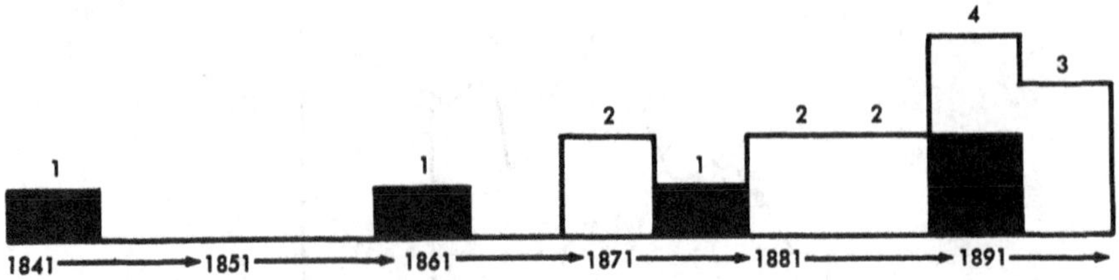

GRAPH V. Grassmannian Analysis Books from 1841 to 1900.

Graphs IV and V present the results of a similar study of Grassmannian analysis publications and books. The works written by Grassmann are indicated by the solid areas. From these graphs it becomes clear that the beginning of the main period of interest in Grassmannian analysis came roughly fifteen years later than the similar period for quaternions (1891 compared to 1876). On the other hand, the number of Grassmannian analysis publications for the period 1896 to 1900 was approaching the number of quaternion publications for the same interval. In regard to Grassmann's per-

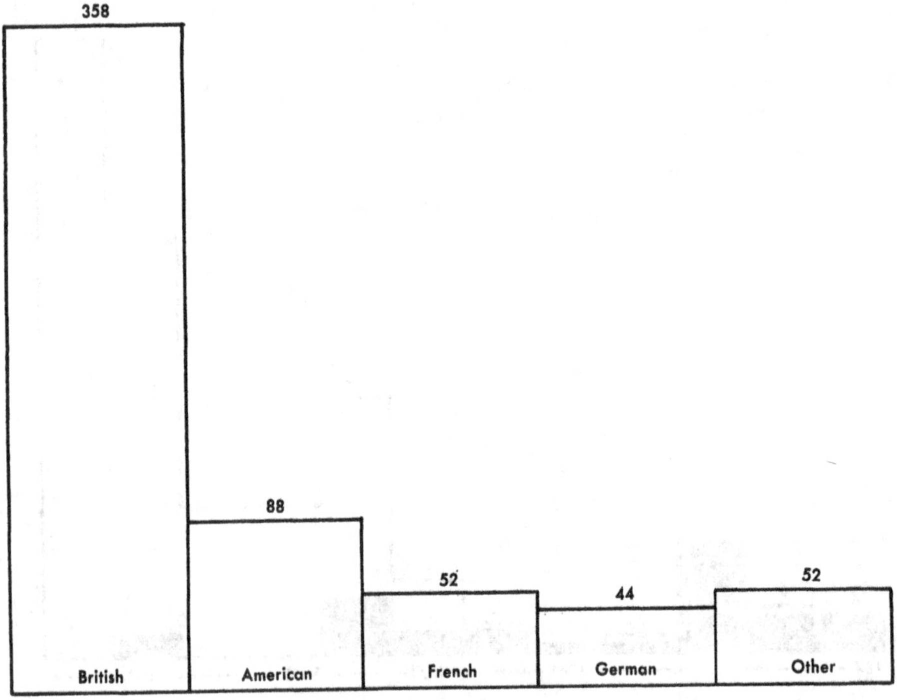

GRAPH VI. Quaternion Publications by Country.

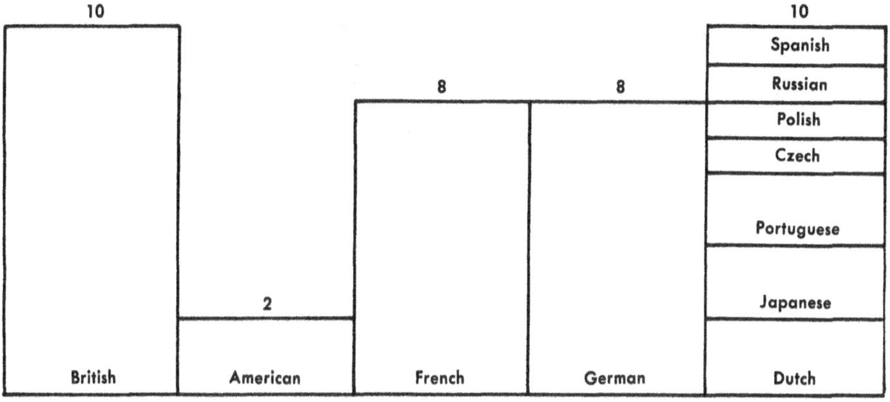

GRAPH VII. Quaternion Books by Country.

sonal contribution it is noteworthy that he published 25 of the 33 (or 76 percent) of the publications up to 1875, and 33 of the total of 217 (or 15 percent) of Grassmannian analysis publications.

A study of the two fields in terms of interest by country is of significance. Thus Graphs VI and VII represent quaternion publications and books respectively as classified into five groups: British, American, French, German, and those of other countries. Quaternion books appeared in ten languages as follows (the number after

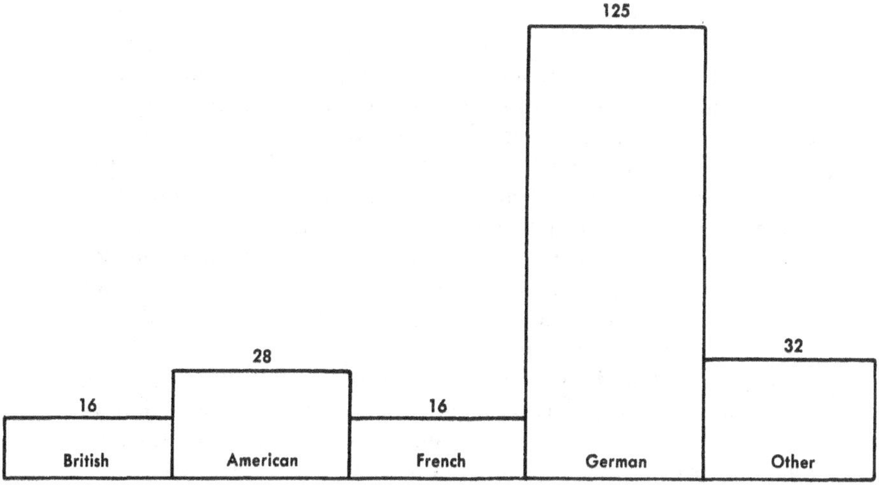

GRAPH VIII. Grassmannian Analysis Publications by Country. (Note that if any given height on this scale indicates x publications, then an equal height on Graph VI indicates 2x publications.)

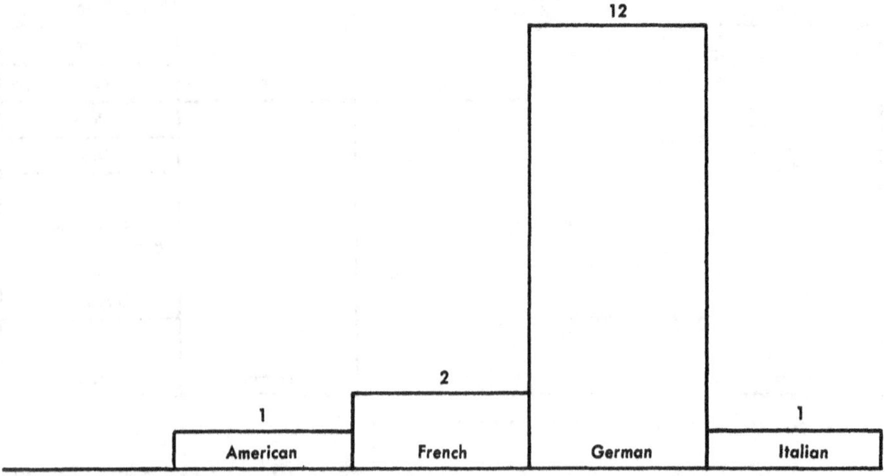

GRAPH IX. Grassmannian Analysis Books by Country.

each language indicates the number of books): English (12, of which 10 were published in Britain and 2 in America), French (8), German (8), Dutch (2), Japanese (2), Portuguese (2), Czechoslovakian (1), Polish (1), Russian (1), and Spanish (1). There were in addition a number of papers in Italian, at least one in Danish, and at least one paper was published in Australia. From this quantitative study it may be inferred that 60 percent of the publications (books and papers) were British, 15 percent were American, 9 percent were French, 8 percent German, with the remaining 8 percent coming from other countries. This should be considered in relation to the fact that 26 percent of the books on quaternions were of British origin, 5 percent of American origin (the British books were of course often used by Americans), 21 percent of French and 21 percent of German origin, and the remaining were in other languages. These statistics point out that interest in quaternions was strongest in Britain but was substantial in America, Germany, and France, and that it extended to most of the then intellectually productive countries of the world.

Graphs VIII and IX represent Grassmannian analysis publications and books respectively that are classified on the same basis of country of publication. Grassmannian analysis books appeared in four countries: 12 were published in Germany, 2 in France, and 1 each in America and Italy. The study revealed that 57 percent of the Grassmannian analysis publications appeared in Germany, 18 percent in America, 10.5 percent in both Britain and France, with a few works appearing in Polish, Italian, Spanish, Russian, and Czechoslovakian.

Thus it appears that interest in Grassmannian analysis was centralized in Germany, with less proportionate interest outside Germany than the interest in quaternions outside Britain. It is noteworthy that for both systems the country in which the most interest developed after the mother country of the analysis was America.[14]

This study may be summarized by the statement that the interest in quaternion analysis was roughly two and one-half times as great as interest in Grassmannian analysis and extended to more countries, with greater interest proportionately developing in countries outside the country in which the system originated. To this may be added the observation that there was substantial interest in quaternions from 1876 to 1900 and that although interest in Grassmannian analysis came somewhat later, it did by the period 1891–1900 attain substantial magnitude.

It is the author's belief that this quantified study tells no more than part of the story. It does however supply a valuable perspective into which developments discussed in later sections may be set.

III. *Peter Guthrie Tait: Advocate and Developer of Quaternions*

The importance of Tait for this history is fourfold. (1) He was the acknowledged leader of the quaternion analysts from 1865 until his death in 1901. Indeed eight books on quaternions (including later editions, translations, and coauthorships) carried his name on the title page. (2) Tait developed quaternion analysis as a tool for research in physical science (as Hamilton had not) and created many new theorems in quaternion analysis which were capable of being translated into modern vector analysis. (3) It was probably through Tait that Maxwell became interested in quaternions. (4) Tait was the most important opponent of modern vector analysis.

Peter Guthrie Tait was born in 1831 near Edinburgh, Scotland. In 1841 he entered Edinburgh Academy where one year earlier the young James Clerk Maxwell had enrolled. Playmates in their youth, the two became fast friends and frequent correspondents in their maturity. Maxwell's 1846 entrance into Edinburgh University was followed by Tait's in 1847, with the order of entry being reversed when Tait left for Cambridge after one year at Edinburgh University, while Maxwell stayed for three. After Tait's graduation in 1852 as Senior Wrangler and First Smith's Prizeman he was elected a Fellow of Peterhouse College, Cambridge, and began writing the first of his many books. This was coauthored by W. J.

Steele (who died before its completion) and was entitled *Dynamics of a Particle.* In 1853 Tait ordered a copy of the just-published *Lectures on Quaternions.* As Tait later wrote to Hamilton: "when I ordered your book, on account of an advertisement in the Athenaeum, I had NO IDEA what it was about. The startling title caught my eye in August '53, and as I was going off to shooting quarters I took it and some scribbling paper with me to beguile the time. . . . However as I told you in my first letter I got easily enough through the first six Lectures. . . ." (1; 126)

On his return to Cambridge Tait did not continue his study of quaternions, primarily because of the labor involved in the book he was writing. In 1854 he accepted the Professorship in Mathematics at Queen's College, Belfast, Ireland. Here he became involved in teaching as well as in experimental work with his colleague Thomas Andrews. He also pursued the study of the "Theories of Heat, Electricity and Light." Finally in August, 1857, his interest in quaternions returned as a result of reading Helmholtz' famous paper on vortex motion. Then as Tait wrote to Hamilton: "I suddenly bethought me of certain formulae I had admired years ago at p. 610 of your Lectures—and which I thought (and still think) likely to serve my purpose exactly." (1; 127) The formulae to which Tait referred were those associated with the operator "$\lhd = i\frac{d}{dx} + j\frac{d}{dy} + k\frac{d}{dz}$."[15] It is clear from the above (and from numerous other statements by Tait) that his interest in quaternions was for their physical applications. From 1857 Tait's interest in quaternions continued to increase, and in 1858 Andrews wrote to Hamilton on Tait's behalf to request Hamilton to allow Tait to correspond with him. Hamilton responded cordially, and a voluminous correspondence ensued. Perhaps fifty letters were written, one of which was ninety-six pages in length. (1; 141)

Hamilton soon realized that Tait was a gifted mathematician and in 1859 encouraged Tait to publish a paper based on a quaternion investigation of the Fresnel wave surface that Tait had carried out. This Tait did, and thus the first of his numerous quaternion papers appeared in 1859. In the same year Tait was encouraged by some Cambridge friends to write a book on quaternions. Since Hamilton had communicated unpublished results to Tait, it was natural for Tait to ask Hamilton's permission to publish such a book, even though Tait intended that it would consist mainly of examples. Hamilton asked Tait to wait for the publication of his *Elements of Quaternions,* which was at that time planned as an elementary work. Tait agreed to this request, and thus his book on quaternions was not published until 1867. It is interesting that at nearly the same time John Herschel (at age 72) was engaged in preparing an elementary work on quaternions, but it was never published. (1; 141–142)

In 1860 Tait was appointed to the Chair of Natural Philosophy at Edinburgh University after he had been selected for this position from a noteworthy group of candidates that included Maxwell. Tait remained at Edinburgh until his death in 1901 and distinquished himself as an excellent lecturer, an important writer of scientific books (twenty-two books were either wholly or partially written by him), and as a productive scientific researcher. He published 365 papers, of which approximately 70 were on quaternions. The commonly accepted view of his importance as a scientist places him below the group consisting of such men as Kelvin, Maxwell, and Stokes, but certainly above the majority of the scientific professors of nineteenth-century Britain.

Soon after arriving at Edinburgh, Tait began writing with William Thomson (known from 1892 as Lord Kelvin) a work on mathematical physics entitled *Treatise on Natural Philosophy*. Originally the authors planned to survey all of physical science, but only the volume on mechanics was completed. This was published in 1867 and became an immediate success; it was often referred to as the *Principia* of the nineteenth century and was compared to the works of Lagrange and Laplace. In 1871 Helmholtz arranged for its translation into German. This was the golden opportunity for quaternions, for their inclusion in such an important work on mathematical physics would have acquainted numerous readers with quaternion methods. Kelvin however was strongly opposed to the introduction of quaternions, and they were not introduced. After Tait's death Kelvin in 1901 wrote to George Chrystal:

> We [Kelvin and Tait] have had a thirty-eight years' war over quaternions. He had been captivated by the originality and extraordinary beauty of Hamilton's genius in this respect, and had accepted, I believe, definitely from Hamilton to take charge of quaternions after his death, which he has most loyally executed. Times without number I offered to let quaternions into Thomson and Tait [the *Treatise*], if he could only show that in any case our work would be helped by their use. You will see that from beginning to end they were never introduced.[16]

Similarly Kelvin wrote in 1892 to R. B. Hayward concerning the latter's *Algebra of Coplanar Vectors and Trigonometry:* "I do think, however, that you would find it would lose nothing by omitting the word 'vector' throughout. It adds nothing to the clearness or simplicity of the geometry, whether of two-dimensions or three-dimensions. Quaternions came from Hamilton after his really good work had been done; and, though beautifully ingenious, have been an unmixed evil to those who have touched them in any way, including Clerk Maxwell." [17] The following quotation indicates that Kelvin's strong opposition to quaternions probably extended to any sort of vector analysis. In 1896 Kelvin wrote to G. F. FitzGerald in regard to FitzGerald's mention of "vectors and symmetrical equations": "Symmetrical equations are good in their place, but 'vector' is a useless survival, or offshoot, from quaternions, and has never been of the slightest use to any creature. Hertz wisely shunted it, but unwisely he adopted temporarily Heaviside's nihilism. He even tended to nihilism in dynamics, as I warned you soon after his death. He would have grown out of all this, I believe, if he had lived." [18] Whatever might have been Kelvin's reasons for rejecting vectors and quaternions, his opposition is illustrative of the feeling of many other physical scientists of the time. Kelvin was probably the most influential British physicist of the late nineteenth century, and his opposition to vector methods, though never put forward in a published form, was almost certainly well known.

Knott pointed out (1; 188) that Tait could, and did, take advantage of the *Treatise on Natural Philosophy* by developing in quaternion form in his other writings some of the areas included in the *Treatise*, to show in this way the compactness that resulted from quaternion treatment. Thus at one point Tait covered in his quaternion book in a matter of lines what had been covered in the jointly authored work in paragraphs or pages. (1; 188)

Tait acted in a surprising manner on one aspect of quaternion development. Although he published numerous papers and books on quaternions, he drew a sharp line between his quaternion works and his other publications. Thus despite the fact that in many of his books he could have used the quaternion method, he did not. One example of this is his long article "Mechanics" for the ninth edition of the *Encyclopaedia Britannica*. Similarly one would expect that Tait in his mathematical physics courses at Edinburgh would have used quaternions wherever possible. This he did not do. One of Tait's students wrote in a biographical sketch of Tait: "Tait, so far as I know, never lectured on the subject [quaternions] at the University of Edinburgh." [19] To do so was in one way unnecessary since Kelland, the Professor of Mathematics, did discuss quaternions in his courses; on the other hand, this may actually make it more surprising that Tait probably made no use of them in his lectures on physics. Nevertheless at least two of Tait's students,

Knott and Macfarlane, became important contributors to quaternion analysis.

Although Tait possessed a high order of ability in pure mathematics and had received incomparable training in quaternions through his correspondence with Hamilton, his strongest interests were in physics rather than pure mathematics. This is reflected in the fact that his change from Belfast to Edinburgh involved more than a change in location: he had been Professor of Mathematics; he was now (and happily so) Professor of Natural Philosophy. Tait referred to this change in professional chair and to a consequence thereof in the preface to *Elementary Treatise on Quaternions* (1867): "The present work was commenced in 1859, while I was Professor of Mathematics, and far more ready at Quaternion analysis that I can now pretend to be. . . . The duties of another Chair, and Sir W. Hamilton's wish that my volume should not appear till after the publication of his *Elements*, interrupted my already extensive preparations." [20]

As it was mentioned before, eight books on quaternions (including later editions, translations, and coauthorships) are credited to Tait. His *Elementary Treatise on Quaternions* of 1867 was republished in a "Second Edition, Enlarged" in 1873, and in a "Third Edition, Much Enlarged" in 1890.[21] Tait's *Treatise* was translated into German by von Scherff (1880) and into French by Plarr (1884). Despite its title the work (especially in the later editions) was not really elementary. The need for an elementary work was fulfilled by the book *Introduction to Quaternions* by Philip Kelland and Tait. This jointly authored work appeared in 1873 and was followed by a second edition in 1882 and a third edition in 1904 with C. G. Knott as editor. Tait essentially wrote only one chapter (the tenth and last) of this book.[22] Nonetheless it was in a real sense a joint work, for as Kelland wrote in his preface, Tait "being my pupil in youth is my teacher in riper years. . . ." [23] Kelland also wrote: "I must confess that my knowledge of Quaternions is due exclusively to him. The first work of Sir Wm. Hamilton, *Lectures on Quaternions,* was very dimly and imperfectly understood by me and I dare say by others, until Professor Tait published his papers on the subject in the *Messenger of Mathematics.*" [24]

Tait's *Elementary Treatise on Quaternions* was written in such a way as to stress the applications of quaternions to physical science. Thus approximately 62 (of 320) pages in the first edition, 93 (of 298) pages of the second edition, and 130 (of 422) pages of the third edition were devoted to physical applications. Furthermore the other chapters were directed towards preparing for the final chapters on physical applications. The most important edition of the *Treatise* for this study is the second edition, since it was probably from this edition that both Gibbs and Heaviside learned quaternions.

Chapter I of Tait's *Treatise* is entitled "Vectors and Their Composition." In this chapter Tait treated vector equality, vector addition and subtraction, and the multiplication of a vector by a scalar quantity. He also treated the differentiation of a vector in terms of a single scalar variable. A noteworthy aspect of this chapter is that the quaternion never enters; the chapter could in fact serve as the first chapter of a modern vector analysis book, for the vector part of a quaternion is added, subtracted, and so on in the same manner as a modern vector. This chapter is only an extreme example of what is true to a certain extent of all of Tait's *Treatise;* similarities between Tait's book and modern vector analysis books abound. And this is to be expected: offspring usually resemble their ancestors. The lineage of the vast majority of the basic ideas and theorems [25] found in modern vector analysis books can be traced back to or through such quaternion books as Tait's *Treatise.*[26] Numerous illustrations of this all-important point will be given in this and the following chapter.

Chapter II of Tait's *Treatise* is entitled "Products and Quotients of Vectors." Since in general the product and the quotient of two vectors is a quaternion, this chapter concentrated on quaternions. Here again, but naturally to a lesser extent, many of the theorems can be translated into modern vector analysis. Many of the similarities derive from the fact shown before: the product of two vectors in quaternion analysis is equal to the negative of their scalar (dot) product plus their vector (cross) product. Thus, for example, Tait's statements that for any two vectors α and β, "$S\alpha\beta = S\beta\alpha$" and "$V\alpha\beta = -V\beta\alpha$" (2; 43) are equivalent to the statements in modern vector analysis that the dot product is commutative and the cross product is anticommutative.

Chapter III is entitled "Interpretations and Transformations of Quaternion Expressions." Among the first formulae in the chapter are those for α and β vectors: "$S\alpha\beta = -T\alpha T\beta \cos\theta$" and "$V\alpha\beta = T\alpha T\beta \sin\theta \cdot \eta$." (2; 49) The quaternion symbol T applied to a vector indicates its length, and the vector η is defined as a unit vector perpendicular to α and β. Thus the counterparts of these formulae in modern vector analysis would be

$$\boldsymbol{\alpha} \cdot \boldsymbol{\beta} = |\boldsymbol{\alpha}|\,|\boldsymbol{\beta}| \cos\theta$$

$$\boldsymbol{\alpha} \times \boldsymbol{\beta} = (|\boldsymbol{\alpha}|\,|\boldsymbol{\beta}| \sin\theta)\boldsymbol{\eta}$$

Then follow numerous examples and a number of other theorems which may also be translated into the now common vector analysis. This chapter concludes with a sketch of biquaternions, which are quaternions in which the four scalar multiples (of 1, *i*, *j*, *k*) may assume imaginary values.

Chapter IV, "Differentiation of Quaternions," is only six pages long and applies the processes developed earlier for vector differentiation to quaternion differentiation.

Tait began Chapter V by stating that

$$\text{“}\alpha S\beta\rho + \alpha_1 S\beta_1\rho + \ldots = \Sigma . \alpha S\beta\rho = \gamma\text{”}$$

which he abbreviated (following Hamilton) as

$$\text{“}\phi\rho = \Sigma . \alpha S\beta\rho\text{” and “}\phi\rho = \gamma\text{”} \quad (2;\ 78)$$

The reader acquainted with the modern treatment of the linear vector function by means of dyadics will see immediately that these equations are essentially (when translated)

$$\boldsymbol{\alpha\beta} . \boldsymbol{\rho} + \boldsymbol{\alpha}_1\boldsymbol{\beta}_1 . \boldsymbol{\rho}_1 + \ldots = \Sigma\boldsymbol{\alpha\beta} . \boldsymbol{\rho} = \gamma$$

or

$$\boldsymbol{\phi} . \boldsymbol{\rho} = \Sigma\boldsymbol{\alpha\beta} . \boldsymbol{\rho} \qquad \text{or} \qquad \boldsymbol{\phi} . \boldsymbol{\rho} = \gamma$$

where $\boldsymbol{\phi}$ is the modern dyadic. Tait thus devoted this chapter to the development of the linear vector function. His form of presentation is somewhat similar to the modern method of presentation by means of dyadics.

The next four chapters deal with the application of quaternion analysis to problems in geometry. For the present purposes all that need be said in regard to these chapters is that in them the linear vector function is developed further and the operator ∇ is introduced, but not developed.

The remaining third of the book consists of two chapters entitled "Kinematics" and "Physical Applications." Tait began the first of these by a discussion of Hamilton's hodograph and followed this by a treatment of the rotation of a rigid body by quaternion methods and of homogeneous strain by means of the linear vector function. It was perhaps in regard to methods of treatment for rotations and strains that quaternion analysis was of most value.

The final long chapter, entitled "Physical Applications," begins with a discussion of some theorems in dynamics. Tait began the chapter by stating: "When any forces act on a rigid body, the force β at the point whose vector is α, &c., then, if the body be slightly displaced, so that α becomes $\alpha + \delta\alpha$, the whole work done is $\Sigma S\beta\delta\alpha$.

This must vanish if the forces are such as to maintain equilibrium. Hence *the condition of equilibrium of a rigid body is* $\Sigma S\beta\delta\alpha = 0$." (2; 222) These equations express exactly the equivalent of the equations $\Sigma\boldsymbol{\beta}.\delta\mathbf{a} = 0$ and $\Sigma\boldsymbol{\beta}.\delta\mathbf{a} = 0$ in modern vector analysis. The quaternionists were well aware of the simplification that the dot product introduces into mechanics. Tait naturally proceeded to more complicated matters such as the treatment of the Foucault pendulum. To illustrate quaternionic applications in optics, Tait treated the Fresnel wave surface, followed this with a treatment of the effects of electric currents on one another and on magnets, and invited the reader to compare the mathematical methods with those of Ampère. The important discussion of the operator $\nabla = i\frac{d}{dx} + j\frac{d}{dy} + k\frac{d}{dz}$ included such theorems as those of Stokes, Green, and Gauss. The book is concluded by miscellaneous applications and problems.

From the above discussion of Tait's *Elementary Treatise on Quaternions* it should be clear that quaternion analysis, especially as presented by Tait, had many similarities to modern vector analysis. Vector addition and subtraction, vector multiplication in both the scalar (dot) and vector (cross) products, vector differentiation, vector algebra, the properties of ∇, and even the linear vector function were present. The exact form was not that of modern vector analysis, and there were some sections in the quaternion treatment that could not be translated into modern vector analysis, but the over-all similarity is beyond question. The function of the above discussion for this history is to indicate the *possibility* of deriving modern vector analysis from quaternion analysis as presented by Tait. The possibility of developing modern vector analysis from Grassmannian analysis was previously discussed. The problem thus clearly emerges: Did modern vector analysis emerge historically from the quaternion or from the Grassmannian methods of analysis?

Tait's contribution to quaternion analysis was not only as expositor of known methods but also as creator of new methods, many of which were later transferred to vector analysis. Tait made important advances in the theory and application of the linear vector function, but more important than this is his development of the operator ∇, which has been called "nabla," "del," and "atled." [27] Hamilton introduced this operator, but as Maxwell wrote: "The extension of this operator [∇] to vector displacements, and most of its further development, are due to Professor Tait." [28] Essentially what

Tait did was to state such important theorems as those of Green, Stokes, and Gauss in quaternion form, show their applications, and develop related theorems.[29] Among the most important papers written by Tait in this regard are "On Green's and Other Allied Theorems"[30] and "On Some Quaternion Integrals."[31] Some of Tait's polemical articles on quaternions will be discussed in a later chapter.

Tait's importance for this history may now be partially summed up. Tait, as the leading advocate and developer of quaternion analysis in the last third of the nineteenth century, changed the direction of emphasis in quaternion analysis toward its usefulness as a tool for physical science. He did this by developing and stressing those parts of the analysis which were most useful for physical science. This transformation of quaternion analysis was very probably a necessary preliminary for the development of the Gibbs-Heaviside modern vector analysis from quaternion analysis. One important example of this is the fact that in Hamilton's works one can find almost no discussion of the operator ∇, which is a fundamental part of modern vector analysis, while in Tait's works the discussion of this operator was probably fuller and better than could be found in any other mathematics books of the time. Historically it turns out that this change in direction in quaternion analysis was decisive for later developments.

IV. *Benjamin Peirce: Advocate of Quaternions in America*

Tait was not the only advocate of quaternions nor was enthusiasm for quaternions confined to the British Isles. One of the earliest and most influential advocates of quaternions was Benjamin Peirce, who did more than anyone else to develop interest in quaternions in the United States. This is important, since this was the homeland of both Gibbs and Wilson. Peirce did no creative work directly within the quaternion tradition; his greatest creative achievement (his "Linear Associative Algebra") is however intimately linked to Hamilton's discoveries and can be viewed as an excellent example of the tendency of mathematicians to proceed from quaternions to important discoveries outside the quaternion tradition. Thus Peirce will be discussed as (1) a figure influential on the history of quaternion analysis and as (2) a figure illustrative of an important tendency that was associated with the quaternion tradition.

Benjamin Peirce is generally considered to have been the first great American mathematician. Dirk J. Struik referred to his

"Linear Associative Algebra" as "the first major original contribution to mathematics produced in the United States."[32] As professor of astronomy and mathematics at Harvard from 1833 until his death in 1880 he exerted a vast influence on mathematics in the United States. It was shown previously that there was a surprisingly large interest in quaternion analysis in America; the greatest single cause of this was Benjamin Peirce's advocacy of Hamilton's system. In the chapter on Hamilton a quotation from Peirce's *Analytical Mechanics* of 1855 was given, which shows the great importance that Peirce attached to this discovery. A second quotation taken from Peirce's last major publication demonstrates that his enthusiasm for quaternions did not diminish with time. Peirce, after having mentioned the symbol for the square root of minus one, wrote:

> This symbol is restricted to a precise signification as the representative of perpendicularity in quaternions, and this wonderful algebra of space is intimately dependent upon the special use of the symbol for its symmetry, elegance, and power. The immortal author of quaternions has shown that there are other significations which may attach to the symbol in other cases. But the strongest use of the symbol is to be found in its magical power of doubling the actual universe, and placing by its side an ideal universe, its exact counterpart, with which it can be compared and contrasted, and, by means of curiously connecting fibres, form with it an organic whole, from which modern analysis has developed her surpassing geometry.[33]

Peirce's enthusiasm for quaternions dates back to at least 1848; in that year he seems to have included this subject in his mathematics courses at Harvard.[34] It is perhaps impossible to determine whether Peirce taught his quaternion course every year, but such seems probable. One student of Peirce, W. E. Byerly, referred to quaternions as Peirce's "favorite subject,"[35] and a number of his students went on to publish in quaternion analysis. Among these were Thomas Hill (who became president of Harvard in 1862), A. Lawrence Lowell (also a president of Harvard), Arnold B. Chase (who became chancellor of Brown), and two of Peirce's sons, Charles Santiago Saunders Peirce and James Mill Peirce. The latter son was professor of mathematics at Harvard from 1869–1906 and continued the teaching of quaternions at Harvard after his father's death.[36]

Benjamin Peirce published little on quaternion analysis; H. A. Newton wrote that Peirce had said in regard to quaternion analysis: "I wish I was young again that I might get such power in using it as only a young man can get."[37] Peirce's greatest contribution to mathematics, his "Linear Associative Algebra," did however stem from his interest in quaternions. This work, which appeared in lithograph in 1870, was published in 1881.[38] It was aimed at developing "so much of the theory of hyper-complex numbers as would enable him to enumerate all inequivalent, pure, non-reciprocal number systems in less than seven units." [39] In this work Peirce developed 162 different algebras. In a broad sense his work illustrates the fact that the discovery of and interest in quaternions led in many different directions; among these was the discovery of other similar systems. A more important result was that it led such men as Peirce to a study of what mathematical systems are possible or, still more generally, to a study of algebraic structure.

Information compiled by Florian Cajori serves to indicate the degree to which Peirce's enthusiasm for quaternions spread to his students and countrymen. Not all the interest stemmed of course from Peirce; nevertheless it seems probable that a large measure of it came from him, for he was the recognized leader in mathematics as well as the first advocate of quaternions in the United States. Cajori supplied information on twenty-two colleges and universities in nineteenth-century America; quaternion methods were taught at no less than twelve of these. This is surprising since the level of mathematical instruction was not at a high level at that time; indeed at some of the schools where quaternions were not taught, no other mathematics courses on a comparable level of complexity were taught.[40]

V. *James Clerk Maxwell: Critic of Quaternions*

The discussion of Maxwell presented in this section serves a twofold function, for this brilliant Scot not only was a very influential figure in the history of vector analysis but he also aptly represents an important development in the physical science of the nineteenth century. This development was the increasing awareness of the need (itself increasing) for a vectorial approach to the solution of physical problems. There is much truth and much relevance in Saloman Bochner's comparison of the space of Euclid with the space of Newton and his successors:

> And yet, inwardly, the Euclidean Space that underlies the *Principia* is mathematically not quite the same as the Euclidean Space that underlies Greek mathematics (and physics) from Thales to Apollonius. The geometry of the Greeks emphasized congruences and similarities between figures, that is, in analytical parlance, orthogonal and homothetic transformations of the underlying space. . . . The Euclidean Space of the *Principia* continued to be all this, but it was also something new in

> addition. Several significant physical entities of the *Principia,* namely, velocities, momenta, and forces are, by mathematical structure, vectors, that is, elements of vector fields, and vectorial composition and decomposition of these entities constitute an innermost scheme of the entire theory.[41]

Thus with Newton a host of vectorial entities (though not the concept of a vector) entered into physical science. The revolution associated with Newton is paralleled by a development that occurred in the nineteenth century; this development is the introduction into physical science of the concept of field, a concept which Einstein and Infeld have described as the "most important invention [in physics] since Newton's time. . . ." [42] With the introduction of this concept the vectorial entities of the space of Newton were multiplied in number and the need for vector analysis was greatly increased. The concept of field as well as the increased need for vector analysis arose in part from developments in mechanics (especially hydrodynamics), in part from advances in optical theory, in part from the creation and elaboration of potential theory, and above all from successes in electrical theory. It should thus come as no surprise that the eventual emergence of vector analysis should intimately be associated with the achievements of the great nineteenth-century theoretical electricians, especially Maxwell. Much is symbolized in the fact that Maxwell first presented his famous equations in the 1860's, written in component notation, whereas in his 1873 *Treatise on Electricity and Magnetism* he wrote them both in component *and* in quaternion notation.

It is perhaps an accident of history that Tait and Maxwell spent a number of years together studying at Edinburgh and at Cambridge. That a close friendship developed between them is less accidental; both were men of brilliant intellect, and more importantly, both tended to take similar approaches in mathematical physics. The approaches of Tait and Maxwell in mathematical physics were alike in that both wished their mathematics to be as close a representation of the physical entities as possible.

Tait left Cambridge in 1854 for Queen's College, Belfast; Maxwell left Cambridge two years later for Marischal College at Aberdeen. In the decade from 1856 to 1865, while Tait pursued quaternion researches with Hamilton and experimental studies with Andrews, Maxwell composed his four great electrical papers. The discoveries published in these papers, especially those concerned with Maxwell's concept of the electromagnetic field, involved mathematical problems treatable by quaternion analysis. In the published forms of these papers quaternions were never used, and it is clear from the correspondence between Tait and Maxwell[43] that Maxwell began to learn quaternion methods only in the 1870's.[44] Nevertheless a few people made the assumption that Maxwell had been aided in developing his ideas by the use of quaternion methods. Though this assumption was not correct, it was probably influential in leading people to study quaternions.

E. T. Whittaker described the developments in physics that led to Maxwell's great electrical papers in the following way:

> In that power to which Gauss attached so much importance, of devising dynamical models and analogies for obscure physical phenomena, perhaps no-one has ever excelled W. Thomson; and to him, jointly with Faraday, is due the credit of having initiated the theory of the electric medium. In one of his earliest papers, written at the age of seventeen when he was a very young freshman at Cambridge, Thomson compared the distribution of electrostatic force, in a region containing electrified conductors, with the distribution of the flow of heat in an infinite solid; the equipotential surfaces in the one case correspond to the isothermal surfaces in the other, and an electric charge corresponds to a source of heat.
>
> It may, perhaps, seem as if the value of such an analogy as this consisted merely in the prospect which it offered of comparing, and thereby extending, the mathematical theories of heat and electricity. But to the physicist its chief interest lay rather in the idea that formulae which relate to the electric field, and which had been deduced from laws of action at a distance, were shown to be identical with formulae relating to the theory of heat, which had been deduced from hypotheses of action between contiguous particles. "This paper," as Maxwell said long afterwards, "first introduced into mathematical science that idea of electrical action carried on by means of a continuous medium, which, though it had been announced by Faraday, and used by him as the guiding idea of his researches, had never been appreciated by other men of science, and was supposed by mathematicians to be inconsistent with the law of electrical action, as established by Coulomb, and built on by Poisson."
>
> In 1846—the year after he had taken his degree as second wrangler at Cambridge—Thomson investigated the analogies of electric phenomena with those of elasticity. For this purpose he examined the equations of equilibrium of an incompressible elastic solid which is in a state of strain; and showed that the distribution of the vector which represents the elastic displacement might be assimilated to the distribution of the electric force in an electrostatic system. This, however, as he went on to show, is not the only analogy which may be perceived with the equations of elasticity; for the elastic displacement may equally well be identified with a vector **a**, defined in terms of the magnetic induction **B** by the relation curl **a** = **B**.
>
> The vector **a** is equivalent to the vector-potential which had been used in the memoirs of Neumann, Weber and Kirchhoff, on the induction of currents; but Thomson arrived at it independently by a different process, and without being at the time aware of the identification.
>
> The results of Thomson's memoir seemed to suggest a picture of the

> propagation of electric or magnetic force: might it not take place in somewhat the same way as changes in the elastic displacement are transmitted through an elastic solid? These suggestions were not at the time pursued further by their author; but they helped to inspire another young Cambridge man to take up the matter a few years later. James Clerk Maxwell, by whom the problem was eventually solved, was born in. . . .[45]

This extensive and important quotation is significant for the history of vector analysis in numerous ways. The most important of these is that Maxwell presented an historical interpretation of what led to Thomson's discoveries, and this interpretation indicates what is very possibly the way in which Maxwell became interested in quaternions. Maxwell's interpretation was given in two papers, the first of which consists of an address to the British Association for 1870 and the second of which is his famous paper entitled "On the Mathematical Classification of Physical Quantities."[46] The main idea developed in both papers is the following: "Of the methods by which the mathematician may make his labours most useful to the student of nature, that which I think is at present most important is the systematic classification of quantities." (5; 218) Maxwell went on to explain why the mathematical classification of physical quantities is important.

> But when the student has become acquainted with several different sciences, he finds the mathematical processes and trains of reasoning on one science resemble those in another so much that his knowledge of the one science may be made a most useful help in the study of the other.
>
> When he examines into the reason of this, he finds that in the two sciences he has been dealing with systems of quantities, in which the mathematical forms of the relations of the quantities are the same in both systems, though the physical nature of the quantities may be utterly different. (5; 218)

Maxwell supplied two historical illustrations of the usefulness of such a classification; the second is the more important. Maxwell wrote: "Another example, by no means so obvious, is that which was originally pointed out by Sir William Thomson, of the analogy between problems in attractions and problems in the steady conduction of heat, by the use of which we are able to make use of many of the results of Fourier for heat in explaining electrical distribution, and of all the results of Poisson in electricity in explaining problems in heat." (6; 258) Maxwell went on to comment that Thomson's discovery depends on the fundamental principle that the mathematical representations of physical entities may be classified; his chief example of such a classification was that found by Hamilton:

> A most important distinction was drawn by Hamilton when he divided the quantities with which he had to do into Scalar quantities, which are completely represented by one numerical quantity, and Vectors, which require three numerical quantities to define them.
>
> The invention of the calculus of Quaternions is a step towards the knowledge of quantities related to space which can only be compared, for its importance, with the invention of triple coordinates by Descartes. The ideas of this calculus, as distinguished from its operations and symbols, are fitted to be of the greatest use in all parts of science. (6; 259)

Maxwell said in effect that if scientists would pay closer attention to the classification of physical quantities, such analogies as those discovered by Kelvin would be obvious. His primary example of such a classification is that which originated with Hamilton, the classification of physical entities into scalars and vectors. The point is that Maxwell seems to have interpreted the discoveries of Thomson (which have been presented in the quotation from Whittaker and cited in part by Maxwell) in the following way: Thomson saw that distributions of electrostatic forces in a region containing electrified conductors produced a vector field and that this vector field was analogous to the vector field associated with the flow of heat in an infinite solid. Associated with these vector fields are the two scalar fields, the lines of equal potential in the case of electrostatic case and the isothermal lines in the case of heat flow. Thus when Thomson saw that both cases involved scalar and vector fields, the next step was nearly obvious. If the above interpretation of Maxwell's ideas is correct, it is probable that this realization led him to an increased interest in quaternion analysis.

In the second of his two papers Maxwell classified vectors into "force" vectors (which are referred to a unit of length) and flux vectors (which are referred to a unit of area).[46] Later in this paper quaternions were discussed briefly with special attention to Hamilton's operator ∇, for the results of which Maxwell proposed a series of names. He gave the definitions "$\nabla = i\frac{d}{dx} + j\frac{d}{dy} + k\frac{d}{dz}$" and "$\nabla^2 = -\left(\frac{d^2}{dx^2} + \frac{d^2}{dy^2} + \frac{d^2}{dz^2}\right)$" (6; 263–264) and then stated: "The discovery of the square root of this operation is due to Hamilton; but most of the applications here given, and the whole development of the theory of this operation, are due to Professor Tait. . . ." (6; 264) He proposed "to call the result of ∇^2 . . . the *Concentration* of the quantity to which it is applied." (6; 264) When ∇ is applied to a scalar function of position P, the result is ∇P. "The quantity ∇P is a vector, indicating the direction in which P decreases most rapidly, and measuring the rate of that decrease. I venture, with much diffidence, to call this the *slope* of *P*." (6; 264) The reader should note that according to quaternion multiplication, if $\sigma = it + ju + kv$, then $\nabla\sigma = -\left(\frac{dt}{dx}+\frac{du}{dy}+\frac{dv}{dz}\right)+i\left(\frac{dv}{dy}-\frac{du}{dz}\right)+j\left(\frac{dt}{dz}-\frac{dv}{dx}\right)+k\left(\frac{du}{dx}-\frac{dt}{dy}\right)$.[47] Maxwell called the scalar part of this product, represented in quaternion language as $S\nabla\sigma$, "the *Convergence* of σ," and the vector part, $V\nabla\sigma$, "the *Curl* or *Version* of the original vector function."[48]

Before publishing this paper, Maxwell had written to Tait to request his views on the suitability of the names: "Here are some rough hewn names. Will you like a good Divinity shape their ends properly so as to make them stick?" Maxwell went on to ask in particular about the names for ∇ applied to a vector function.

> The scalar part I would call the Convergence of the vector function, and the vector part I would call the Twist of the vector function. Here the word twist has nothing to do with a screw or helix. If the word *turn* or *version* would do they would be better than twist, for twist suggests a screw. Twirl is free from the screw notion and is sufficiently racy. Perhaps it is too dynamical for pure mathematicians, so for Cayley's sake I might say Curl (after the fashion of Scroll).[49]

The letter quoted above is the first letter from the Tait-Maxwell correspondence in which Maxwell showed some knowledge of quaternions. It is dated November 7, 1870. The first paper in which Maxwell included a discussion of quaternion ideas is in the aforementioned "Address to the Mathematical and Physical Sections of the British Association," which was read on September 15, 1870. It is thus probable that Maxwell's interest in quaternions originated at about this time. At least as early as 1867 Tait had recommended quaternions to Maxwell. Tait wrote on December 13, 1867, to Maxwell: "If you read the last 20 or 30 pages of my book [*Treatise on Quaternions* (1st ed., 1867)] I think you will see that 4ions are worth getting up, for there it is shown that they go into that $\triangleleft$ business like greased lightning. Unfortunately I cannot find time to work steadily at them. . . ."[50]

Maxwell's letter to Tait of November 7, 1870, was followed by another letter dated November 14; Maxwell's message therein must have been good news for Tait: "With regard to my dabbling in Hamilton I want to leaven my book [*Treatise on Electricity and Magnetism*] with Hamiltonian ideas without casting the operations into Hamiltonian form for which neither I nor I think the public are ripe. Now the value of Hamilton's idea of a vector is unspeakable, and so are those of the addition and multiplication of vectors." (1; 144) Maxwell's letter of October 9, 1872, also brought good news to Tait. In the previous year Maxwell had been chosen as the first Professor of Experimental Physics at Cambridge and Director of the Cavendish Laboratory. Maxwell wrote to Tait, "I am going to try, as I have already tried, to sow 4nion seed at Cambridge."[51]

The Maxwell letter to Tait of December 21, 1871, contained perhaps Maxwell's strongest statement in favor of quaternions. Maxwell wrote: "Impress on T. [Thomson] that $\left(\frac{d}{dx}\right)^2 + \left(\frac{d}{dy}\right)^2 + \left(\frac{d}{dz}\right)^2 = -\nabla^2$ and not $+\nabla^2$ as he vainly asserts is now commonly believed among us. Also how much better and easier he would have done his solenoidal and lamellar business if in addition to what we know is in his head he had had say, 20 years ago, Qns. to hunt for Cartesians instead of vice versa. The one is a flaming sword that burns every way; the other is a ram, pushing westward and northward and (downward?)." (1; 150)

A number of interesting points are contained in Maxwell's letter of November 2, 1871; he wrote:

> But try and do the 4nions. The unbelievers are rampant. They say "show me something done by 4nions which has not been done by old plans. At the best it must rank with abbreviated notations."
>
> You should reply to this, no doubt you will.
>
> But the virtue of the 4nions lies not so much as yet in solving hard questions, as in enabling us to see the meaning of the question and of its solution, instead of setting up the question in x y z, sending it to the analytical engine, and when the solution is sent home translating it back from x y z so that it may appear as A, B, C to the vulgar. (1; 101)

The last point was emphasized in an unsigned article by Maxwell entitled "Quaternions" and published in *Nature* in 1873.[52] The article is actually a review of Kelland's and Tait's *Introduction to Quaternions* published in 1873. Maxwell began the article by discussing the popular image of the mathematician as a calculator. He then wrote: "But though much of the routine work of a mathematician is calculation, his proper work—that which constitutes him a mathematician—is the invention of methods. . . . But the methods on which a mathematician is content to hang his reputation are generally those which he fancies will save him and all who come after him the labour of thinking about what has cost himself so much thought."[53] A common argument for quaternions had been that they were just such a labor-saving device, but Maxwell's position was quite different from this.

> Now Quaternions, or the doctrine of Vectors, is a mathematical method, but it is a method of thinking, and not, at least for the present generation, a method of saving thought. . . . It calls upon us at every step to form a mental image of the geometrical features represented by the symbols, so that in studying geometry by this method we have our minds

> engaged with geometrical ideas, and are not permitted to fancy ourselves geometers when we are only arithmeticians.[53]

Maxwell concluded the review with praise for Kelland's expository abilities and with stress on the importance of the treatment of the linear vector function given in the last chapter (written by Tait).

The theme that dominates this review as well as many of Maxwell's other statements on quaternions is that by means of the vectorial approach the physicist attains to a direct mathematical representation of physical entities and is thus aided in seeing the physics involved in the mathematics. It is probable that Maxwell was led to stress this point by pondering the great successes of Faraday in physics, which were obtained without the use of formal mathematics. Faraday's success presented a conundrum to the mathematical physicists of the latter half of the nineteenth century, a conundrum which Maxwell tried to resolve when he wrote: "The way in which Faraday made use of his lines of force in co-ordinating the phenomena of magneto-electric induction shews him to have been in reality a mathematician of a very high order—one from whom mathematicians of the future may derive valuable and fertile methods." (4,II; 360) Thus Maxwell's enthusiasm for quaternions stemmed from viewing them as a method that allows the physicist to keep the physical entities before his mind during calculations, rather than from viewing them as a method for reducing the labor of thought.

The most important of Maxwell's published works for this history is his *Treatise on Electricity and Magnetism* (1873).[54] This is the most important work not only because it gives the fullest conception of Maxwell's view of quaternions but also because historically it was influential in leading people to discuss quaternion methods.

Maxwell began this work with a mainly mathematical "Preliminary" chapter. After mentioning Descartes' discovery of co-ordinate methods, Maxwell wrote:

> But for many purposes of physical reasoning, as distinguished from calculation, it is desirable to avoid explicitly introducing the Cartesian coordinates, and to fix the mind at once on a point of space instead of its three coordinates, and on the magnitude and direction of a force instead of its three components. This mode of contemplating geometrical and physical quantities is more primitive and more natural than the other, although the ideas connected with it did not receive their full development till Hamilton made the next great step in dealing with space, by the invention of his Calculus of Quaternions.
>
> As the methods of Des Cartes are still the most familiar to students of science, and as they are really the most useful for purposes of calculation, we shall express all our results in the Cartesian form. I am con-

vinced, however, that the introduction of the ideas, as distinguished from the operations and methods of Quaternions, will be of great use to us in the study of all parts of our subject, and especially in electrodynamics, where we have to deal with a number of physical quantities, the relations of which to each other can be expressed far more simply by a few expressions of Hamilton's, than by the ordinary equations.

11. One of the most important features of Hamilton's method is the division of quantities into Scalars and Vectors. (3,I; 9–10)

After a discussion of this division of quantities into vectors and scalars, Maxwell wrote:

There are physical quantities of another kind which are related to directions in space, but which are not vectors. Stresses and strains in solid bodies are examples of these, and so are some of the properties of bodies considered in the theory of elasticity and in the theory of double refraction. Quantities of this class require for their definition *nine* numerical specifications. They are expressed in the language of quaternions by linear and vector functions of a vector. (3,I; 10)

Maxwell then stated that he would use German capital letters to represent vectors. The reason for this innovation was that he needed the Greek letters (which were used to designate vectors by Hamilton and Tait) for other uses in the book. After a discussion of line and surface integrals and the introduction of the operator ∇ and other mathematical preliminaries, Maxwell presented in co-ordinate form the theorems of Gauss and Stokes (Maxwell did not use these names). Maxwell mentioned that he would use the right-handed system of co-ordinates axes and pointed out that in this he differed from the practice of Hamilton but agreed with that employed by Thomson and Tait in their *Treatise on Natural Philosophy* and by Tait in his quaternion publications. (3,I; 25–26) This section was concluded with a translation of Stokes' and Gauss' theorems into quaternion notation and with an explanation of his terms (in the first edition) "concentration," "curl" or "version," "convergence," and "slope." [55] In later editions "curl" or "version" was dropped and ("with great diffidence") "rotation" substituted. "Slope" also was dropped and "space-variation" substituted. (3,I; 30–31)

In the text of the book quaternion notations were used in a substantial number of cases. These may be classified into three groups: to the first belong those cases in which at the end of a section Maxwell stated that he would give the quaternion form of an equation; to the second group belong those cases in which quaternion notation was used in such an elementary form that little importance can be attributed to it; and to the third class belong the most important cases, those in which Maxwell used quaternion expressions almost in such a way as to integrate them into the development.

The most important example of the first group (of which there are perhaps fifteen cases) comes in the chapter "General Equations of the Electromagnetic Field." The last section of this chapter was entitled "Quaternion Expressions for the Electromagnetic Equations." Herein Maxwell wrote: "In this treatise we have endeavoured to avoid any process demanding from the reader a knowledge of the Calculus of Quaternions. At the same time we have not scrupled to introduce the idea of a vector when it was necessary to do so." (3,II; 257) Maxwell then wrote in quaternion notation the most important equations developed in his book. (3,II; 257–259)

Examples of the second kind of use that Maxwell made of quaternion notation are very numerous, especially in the second volume. Repeatedly he wrote vectorial quantities as vectors; this was done especially to stress the nature of the quantity represented, and also as an abbreviation. At other times one vector was given as a scalar multiple of another vector. Similarly he made frequent use of the symbol ∇ but wrote out the full Cartesian form of his equations when presenting proofs.

Cases from the third class are somewhat rare: examples are the sections in which he used the quaternion scalar and vector products.[56] The vector and scalar products were used sufficiently frequently that readers probably noticed their importance. The full quaternion product was never employed, but it at times influenced expressions that were used. Thus in Cartesian analysis ∇^2 can be defined as either plus or minus $\left(\frac{d^2}{dx^2}+\frac{d^2}{dy^2}+\frac{d^2}{dz^2}\right)$. Kelvin used the positive sign, whereas Maxwell used the negative sign and noted: "The negative sign is employed here in order to make our expressions consistent with those in which Quaternions are employed." (3,II; 255)

The following three quotations are of considerable importance.

> The theory of the complete system of equations of resistance and of conductivity is that of linear functions of three variables, and it is exemplified in the theory of Strains, and in other parts of physics. The most appropriate method of treating it is that by which Hamilton and Tait treat a linear and vector function of a vector. We shall not, however, expressly introduce Quaternion notation. (3,I; 422)
>
> This analysis of the forces acting between two small magnets was first given in terms of the Quaternion Analysis by Professor Tait in the *Quarterly Math. Journ.* for Jan. 1860. See also his work on *Quaternions,* Arts. 442–443, 2nd Edition.[57]

Finally, in Maxwell's chapter on "Ampère's Investigation of the Mutual Action of Electric Currents" he stated:

> The only experimental fact which we have made use of in this investigation is the fact established by Ampère that the action of a closed circuit on any portion of another circuit is perpendicular to the direction of the latter. Every other part of the investigation depends on purely mathematical considerations depending on the properties of lines in space. The reasoning therefore may be presented in a much more condensed and appropriate form by the use of the ideas and language of the mathematical method specially adapted to the expression of such geometrical relations—the *Quaternions* of Hamilton.
>
> This has been done by Professor Tait in the *Quarterly Journal of Mathematics*, 1866, and in his treatise on *Quaternions*, § 399, for Ampère's original investigation, and the student can easily adapt the same method to the somewhat more general investigation given here. (3,II; 171–172)

Maxwell's use of quaternion expressions and notations was sufficiently frequent and the above quotations sufficiently strong in praise of quaternions that it seems probable that not a few readers of his *Treatise* left it with the impression that Maxwell had rather strongly recommended the study of quaternions. To those readers who wondered why quaternions were not more frequently used, a perfectly natural explanation was available: Maxwell had in fact wished to use quaternions more frequently but had not done so since few readers of his *Treatise* were versed in such methods. Similarly Maxwell's one explicit statement (he had recommended quaternion "ideas, as distinguished from the operations and methods of Quaternions" [3,I; 9]) was easily missed or easily misinterpreted. The conjecture that readers of the *Treatise* may have been led thereby to study quaternions gains support from evidence that strongly indicates that both Gibbs and Heaviside did in fact go from the *Treatise* of Maxwell to the works of Hamilton and Tait. Even if this conjecture is incorrect, still it is very possible that readers of Maxwell would tend to become readers of Tait in order to avail themselves of the latter's treatment of the linear vector function and the operator ∇.

A letter to Tait written by Maxwell in 1878 gives the clearest picture of Maxwell's views on quaternions. The statements contained therein are sufficient to explain completely why Maxwell took the course that he did in his *Treatise*. Maxwell wrote:

> Here is another question. May one plough with an ox and an ass together? The like of you may write everything and prove everything in pure 4nions, but in the transition period the bilingual method may help to introduce and explain the more perfect.

> But even when that which is perfect is come that which builds over three axes will be useful for purposes of calculation by the Cassios of the future.
>
> Now in a bilingual treatise it is troublesome, to say the least, to find that the square of *AB* is always positive in Cartesians and always negative in 4nions, and that when the thing is mentioned incidentally you do not know which language is being spoken.
>
> Are the Cartesians to be denied the idea of a vector as a sensible thing in real life till they can recognise in a metre scale one of a peculiar system of square roots of −1?
>
> It is also awkward when you are discussing, say, kinetic energy to find that to ensure its being +*ve* you must stick a − sign to it, and that when you are proving a minimum in certain cases the whole appearance of the proof should be tending towards a maximum.
>
> What do you recommend for El. and Mag. to say in such cases?
>
> Do you know Grassmann's Ausdehnungslehre? Spottiswoode spoke of it in Dublin as something above and beyond 4nions. I have not seen it, but Sir W. Hamilton of Edinburgh used to say that the greater the extension the smaller the intention. (1; 151–152)

A. P. Wills has speculated that if Maxwell had read Grassmann, he might have been led to adopt its methods.[58] Maxwell almost certainly did not read Grassmann's *Ausdehnungslehre,* since Maxwell died on November 5, 1879. In relation to this question it is interesting to note that Grassmann was not unknown to Maxwell even before 1878, although probably Maxwell had never seen his books. Twice in his *Treatise on Electricity and Magnetism* Maxwell referred to an idea presented in one of Grassmann's papers on electricity.[59] In this paper Grassmann mentioned his *Ausdehnungslehre* and gave a few of the elementary ideas contained in it. Also in three of Maxwell's early (1855–1860) optical papers some discussion was given concerning one of Grassmann's optical papers [60] in which also he had referred to his *Ausdehnungslehre.* Of this paper by Grassmann, Maxwell wrote in 1860: "It appears therefore that if colours are represented in quantity and quality by the magnitude and direction of straight lines, the rule for the composition of colours is identical with that for the composition of forces in mechanics. This analogy has been well brought out by Professor Grassmann in . . . ." (4,I; 418–419)

Of the four books written by Maxwell after 1873 only one mentioned vectors; this is his elementary work on mechanics published in 1876 and entitled *Matter and Motion.* Herein Maxwell included a short section on the idea of and the addition and subtraction of vectors. Quaternions were not mentioned.

Maxwell's strange, unique, and important place in this history may now be summed up. Maxwell favored quaternion analysis for the naturalness of its representations of physical entities and for the abbreviations stemming from this. Most of all he favored quaternions because the physical entities were kept before the eye of the mathematician. He was particularly impressed by the ∇ operator and the linear vector function. On the other hand, Maxwell in general disliked quaternion "methods" (as opposed to quaternion "ideas"); thus for example he was troubled by the nonhomogeneity of the quaternion or full vector product and by the fact that the square of a vector was negative, which in the case of the velocity vector made kinetic energy negative. The aspects of quaternion analysis that Maxwell liked were clearly brought out in his great work on electricity; the aspects that he did not like were indicated only by the fact that Maxwell did not include them.

Thus Maxwell's importance for this history is twofold: (1) he associated vectorial ideas with electricity in such a way that this linkage was maintained and (2) he to some extent outlined the form that a suitable vector analysis should take. Such a vector analysis was not known to Maxwell, but soon after his death and probably in part due to his inspiration such a system was created by Gibbs and Heaviside. Thus, justification may be seen for Macfarlane's statement that in Maxwell's departure from quaternion orthodoxy "we have the origin of the school of vector-analysts as opposed to the pure quaternionists."[61]

VI. *William Kingdon Clifford: Transition Figure*

Maxwell died in late 1879 at age forty-eight; earlier in the same year British science lost another of its most brilliant representatives when William Kingdon Clifford died at age thirty-four. Clifford had graduated from Cambridge in 1867 as Second Wrangler, and was elected Professor of Applied Mathematics and Mechanics at University College, London, in 1871.

Clifford's significance for this history is twofold: (1) he was one of the few mathematicians of the time who knew both quaternion and Grassmannian analysis and (2) he wrote a work which is in a sense transitional from quaternion analysis to vector analysis.

In Clifford's writings the first mention of Grassmann came in 1868 (7; 114); his acquaintance with Grassmannian analysis is probably due to Hankel's *Theorie der complexen Zahlensysteme* of 1867. His interest in quaternions probably stems from 1867 or earlier, but his first mention of quaternions came in an 1873 paper.[62]

It is clear from his writings that Clifford's interest was greater in the Grassmannian than in the Hamiltonian system. In a paper of 1878 published in the *American Journal of Mathematics Pure and*

Applied and entitled "Applications of Grassmann's Extensive Algebra" Clifford wrote:

> I propose to communicate in a brief form some applications of Grassmann's theory which it seems unlikely that I shall find time to set forth at proper length, though I have waited long for it. Until recently I was unacquainted with the *Ausdehnungslehre*, and knew only so much of it as is contained in the author's geometrical papers in *Crelle's Journal* and in *Hankel's Lectures on Complex Numbers*. I may, perhaps, therefore be permitted to express my profound admiration of that extraordinary work, and my conviction that its principles will exercise a vast influence upon the future of mathematical science. (7; 266)

Clifford delivered in 1877 a series of lectures on quaternions at University College. Clifford's mathematical interests were more in pure mathematics (especially geometry) than in applied mathematics; this becomes clear when one looks at the notes from his lectures on quaternions.[63]

The most important of Clifford's writings for this history is his *Elements of Dynamic*, which was intended as the first in a series of elementary texts. The series was never completed, for Clifford died in 1879. In this work Clifford introduced vectors (or "steps," as he sometimes called them) and early in the work explained such ideas as the addition of vectors. Near the middle of the book, in a section entitled "Product of Two Vectors," Clifford wrote:

> On account of the importance of the theorem of moments, we shall present it under yet another aspect. The area of the parallelogram *abdc* may be supposed to be generated by the motion of *ab* over the step *ac*, or by the motion of *ac* over the step *ab*. Hence it seems natural to speak of it as the *product* of the two steps *ab*, *ac*. We have been accustomed to identify a rectangle with the product of its two sides, when their lengths only are taken into account; we shall now make just such an extension of the meaning of a product as we formerly made of the meaning of a sum, and still regard the parallelogram contained by two steps as their product, when their directions are taken into account. The magnitude of this product is *ab*.*ac* sin *bac*; like any other area, it is to be regarded as a directed quantity.
>
> Suppose, however, that one of the two steps, say *ac*, represents an area perpendicular to it; then to multiply this by *ab*, we must naturally make that area take the step of translation *ab*. In so doing it will generate a volume, which may be regarded as the product of *ac* and *ab*. But the magnitude of this volume is *ab* multiplied by the area into the sine of the angle that it makes with *ab*, that is, into the *cosine* of the angle that *ac* makes with *ab*. This kind of product therefore has the magnitude *ab*.*ac* cos *bac*; being a volume, it can only be greater or less; that is, it is a *scalar* quantity.
>
> We are thus led to two different kinds of product of two vectors *ab*, *ac*; a *vector product*, which may be written *V*.*ab*.*ac*, and which is the area of the parallelogram of which they are two sides, being both regarded as

> steps; and a *scalar product,* which may be written $S.ab.ac$, and which is the volume traced out by an area represented by one, when made to take the step represented by the other. (8; 94–95)

After showing that both these products are distributive, Clifford continued:

> But there is a very important difference between a vector product and a product of two scalar quantities. Namely, the *sign* of an area depends upon the way it is gone round; an area gone round counter-clockwise is positive, gone round clockwise is negative. Now if $V.ab.ac =$ area *abcd*, we must have by symmetry $V.ac.ab =$ area *acdb*, and therefore $V.ac.ab = -V.ab.ac$, or $V\gamma\beta - V\beta\gamma$. Hence *the sign of a vector product is changed by inverting the order of the terms.* It is agreed upon that $V\alpha\beta$ shall be a vector facing to that side from which the rotation from α to β appears to counter-clockwise. (8; 96).

This passage seems to be a definition of the modern scalar and vector products. Clifford's definition of the vector product of two vectors is essentially equivalent to the modern definition and is in one sense equivalent to the quaternion vector product. However, at no point in the above passage did Clifford indicate whether the scalar product was to be taken as positive or negative; he referred only to the magnitude of the product. Nearly one hundred pages later he returned to the scalar product and defined it as "the negative sum of the products of their components along the axes." (8; 186) Clifford thus used the scalar product in the quaternionic sense, not in the modern sense.

In Clifford's posthumously published *Common Sense of the Exact Sciences* the modern definition of the scalar product was given. This appeared in chapter IV, section 16, of the book.[64] Before Clifford died, he had written some of the chapters in this book, and Karl Pearson wrote the remainder. In his preface Pearson wrote: "For the latter half of Chapter III and for the whole of Chapter IV (pp. 116–226) I am alone responsible."[65] Thus the appearance of the modern scalar product in this book is to be credited, not to Clifford, but to Pearson. It is possible that in the year between the publication of his *Elements of Dynamic* (1878) and his death in 1879 Clifford might have changed his definition of the scalar product and that this fact was communicated to Pearson. However, it seems far more probable that Pearson based his definition on the first, rather than the second, discussion of the scalar product as given by Clifford in his *Elements of Dynamic* and hence wrote the product in its modern form only because of a misunderstanding! That this is probable is further indicated by the great similarity of the discussion of the scalar product given by Pearson as compared with the first discussion of the two given by Clifford in his *Elements of Dynamic.* The confusing nature of Clifford's first definition adds to this probability.

Later in his *Elements of Dynamic* Clifford made use of the operator ∇ (without introducing this symbol) and introduced the term *divergence* as the negative of Maxwell's *convergence.* (8; 209) He also employed the linear vector operation as his mathematical method in his section on strains. The incomplete second volume of this work, published posthumously in 1887, followed the same lines of development. It was perhaps natural that in an elementary text Clifford would give few historical references. In fact the index to the two volumes does not refer the reader to a single helpful historical reference in regard to the origin of Clifford's vectorial methods.[66] Because of this fact Tait came close to accusing Clifford of plagiarism.

Writing a review of Clifford's *Common Sense of the Exact Sciences,* Tait stated: "Thus, especially in matters connected with the development of recent mathematical and kinematical methods, his statements were by no means satisfactory (from the historical point of view) to those who recognized, as their own, some of the best 'nuggets' that shine here and there in his pages. His *Kinematic* [the subtitle of Clifford's *Elements of Dynamic*] was, throughout, specially open to this objection. . . ." (1; 272–273) In 1878 Tait reviewed Clifford's *Elements of Dynamic;* Tait wrote: "Though this preliminary volume contains only a small installment of the subject, the mode of treatment to be adopted by Prof. Clifford is made quite obvious. . . . For, although (so far as we have seen) the word quaternion is not once mentioned in the book, the analysis is in great part purely quaternionic. . . ." (1; 270) Tait went on to complain that quaternionic notations and full quaternionic methods should have been used more extensively, although he suggested that Clifford did not do this because students had already been offered such things as ∇ and might "refuse altogether to bite again." (1; 272)

The significance of Clifford in the history of vector analysis may be best understood by considering him as a transition figure. Writing at a time after Grassmann and Hamilton had created their systems and before Gibbs and Heaviside created the modern system of vector analysis, Clifford came to appreciate the benefits to be derived from the use of vector methods, especially in mechanics. Moreover in his *Elements of Dynamic* he introduced the practice of defining separately his vector and scalar products. Considered against the background of the quaternion tradition, this was a major conceptual innovation, since quaternionists never viewed $V\alpha\beta$

and $S\alpha\beta$ as two separate products of the two vectors α and β but only as two *parts* of the full quaternion product $\alpha\beta$. Here and elsewhere in his book we see Clifford selecting and altering parts of the quaternion system and forming thereby the rudiments of a new system of vector analysis. This process of selection and alteration begun by Clifford was carried to completion by Gibbs and Heaviside. Neither of these men seem to have been influenced by Clifford, nor were later vector analysts. This may be attributed to the fact that Clifford died in 1879, and left only an unfinished presentation of the elements he had worked out. If Clifford had lived, the history of vector analysis might be quite different.

As a conclusion to this chapter the writings of Tait, Maxwell, and Clifford may be briefly compared. It is obvious that certain similarities of view run through these writings: all three authors were, for example, convinced that the vectorial approach to physical problems presented many advantages. One similarity which is less than obvious is especially important. If we read the statements of these three authors with the perspective of the present, we see that from the writings of each the idea comes forth that the scalar product and the vector product are of paramount importance and that the quaternion product is of limited use. In Clifford this is clearly evident from the form of presentation used in his *Elements of Dynamic.* In Maxwell the idea is expressed metaphorically—"May one plough with an ox and an ass together?"—as well as indirectly—the full quaternion product is rarely found in his *Treatise on Electricity and Magnetism.* In Tait's *Treatise on Quaternions* this idea may also be found, but only by one who (unlike Tait) has eyes to see it.

Such a man was Gibbs, who in an 1891 paper pointed out that the careful reader of Tait's *Treatise* will find that what is essential to the developments and what occurs most frequently in the treatment of actual problems is, not the quaternion product $\alpha\beta$, but rather the separate products $S\alpha\beta$ and $V\alpha\beta$.[67] In making this acute observation, which he presented in a paper arguing for the advantages of his system of vector analysis over the quaternion system, Gibbs perhaps revealed more than he intended: his debt to Tait was probably greater than he realized. It is no accident that Gibbs read Tait's *Treatise* shortly before he created the modern system of vector analysis. And it is perhaps not too much to say that although Tait felt that he was not in accord with the ideas of Maxwell and Clifford, the three were in fact moving in the same direction.

Notes

[1] Cargill Gilston Knott, *Life and Scientific Work of Peter Guthrie Tait* (Cambridge, England, 1911).

[2] Peter Guthrie Tait, *Elementary Treatise on Quaternions*, 2nd ed. (Oxford, 1873).

[3] James Clerk Maxwell, *Treatise on Electricity and Magnetism*, 2 vols., 3rd ed. (New York, 1954).

[4] James Clerk Maxwell, *The Scientific Papers of James Clerk Maxwell*, 2 vols. bound as 1, ed. W. D. Niven (New York, 1965).

[5] James Clerk Maxwell, "Address to the Mathematical and Physical Sections of the British Association" in (4,II; 215–229). Originally published in *British Association for the Advancement of Science Report, 40* (1870).

[6] James Clerk Maxwell, "On the Mathematical Classification of Physical Quantities" in (4,II; 257–266). Originally published in *Proceedings of the London Mathematical Society*, 3 (1871), 224–232.

[7] William Kingdon Clifford, *Mathematical Papers by William Kingdon Clifford*, ed. R. Tucker (London, 1882).

[8] William Kingdon Clifford, *Elements of Dynamic: Part I. Kinematic* (London, 1878). A second incomplete volume was published posthumously, ed. R. Tucker (London, 1887).

[9] The following note is aimed at discussing the main techniques and assumptions involved in this study. I have titled the various sections for clarity and ease of reference. Definition of the term "publication": A paper or book that appeared during the interval 1841 to 1900. Definition of the term "book": A publication of more than fifty pages which did not appear in a journal. Concerning the question of circulation (for example, in regard to theses that seem to have been published) no work was included unless it was listed in one of the standard mathematical bibliographies or in the catalogue of at least one of the major libraries of the world. Translations and later editions were counted as separate books. Definition of the term "subject": The publications have been classified into Grassmannian and Hamiltonian or quaternionic. Works were included if they included some discussion or use of the ideas or methods of either or *both* of these traditions. This classification was in most cases based on an examination of the publication itself; in those cases (few in number) where this was impossible, descriptions of the work, such as those in *Fortschritte der Mathematik*, were used. Definition of the term "country of publication": The country in which the book or journal article was published. It has been found more fruitful to use this classification rather than one based on language of publication, since, for example, the latter basis would not allow an analysis of the different levels of interest in Britain and America. By a German publication I mean a publication that appeared in a German-speaking country. By a British publication I mean a publication appearing in an English-speaking country, but not the United States. Sources of items classified: The major source was Alexander Macfarlane, *Bibliography of Quaternions and Allied Systems of Mathematics* (Dublin, 1904) and supplements thereto published in the *Bulletin of the International Association for Promoting the Study of Quaternions and Allied Systems of Mathematics*. Supplements appeared in the issues of this journal for 1905, 1908, 1909, 1910, 1912, and 1913. This bibliograph-

ical source, compiled by the mathematician Alexander Macfarlane and his fellow members of the Association, was probably extremely complete, since the members themselves had written many of the articles included. From my experience I would estimate that not less than 97 percent of the relevant articles were included. Technique used: I have based the study on a randomly selected sample consisting of 334 publications which constitute 25 percent of the 1338 items listed in the above bibliography and its supplements.

[10] These numbers were arrived at in the following manner. In the 25 percent sample there were 146 quaternion publications, of which 35 came in the period from 1841 to 1865. There were 54 Grassmannian publications in the sample, of which 8 were from the period from 1841 to 1875. Thus the sample would predict: total quaternion publications $= 4 \times 35 + 4 \times 111 = 584$. Total Grassmannian publications $= 4 \times 8 + 4 \times 46 = 216$. It was convenient to determine by actual count the publications in the first tradition for the period from 1841 to 1865 and for the second tradition from 1841 to 1875. The numbers obtained were 150 and 33 for the quaternion and Grassmannian traditions respectively. An accuracy check on the sampling technique is provided by comparing these figures. Thus for the quaternion tradition and the period 1841 to 1865 the sample predicted ten too few publications. This is an error of 6.7 percent. For the Grassmannian tradition the sample predicted one publication less than the actual number for the period from 1841 to 1875. This is an error of 3.0 percent. I have used the numbers obtained by actual count for the first period in each tradition and used the numbers predicted from the sample for the later period in each tradition. Thus I have concluded that there were $150 + 4 \times 111 = 594$ quaternion publications and $33 + 4 \times 46 = 217$ Grassmannian publications.

[11] If the four extremely long books by Hamilton are not included, the average length of quaternion books became 208 pages. If the five long Grassmann books are not included, the average length of Grassmannian books became 181 pages. It may be noted that Hamilton and Grassmann each wrote only two books; the higher number given above is accounted for by the fact that translations and later editions were included.

[12] H. S. White, "Forty Years' Fluctuations in Mathematical Research" in *Science* (New Series), *42* (1915), 105–113.

[13] Graph III is from White, *art. cit.*, 106. Unfortunately White did not give any indication as to the actual number of publications in any year. The graph is thus useful only for comparison of its form with that of Graphs I and IV.

[14] It is to be noted that the accuracy of the sample as a criterion of inference on any given point is a function of the size of the sample. Thus for example the degree of accuracy may not be great for inferences in regard to the number of Grassmannian analysis publications written in any country except Germany.

[15] William Rowan Hamilton, *Lectures on Quaternions* (Dublin, 1853), 610.

[16] As quoted in G. Chrystal, "Professor Tait" in *Nature, 64* (1901), 306.

[17] As quoted in Silvanus P. Thompson, *The Life of William Thomson,* vol. II (London, 1910), 1138.

[18] *Ibid.,* 1070. Hertz used vectors in a very elementary form in his *Principles of Mechanics* of 1894.

[19] Alexander Macfarlane, "Peter Guthrie Tait" in *Ten British Physicists* (New York, 1919), 45. See also (1; 21–22).

[20] Peter Guthrie Tait, *Elementary Treatise on Quaternions* (Oxford, 1867), v.

[21] These editions were both published at Cambridge. R. W. David, the present manager of Cambridge University Press, has generously supplied me with the following information. The second edition probably consisted of 1000 copies, since

796 copies were still in stock at the Press in 1875. The third edition consisted of 750 copies, and this edition went out of print in 1910. This should be compared to the fact that 500 copies of Hamilton's *Elements* were printed.

[22] Philip Kelland and Peter Guthrie Tait, *Introduction to Quaternions*, 2nd ed. (London, 1882), v.

[23] *Ibid.*, x.

[24] *Ibid.*, ix.

[25] Note that for the purposes of this study I have defined "modern vector analysis" so as to include such areas as vector addition, subtraction, and multiplication; elementary vector algebra; vector differentiation and integration; and the properties of the operator ∇, along with such famous transformation theorems as those of Stokes and Gauss. Dyadics and the linear vector function are not included in this definition, but some discussion will be given of their place in this history.

[26] Of course many of the theorems were first stated by Grassmann, but we will show that this was not in most cases the source from which they entered modern vector analysis.

[27] Nabla was the name suggested to Tait by Robertson Smith because of the similarity of the symbol to an Assyrian harp. See (1; 143). Maxwell used the word only once in his published writings, and that was in a poem, "To the Chief Musician upon Nabla, A Tyndallic Ode." The "Chief Musician upon Nabla" was Tait. The poem was published in *Nature* and is given in (1; 171–174). The name del first appeared in print in Edwin Bidwell Wilson's *Vector Analysis, A Text-Book for the Use of Students of Mathematics and Physics Founded upon the Lectures of J. Willard Gibbs* (New York, 1901), 138. The genealogy of atled (which is Δ, delta inverted to form ∇, atled) is unknown to me. It must have been used as early as 1870, for in that year Maxwell asked Tait in a letter if atled was Tait's name for ∇. See (1; 143).

[28] (3,I; 29). On Tait's contribution see also (1; 142–149) and Alexander Macfarlane, "Peter Guthrie Tait" in *Physical Review, 15* (1902), 56.

[29] The history of these theorems has never (to my knowledge) been written. It essentially lies outside the province of the history of vector analysis, for the theorems were all developed originally for Cartesian analysis, and by people who did not work with vectors. Some comments may however be made. Gauss' Theorem (often called the Divergence Theorem) is attributed (by Hermann Rothe, "Systeme Geometrischen Analyse, Erster Teil" in *Encyklopädie der mathematischen Wissenschaften*, vol. III, pt I, 1345) to Gauss in his "Allgemeine Lehrsätze in Beziehung auf die im verkehrten Verhältnisse des Quadrates der Entfernung wirkenden Anziehungs-und Abstossungskräfte" in *Resultate aus den Beobachtungen des magnetischen Vereins im Jahre 1839*, (Leipzig, 1840), 1–51, with special attention to pages 34 to 35. This was translated into English and published in Richard Taylor's *Scientific Memoirs, 3* (London, 1843), 153–196. James Clerk Maxwell in (3,I; 125) stated that Gauss' Theorem ". . . seems to have been first given by Ostrogradsky in a paper read in 1828, but published in 1831 in the Mem. de l'Acad. de St. Petersburg, T. I. p. 39." This note is not contained in the first edition of his *Treatise*. This fact is doubly interesting as possibly indicating where Maxwell first found the theorem. Oliver Dimon Kellogg, *Foundations of Potential Theory*, (New York, 1953), 38, wrote the following in regard to Gauss' Theorem: "A similar reduction of triple integrals to double integrals was employed by LAGRANGE: *Nouvelles recherches sur la nature et la propagation du son*, t. II, 1760–61, 45; *Œuvres*, t. I, p. 263. The double integrals are given in more definite form by GAUSS, *Theoria attractionis corporum sphaeroidicorum ellipticorum homogeneorum methodo novo tractata*, Commentationes societatis regiae scientiarum Gottingensis recentiores,

vol. II, 1813, 2–5; *Werke*, Bd. V, pp. 5–7 [.] A systematic use of integral identities equivalent to the divergence theorem was made by George Green in his *Essay on the Application of Mathematical Analysis to the Theory of Electricity and Magnetism,* Nottingham, 1828." The history of Stokes' Theorem is clear but very complicated. It was first given by Stokes without proof—as was necessary—since it was given as an examination question for the Smith's Prize Examination of that year! Among the candidates for the prize was Maxwell, who later traced to Stokes the origin of the theorem, which by 1870 was frequently used. On this see George Gabriel Stokes, *Mathematical and Physical Papers,* vol. V (Cambridge, England, 1905), 320–321. See also the important historical footnote which indicates that Kelvin in a letter of 1850 was the first who actually stated the theorem, although others as Ampère had employed "the same kind of analysis . . . in particular cases." See also Max Bacharach, *Abriss der Geschichte der Potentialtheorie* (Göttingen, 1883).

30 Peter Guthrie Tait, "On Green's and Other Allied Theorems" in *Transactions of the Royal Society of Edinburgh, 26* (1870), 169–184. Published also in Peter Guthrie Tait, *Scientific Papers,* vol. I (Cambridge, England, 1898), 136–150.

31 Peter Guthrie Tait, "On Some Quaternion Integrals" in *Proceedings of the Royal Society of Edinburgh, 7* (1870), 318–320, and (1872), 784–788. Published also in Tait, *Scientific Papers,* vol. I, 159–163.

32 Dirk J. Struik, *Yankee Science in the Making,* rev. ed. (New York, 1962), 415.

33 Benjamin Peirce, "Linear Associative Algebra," in *American Journal of Mathematics, 4* (1881), 216–217.

34 Florian Cajori, *The Teaching and History of Mathematics in the United States* (Washington, 1890), 137.

35 In a "Reminiscences" included in Raymond Clare Archibald, *Benjamin Peirce* (Oberlin, 1925), 6.

36 On Hill, Lowell, Chase, and the two Peirce sons see Archibald, *Benjamin Peirce.* On the teaching of quaternions at Harvard see Cajori, *Math in the U.S.,* 127–151.

37 H. A. Newton, "Benjamin Peirce" (Obituary Notice) in *American Journal of Science,* 3rd Ser., 22 (1881), 74.

38 Benjamin Peirce, "Linear Associative Algebra" in *American Journal of Mathematics, 4* (1881), 97–229.

39 As described by H. E. Hawkes and as quoted by Archibald, *Peirce,* p. 16. H. A. Newton's description of this work is too long to quote but too important not to mention. See Newton, "Peirce" in *American Journal of Science,* 3rd Ser., 22 (1881), 167–178.

40 Quaternion methods were taught at Johns Hopkins, Wisconsin, Michigan, Harvard, Princeton (then the College of New Jersey), Dartmouth, Cornell, Virginia, South Carolina, Alabama, Tennessee, and Texas. Cajori gave no information as to whether courses in quaternions were (or were not) taught at Yale, Bowdoin, Georgetown, Virginia Military Institute, North Carolina, Mississippi, Kentucky, Tulane, Washington (of Saint Louis), and the United States Military Academy. It is to be noted that the first group contains the majority of the better schools at that time. See Florian Cajori, *op. cit.* Cajori's main concern in this book was the history of mathematics teaching; nonetheless his book includes much information relevant to the topic at hand.

41 Saloman Bochner, "The Role of Mathematics in the Rise of Mechanics" in *American Scientist, 50* (1962), 301–302.

42 Albert Einstein and Leopold Infeld, *The Evolution of Physics* (New York, 1961), 244.

[43] Many of the most important letters in the correspondence for this history were published in Knott's book, cited in note (1) above. Knott's reason for publishing so many of Maxwell's letters is in large part that most of Tait's letters were lost, and hence the only way of indicating their content was by publishing Maxwell's replies.

[44] See for example (1; 144), where letters from Maxwell to Tait of 1870 are quoted and in which Maxwell admitted he was "unlearned" in quaternions but was "dabbling" in Hamilton.

[45] Edmund Whittaker, *A History of the Theories of Aether and Electricity,* vol. I (London, 1958), 241–242. It should be noted that Whittaker slipped into a number of anachronisms, as for example when he wrote that Thomson in 1846 found "curl **a** = **B**." This anachronism would have been particularly displeasing to Thomson, whose strong feelings against vectors have been discussed.

[46] (6; 261). The reader may note that quotations have been drawn from both papers cited in notes (5) and (6) above. Rhetorically this is useful and not misleading historically. According to Clifford, Maxwell was the first to distinguish between "force" and "flux" vectors; see (7; 497).

[47] For this see Tait, *Treatise* (1st ed.), 268, or Hamilton, *Lectures,* 610. Note that neither Hamilton nor Tait nor Maxwell use the now common δ for partial differentiation.

[48] (6; 265). Note that since $S\nabla\sigma$ (in quaternion language) is equal to the modern $-\nabla \cdot \boldsymbol{\gamma}$, Maxwell's name "convergence" required alteration. The modern name "divergence" for $\nabla \cdot \boldsymbol{\sigma}$ was given by William Kingdon Clifford in (8; 209).

[49] (1; 143). The following discussion of Maxwell's ideas as expressed in his letters has been aided by the availability of unpublished parts of Maxwell's correspondence. These letters are in the archives of Cambridge University Library. The officials of the Library permitted Professor Derek J. de Solla Price to microfilm the collection. I have seen the letters in a copy of this microfilm in the possession of Professor Erwin N. Hiebert of the University of Wisconsin.

[50] From the unpublished portion of the Tait-Maxwell correspondence.

[51] Quaternions were probably not taught at Cambridge however. On this point see A. R. Forsyth, "Old Tripos Days at Cambridge" in *Mathematical Gazette, 19* (1935), 164 and 172. Forsyth wrote: "Occasional attention, by individual students who had been pupils of Tait at Edinburgh or pupils of Barker at Manchester, was paid to quaternions: never as a non-commutative algebra: always with an introductory testimonial from Natural Philosophy." (p. 172).

[52] [James Clerk Maxwell], "Quaternions" in *Nature,* 9 (1873), 137–138. In attributing this review to Maxwell I have followed Knott (1; 115) and Macfarlane *Bibliography of Quaternions*). The content is of such a nature as to put the question beyond a reasonable doubt. This paper was not included in Maxwell's *Scientific Papers* (cited in note [4] above).

[53] *Ibid.,* 137.

[54] James Clerk Maxwell, *Treatise on Electricity and Magnetism,* 2 vols. (Oxford, 1873). A second (posthumous, but with some revisions by Maxwell) edition appeared in 1881, and a third edition in 1891 (the book cited in note [3] above is a reprint of the 3rd. ed.). Quotations from the *Treatise* will be referred to the third edition. All quotations may be assumed to be identical in all editions unless the contrary is noted.

[55] Maxwell, *Treatise,* I (1st ed.), 28–29.

[56] For the scalar product see (3,II; 256 and 274). For the vector product see (3,II; 240, 244, and 305).

[57] (3,II; 13). In the first edition of Maxwell's *Treatise* the reference was to the first edition of Tait's *Quaternions.*

[58] A. P. Wills in the "Historical Introduction" to his *Vector Analysis with an Introduction to Tensor Analysis* (New York, 1958), xxvi.

[59] (3,II; 174 and 319). Grassmann's paper is his "Neue Theorie der Electrodynamik" in Poggendorff's *Annalen der Physik, 74* (1845), 1–18.

[60] Maxwell referred to Grassmann on pages 125, 141–142, 152, 414, and 418–419 in the first volume of the work cited in note (4) above. Grassmann's paper is "Zur Theorie der Farbenmischung" in Poggendorff's *Annalen der Physik,* 99 (1853), 69–84. This was translated into English and published in the *Philosophical Magazine,* 7 (1854), 254–264.

[61] Alexander Macfarlane, "James Clerk Maxwell" in *Ten British Physicists* (New York, 1919), 18.

[62] William Kingdon Clifford, "Preliminary Sketch of Biquaternions" in (7; 181–200). Clifford's biquaternions are not the same as Hamilton's biquaternions. They are an offshoot of Hamilton's quaternion system and Ball's theory of screws and Grassmann's system.

[63] The content of these lectures was preserved in notes on the lectures made by a student and published in (7; 478–515).

[64] William Kingdon Clifford, *Common Sense of the Exact Sciences,* ed. Karl Pearson (New York, 1894). The book first appeared in 1885; the above is a reprint.

[65] *Ibid.,* vii.

[66] The index for both volumes is contained in the second volume. Grassmann is never mentioned; Tait is referred to three times, always in regard to the *Treatise on Natural Philosophy,* which Tait wrote with Kelvin. A single reference to Hamilton appeared, and this is of no importance.

[67] On this see *The Scientific Papers of J. Willard Gibbs,* vol. ii (New York, 1961), 162.

CHAPTER FIVE

Gibbs and Heaviside and the Development of the Modern System of Vector Analysis

I. *Introduction*

To show that Josiah Willard Gibbs and Oliver Heaviside independently and nearly simultaneously created what is essentially the modern system of vector analysis is the chief concern of this chapter. It will be argued that these two men, motivated by an interest in electrical theory and inspired particularly by Maxwell, forged modern vector analysis from quaternion (not Grassmannian) elements. Later chapters will show that Gibbs and Heaviside were spared the fate of Grassmann and Hamilton; within twenty-five years their system of vector analysis was widely appreciated and much used. Since Gibbs' work came slightly earlier than Heaviside's, he will be considered first.

II. *Josiah Willard Gibbs*

In the preface to the third edition of his *Treatise on Quaternions* (1890) Peter Guthrie Tait expressed his disappointment at "how little progress has recently been made with the development of Quaternions."[7] He went on to remark: "Even Prof. Willard Gibbs must be ranked as one of the retarders of Quaternion progress, in virtue of his pamphlet on *Vector Analysis;* a sort of hermaphrodite monster, compounded of the notations of Hamilton and of Grassmann."[7] Tait's remark about Gibbs is correct; Gibbs did retard quaternion progress, for his "pamphlet" *Elements of Vector Analysis* marks the beginning of modern vector analysis.

Many of Tait's readers must have wondered who this "retarder of Quaternion progress" was. Though Gibbs had by 1890 made major scientific discoveries, few scientists had noticed them in the pages of the *Transactions of the Connecticut Academy*. Gibbs' fate was the not uncommon fate of the theoretician: slow recognition. Recognition did come in time, and now Gibbs is frequently placed in a class with Gauss, Faraday, Maxwell, Helmholtz, and Einstein. However when he presented his vectorial system in the 1880's, he lacked that "capital" of reputation which Hamilton had attained so early in his life and which favorably influenced the reception of his ideas. It will thus be important to survey the events in Gibbs' life that led up to the creation of his system and influenced its reception.

Josiah Willard Gibbs was born in 1839: his father was at that time a professor of sacred literature at Yale University.[8] Gibbs graduated from Yale in 1858, after he had compiled a distinguished record as a student. His training in mathematics was good, mainly because of the presence of H. A. Newton on the faculty. Immediately after graduation he enrolled for advanced work in engineering and attained in 1863 the first doctorate in engineering given in the United States. (1; 32) After remaining at Yale as tutor until 1866, Gibbs journeyed to Europe for three years of study divided between Paris, Berlin, and Heidelberg. Not a great deal of information is preserved concerning his areas of concentration during these years, but it is clear that his main interests were theoretical science and mathematics rather than applied science. It is known that at this time he became acquainted with Möbius' work in geometry, but probably not with the systems of Grassmann or Hamilton. (1; 43) Gibbs returned to New Haven in 1869 and two years later was made professor of mathematical physics at Yale, a position he held until his death.

His main scientific interests in his first year of teaching after his return seem to have been mechanics and optics. (1; 61–62) His interest in thermodynamics increased at this time, and his research in this area led to the publication of three papers, the last being his now classic "On the Equilibrium of Heterogeneous Substances," published in 1876 and 1878 in volume III of the *Transactions of the Connecticut Academy*. This work of over three hundred pages was of immense importance. When scientists finally realized its scope and significance, they praised it as one of the greatest contributions of the century. However, for the present purpose the focus must be on the immeidate reception of the paper. Although Maxwell had recognized its importance, few others did before the early 1890's, when Ostwald arranged for a German translation and publication of the work.[9] Thus when in the 1880's Gibbs began to make known his views on vector analysis, his fame as a scientist was not great.

There is good evidence for concluding that Gibbs did not know Maxwell's ideas on electricity until the appearance of the latter's *Treatise on Electricity and Magnetism* in 1873. (1; 62) His increasing interest in this area led him to give a course in electricity and magnetism in 1877 (1; 62), and eventually to publish papers in the Maxwellian tradition.

III. *Gibbs' Early Work in Vector Analysis*

The historian is extremely fortunate that the draft of an 1888 letter written by Gibbs to Victor Schlegel has been preserved, for this letter answers a host of interesting questions. The introductory paragraph of the letter was followed by:

> Your apt characterization of my Vector Analysis in the Fortsch. Math. suggests that you may be intersd [interested] to know the precise relation of that pamphlet to the work of Ham. & Grass. with respect to its composition.
>
> My first acquaintance with quaternions was in reading Maxwell's E. & M. [*Electricity and Magnetism*] where Quaternion notations are considerably used. I became convinced that to master those subjects, it was necessary for me to commence by mastering those methods. At the same time I saw, that although the methods were called quaternionic the idea of the quaternion was quite foreign to the subject. In regard to the products of vectors, I saw that there were two important functions (or products) called the vector part & the scalar part of the product, but that the union of the two to form what was called the (whole) product did not advance the theory as an instrument of geom. investigation. Again with respect to the operator ∇ as applied to a vector I saw that the vector part & the scalar part of the result represented important operations, but their union (generally to be separated afterwards) did not seem a valuable idea. This is indeed only a repetition of my first observation, since the operator is defined by means of the multiplication of vectors, & a change in the idea of that multiplication would involve the change in the use to the operator ∇.
>
> I therefore began to work out ab initio, the algebra of the two kinds of multiplication, the three differential operations ∇ applied to a scalar, & the two operations to a vector, & those functions or rather integrating operators wh [which] (under certain limitations) are the inverse of the said differential operators, & wh play the leading roles in many departments of Math. Phys. To these subjects was added that of lin. vec. functions wh is also prominent in Maxwell's E. & M.
>
> This I ultimately printed but never published, although I distributed a good many copies among such persons as I thought might possibly take an interest in it. My delay & hesitation in this respect was principally

due to difficulty in making up my mind in respect to details of notation, matters trifling in themselves, but in wh it is undesirable to make unnecessary changes.

My acquaintance with Grassmann's work was also due to the subject of E. [electricity] & in particular to the note wh he published in Crelle's Jour. in 1877 calling attention to the fact that the law of the mutual action of two elements of current wh Clausius had just published had been given in 1845 by himself. I was the more interested in the subject as I had myself (before seeing Clausius' work) come to regard the same as the simplest expression for the mechanical action, & probably for the same reason as Grassmann, because that law is so very simply expressed by means of the external product.

At all events I saw that the methods wh I was using, while nearly those of Hamilton, were almost exactly those of Grassmann. I procured the two Ed. of the Ausd. but I cannot say that I found them easy reading. In fact I have never had the perseverance to get through with either of them, & have perhaps got more ideas from his miscellaneous memoirs than from those works.

I am not however conscious that Grassmann's writings exerted any particular influence on my V-A, although I was glad enough in the introductory paragraph to shelter myself behind one or two distinguished names [Grassmann and Clifford] in making changes of notation wh I felt would be distasteful to quaternionists. In fact if you read that pamphlet carefully you will see that it all follows with the inexorable logic of algebra from the problem wh I had set myself long before my acq. with Grass.

I have no doubt that you consider, as I do, the methods of Grassmann to be superior to those of Hamilton. It thus seemed to me that it might [be] interesting to you to know how commencing with some knowledge of Ham's methods & influenced simply by a desire to obtain the simplest algebra for the expression of the relations of Geom. Phys. &c I was led essentially to Grassmann's algebra of vectors, independently of any influence from him or any one else.[10]

This letter is so rich in information that from it we may establish Gibbs' relationship to each of the important figures in the earlier history of vector analysis.

Concerning Gibbs' relationship to Maxwell: we learn that interest in electricity and magnetism led Gibbs to Maxwell's *Treatise;* here he found quaternions "considerably used" and inferred their usefulness. Thus from Maxwell he went to the quaternionists. Maxwell's gentle criticisms and discriminating employment of quaternion methods probably did not go unnoticed by Gibbs. Or perhaps the practices of the quaternionists revealed to him what their theoretical and rhetorical statements could not say: the quaternion is not needed. In any case, what Maxwell wanted in a vector analysis, Gibbs produced.

Concerning Gibbs' relationship to Hamilton and Tait: we learn that Gibbs commenced his search for a vector analysis "with some knowledge of Ham[ilton]'s methods . . ." and ended up with methods that were "nearly those of Hamilton. . . ." Stronger statements showing historical continuity are seldom found. It is clear that Gibbs was no more than a minor heretic in relation to quaternion orthodoxy. Gibbs' achievement was great, but so was his debt to Hamilton and his followers. It will later be argued that Tait was especially influential on Gibbs.

Concerning Gibbs' relationship to Grassmann: we learn that Gibbs created his system "independently of any influence from him or any one else." Obviously Gibbs' "any one else" was not meant to include Maxwell and the quaternionists. Gibbs also stated that he was not "conscious that Grassmann exerted any particular influence on my V-A. . . ." This is to be expected since Gibbs had begun searching for a new vector system "long before my acq[uaintance] with Grass[mann]." When (1877 or later) Gibbs finally began to read Grassmann, he found a kindred spirit. Although Gibbs admitted he had never been able to read through either of Grassmann's books, he did recognize Grassmann's priority and warmly praised his ideas on numerous occasions. Gibbs' statements concerning his independence of Grassmann were fully accepted by Victor Schlegel, who as Grassmann's leading disciple was perhaps least disposed to accept them. In his reply to Gibbs' letter Schlegel stated: "I realized from your letter how you attained to ideas similar to Grassmann's but independently of him; I have repeatedly witnessed this phenomenon: when the time has come—when the development of science demands a definite discovery—then this discovery will be made by a number of investigators." [11]

Concerning Gibbs' relationship to Clifford: we learn that it was limited to the use of Clifford's name "as a shelter" in the introductory paragraph of his vector analysis book.

Thus from this letter we can obtain a general outline of Gibbs' early work in vector analysis. There are however a number of facts that will give a fuller picture of Gibbs' activities.

In 1879 Gibbs gave a course in vector analysis with applications to electricity and magnetism,[12] and in 1881 he arranged for the private printing of the first half of his *Elements of Vector Analysis;* the second half appeared in 1884. In an effort to make his system known, Gibbs sent out copies of this work to more than 130 scientists and mathematicians. (1; 247) Many of the leading scientists of the day received copies, for example, Michelson, Newcomb, J. J. Thomson, Rayleigh, FitzGerald, Stokes, Kelvin, Cayley, Tait, Sylvester, G. H. Darwin, Heaviside, Helmholtz, Clausius, Kirchhoff,

Lorentz, Weber, Felix Klein, and Schlegel. (1; 236–247) Though the work was not given the advertisement that a regular publication would have had, such a selective distribution must have aided in making it known.

IV. *Gibbs'* Elements of Vector Analysis

Some idea of the form of Gibbs' *Elements of Vector Analysis* [13] may be obtained from Gibbs' introductory paragraph:

> The fundamental principles of the following analysis are such as are familiar under a slightly different form to students of quaternions. The manner in which the subject is developed is somewhat different from that followed in treatises on quaternions, since the object of the writer does not require any use of the conception of the quaternion, being simply to give a suitable notation for those relations between vectors, or between vectors and scalars, which seem most important, and which lend themselves most readily to analytical transformations, and to explain some of these transformations. As a precedent for such a departure from quaternionic usage, Clifford's *Kinematic* may be cited. In this connection, the name of Grassmann may also be mentioned, to whose system the following method attaches itself in some respects more closely than to that of Hamilton. (2; 17)

Although Gibbs mentioned only Clifford and Grassmann in the introductory paragraph, the previously cited letter makes it clear that his chief debt was not to either Clifford or Grassmann but to the quaternionists. In the discussion of Gibbs' book this point will be illustrated; specifically it will be suggested that Gibbs was strongly influenced by the content and form of presentation found in the second edition of Tait's *Treatise on Quaternions*.

Chapter I, "Concerning the Algebra of Vectors," began with such definitions as "vector," "scalar," and "vector analysis." In much of the symbolism introduced, Gibbs followed the quaternion traditions; for example, Gibbs represented vectors by Greek letters, their components by means of i, j, and k. Where Hamilton had written AB (for a vector) and Tait $\overline{AB}$, Gibbs wrote $\overline{AB}$. Gibbs, like Tait and Maxwell, but unlike Hamilton, used the right-handed co-ordinate system.

In dealing with vector products Gibbs introduced the "direct product," written $\alpha . \beta$, and the "skew product," written $\alpha \times \beta$. These are the now-current scalar (dot) and vector (cross) products. The relation between these products and the quaternion system may be expressed by writing $\alpha . \beta \equiv -S\alpha\beta$ and $\alpha \times \beta \equiv V\alpha\beta$. Gibbs of course never combined these products in any way analogous to the quaternion equation $\alpha\beta = S\alpha\beta + V\alpha\beta$. To illustrate further the closeness of Gibbs' booklet to Tait's *Treatise on Quaternions* (2nd ed.), various equivalent equations from Chapter I of Gibbs and from Tait have been listed below:

Gibbs (2; 21) $\alpha.\beta = \beta.\alpha$

Tait (3; 43) $S\alpha\beta = S\beta\alpha$

Gibbs (2; 21) $\alpha \times \beta = -\beta \times \alpha$

Tait (3; 43) $V\alpha\beta = -V\beta\alpha$

Gibbs (2; 22) $\alpha.\beta = xx' + yy' + zz'$ and $\alpha \times \beta = (yz' - zy')i + (zx' - xz')j + (xy' - yx')k$

Tait (3; 43) $\alpha\beta = -(xx' + yy' + zz') + (yz' - zy')i + (zx' - xz')j + (xy' - yx')k$

Gibbs (2; 23) $\alpha \times [\beta \times \gamma] = (\alpha.\gamma)\beta - (\alpha.\beta)\gamma$

Tait (3; 44) $V\alpha V\beta\gamma = \gamma S\alpha\beta - \beta S\gamma\alpha$

Gibbs (2; 24) $[\alpha \times \beta] \times [\gamma \times \delta] = (\alpha.\gamma \times \delta)\beta - (\beta.\gamma \times \delta)\alpha$

Tait (3; 44) $VV\alpha\beta V\gamma\delta = -\beta S\alpha V\gamma\delta + \alpha S\beta V\gamma\delta$

Many more such pairs of equations could be given, but the above should be sufficient to suggest strongly that the two works are quite similar in the symbolisms employed and mathematical ideas expressed. The similarities are probably most easily explained by the conjecture that Gibbs learned quaternionic analysis from Tait and hence when he came to write a work with similar mathematical content, it was natural for him to write in the "language" that he had learned from Tait. Unfortunately for the historian, Gibbs, having written his booklet primarily as a teaching aid, made no attempt whatever to indicate the origins of the various theorems.

Chapter I concluded with a treatment of methods for solving vectorial equations, and chapter II was entitled "Concerning the Differential and Integral Calculus of Vectors." Herein Gibbs introduced the operator ∇, proved the related transformation theorems, and gave an extended treatment of the mathematics of potential theory.[14] The part of Gibbs' booklet printed in 1881 terminated near the end of chapter II; the remainder was printed in 1884.

Chapters III and IV centered on linear vector functions, that is, vector functions of such a nature that a function of the sum of any two vectors is equal to the sum of the functions of the vectors. To treat linear vector functions Gibbs introduced the terms and concepts "dyad" and "dyadic." A dyad is an expression of the form $\alpha\lambda$, where α and λ are vectors; and a dyadic refers to the sum of a number of dyads. In these two chapters Gibbs followed the general scheme for the treatment of linear vector functions that had been worked out by Hamilton and Tait. Because of this the quaternionist C. G. Knott was able to argue:

> but when we bear in mind that Professor Gibbs deliberately set out to construct a system free from the fancied blemish of the quaternion and yet did not scruple to introduce in its stead an indeterminate product [the dyad] which is without any geometric significance whatever, and when we find on careful comparison that *practically* the dyadic system is simply a modification of quaternion methods, in large measure a mere difference of notation, we can find no satisfactory reason for a man of Professor Gibbs's great powers leaving quaternionic paths to invent new notations, new names for old things, and an indeterminate purely symbolic product to take the place of the determinate real quaternion.[15]

Although Knott went somewhat too far in this statement, it is typical of a type of argument that proponents of the quaternion tradition were to bring against Gibbs. The same conclusion as to the relation of the systems is implicit in the tables of comparative notation given by later authors.[16] The main physical applications of linear vector functions were to the treatment of rotations and strains, which Gibbs, like Tait, covered in some detail. It was primarily in these sections that Gibbs went beyond the results obtained by the quaternionists.[17] The concluding brief chapter of his booklet dealt with transcendental functions of dyadics, and to this was appended a short note on bivector analysis.[18]

That Gibbs' book was not highly original was the view of most of his contemporaries, and in this view they did not err. E. B. Wilson, Gibbs' student and the author of an important book on vector analysis "founded upon the lectures of" Gibbs, wrote the following in a 1936 article entitled "The Contributions of Gibbs to Vector Analysis and Multiple Algebra": "If I have not claimed very much for his contributions in this field I have but followed what I believe to be his own matured judgment. Had I written thirty years ago I should have claimed much more for him than he would have admitted."[19] Leigh Page seems to have come to the same conclusion, for he is quoted as writing that "the value of his work lies more in the formulation of a convenient and significant notation than in the development of a new mathematical method."[20] Heaviside referred to Gibbs' booklet as "an able and in some respects original little treatise on vector analysis. . . ." (4; 138) If

Gibbs cannot be given great credit for originality of methods, yet he deserves praise for the sensitivity of his judgment as to what deletions and alterations should be made in the quaternionic system in order to make a viable system. His symbolism, now in the main accepted, also constitutes a significant contribution.

Gibbs' work is, historically considered, one of the two sources from which modern vector analysis came into existence. Though his booklet was difficult to read because of its compactness (Heaviside called it a "condensed synopsis"), it could and did form a basis for later writers.

V. *Gibbs' Other Work Pertaining to Vector Analysis*

Gibbs' *Elements of Vector Analysis* was by no means his only contribution to vector analysis. Although Grassmann, Hamilton, Tait, and Heaviside did almost no teaching of their ideas on vector analysis, this was not the case with Gibbs. During the 1880's Gibbs taught vector analysis periodically and in the 1890's taught the course every year.[21] At least in later years the course consisted of ninety lectures.[22]

Gibbs also published a number of articles which have relevance to the history of vector analysis. It is primarily from these that Gibbs' view of Grassmann may be determined. Though Gibbs created his system independently of Grassmann's, the two systems had much in common, and it was natural for Gibbs to argue for the power and importance of Grassmann's work. Moreover Gibbs felt that Grassmann's system, when viewed as a work in multiple algebra, had much of value for the pure mathematician.[23] One of Gibbs' most famous papers was entitled "On Multiple Algebra." It was given as an "Address before the Section of Mathematics and Astronomy of the American Association for the Advancement of Science, by the Vice-President" and was published in 1886.[24] Of all Gibbs' writings this essay gives the best picture of his conception of the place of vector analysis within the wider fields of algebra and mathematics in general. Broadly considered, the article is a plea for greater interest in multiple algebra. Gibbs began with a brief treatment of the history of multiple algebra and considered such men as Möbius, Grassmann, Hamilton, Saint-Venant, Cauchy, Cayley, Hankel, Benjamin Peirce, and Sylvester. By this analysis Gibbs hoped among other things

> to illustrate the fact, which I think is a general one, that the modern geometry is not only tending to results which are appropriately expressed in multiple algebra, but that it is actually striving to clothe itself

in forms which are remarkably similar to the notations of multiple algebra, only less simple and general and far less amenable to analytical treatment, and therefore, that a certain logical necessity calls for throwing off the yoke under which analytical geometry has so long labored. (2; 103)

The strongest praise and fullest treatment was bestowed upon Grassmann. Considering in detail a number of the products defined by Grassmann, Gibbs discussed their significance and argued that Grassmann's system provides a rich and encompassing point of view. Gibbs concluded the paper with a discussion of applications of multiple algebra to physical science. Thus he stated:

First of all, geometry, and the geometrical sciences which treat of things having position in space, kinematics, mechanics, astronomy[,] physics, crystallography, seem to demand a method of this kind, for position in space is essentially a multiple quantity and can only be represented by simple quantities in an arbitrary and cumbersome manner. For this reason, and because our spatial intuitions are more developed than those of any other class of mathematical relations, these subjects are especially adapted to introduce the student to the methods of multiple algebra. (2; 113)

Proceeding then to specifics, Gibbs noted that through Maxwell electricity and magnetism had become associated with the methods of multiple algebra but that astronomy had so far remained aloof. (2; 114) Having noted that for geometrical applications multiple algebra will generally take the form of a point or a vector analysis, Gibbs stated that "in mechanics, kinematics, astronomy, physics, or crystallography, Grassmann's point analysis will rarely be wanted." (2; 115) However arguments were given on behalf of the usefulness of point analysis for investigations in pure mathematics. The concluding paragraph and the concluding sentence in particular have now become classic.

But I do not so much desire to call your attention to the diversity of the applications of multiple algebra, as to the simplicity and unity of its principles. The student of multiple algebra suddenly finds himself freed from various restrictions to which he has been accustomed. To many, doubtless, this liberty seems like an invitation to license. Here is a boundless field in which caprice may riot. It is not strange if some look with distrust for the result of such an experiment. But the farther we advance, the more evident it becomes that this too is a realm subject to law. The more we study the subject, the more we find all that is most useful and beautiful attaching itself to a few central principles. We begin by studying *multiple algebras;* we end, I think, by studying MULTIPLE ALGEBRA. (2; 117)

Gibbs' paper may be taken as symbolic of the ever-increasing interest expressed at that time by pure and applied mathematicians in multiple algebra. The increasing interest in vector analysis was a part, in one way the most influential part, of that trend. Gibbs was able, as some later writers were not, to view vector analysis in the broad perspective offered by multiple algebra. That Gibbs was deeply interested in multiple algebra is shown by the facts that every two or three years he gave a course in multiple algebra[25] and that he planned to publish additional writings on multiple algebra and had actually done research in this regard. (1; 118)

In two of his five papers on the electromagnetic theory of light Gibbs made use of vectorial methods. However he did not in these papers stress the importance of vector methods; in fact anyone who had read Maxwell's *Treatise* would have no difficulty with the methods used. Gibbs did publish one highly mathematical paper aimed at demonstrating the usefulness of vector methods to a group of scientists—the astronomers. This is his "On the Determination of Elliptic Orbits from Three Complete Observations," published in 1889 in the *Memoirs of the National Academy of Sciences*[26] and later translated into German and included in the third edition of Klinkerfuss' *Theoretische Astronomie.* (1; 137) Gibbs discussed the aim of this rather long paper in a letter to Hugo Buchholz, who edited the third edition of this book:

> The object of my paper was to show to astronomers, who are rather conservative . . . the advantage in the use of vector notations, which I had learned in Physics from Maxwell. This object could be best obtained, not by showing, as I might have done, that much in the classic methods could be conveniently and perspicuously represented by vector notations, but rather by showing that these notations so simplify the subject, that it is easy to construct a method for the complete solution of the problem. (2; 149)

In this paper Gibbs gave a sketch of the very elementary parts of his vector analysis (little more than the direct and skew products), and with these methods he developed an improved technique for the determination of an orbit from three complete observations. Shortly after the paper was published, A. W. Phillips and W. Beebe published a paper in which they tested (successfully) his method on Swift's Comet.[27] It has been established that Gibbs did more work on this problem, though he did not publish his results. (1; 137) Gibbs sent out more reprints (199) of this article than of any other article, with the exception of his "On Multiple Algebra," of which 276 copies were distributed. (1; 247) The only remaining papers by Gibbs which bear on the history of vector analysis are the four polemical papers published in *Nature* from 1891 to 1893. These will be discussed in detail in the following chapter.

Gibbs, as mentioned before, was an ardent admirer of Grassmann's works. This led him in 1887 to write a letter to Grassmann's son, Hermann Jr., containing the following:

> Permit me to take this opportunity to express my hope that the publication of another edition of the Ausdehnungslehre of 1862, will not be long delayed. Using that treatise in one of my classes two years ago, I had great difficulty in collecting 3 or 4 copies by borrowing, &c.
>
> Another matter has long been on my mind. Your honored Father, in the preface to the first Ausdehnungslehre, mentions a work on the *tides* in wh he used the principles of the Ausdehnungslehre prior to its publication. If the manuscript of that work is in existence, it seems to me that its publication would be an important contribution to the history of the development of mathematical ideas in this century. (1; 114)

Hermann Grassmann Jr. replied in 1888 that the thesis on the tides had been located but there might be difficulty in finding a publisher. (1; 115) This led Gibbs to write to Thomas Craig, the editor of the *American Journal of Mathematics,* to ask if he would consider publishing the work. The following interesting passage is from this letter:

> I believe that a Kampf ums Dasein is just commencing between the different methods and notations of multiple algebra, especially between the ideas of Grassmann & of Hamilton. The most important question is of course that of merit, but with this questions of priority are inextricably entangled, & will be certain to be the more discussed, since there are so many persons who can judge of priority to one who can judge of merit. Those who are to discuss these subjects ought to have the documents before them, & not be handicapped as I was two years ago, for want of them. (1; 115)

Gibbs' prophecy, that a "struggle for existence" between the different methods was imminent, was certainly fulfilled. Although Craig consented to publish the work, this became unnecessary when it was decided that Grassmann's collected works should be published. In this regard Hermann Grassmann Jr. wrote to Gibbs in 1893:

> For the completion of a collected edition of the mathematical and physical works of my father we have to thank primarily the energetic labors of Professor F. Klein of Gottingen, who through his wide connections has supplied the necessary driving force for the undertaking. However the credit for stirring up interest in the matter is exclusively yours. (1; 115–116)

As a final note it may be mentioned that in 1899 Max Abraham (one of the advocates of vector analysis in Germany) requested Gibbs to write an article on vector analysis for the *Encyklopädie der mathematischen Wissenschaften,* a request which evidently Gibbs declined to fulfill. (1; 231)

By the time that Gibbs died in 1903 he had made numerous important contributions to vector analysis. This was possible not only because of his creative powers in mathematics but also by his great ability to seize upon the most significant traditions in the mathematics and physical science of the time, and thence to construct a system in harmony with the demands of the time.

It is striking that a historical phenomenon of the 1840's occurred again in the 1880's. In the 1840's Hamilton and Grassmann independently and simultaneously created vectorial systems with a number of similarities. In the 1880's another scientist, geographically an ocean away from Gibbs, was covering much the same mathematical terrain as Gibbs. This was Oliver Heaviside.

VI. *Oliver Heaviside*

Oliver Heaviside was born May 18, 1850, in London to a family with little financial security.[28] Heaviside's formal education came to an end in 1866; hence he lacked university training. In 1868, probably through the influence of his uncle Sir Charles Wheatstone, Heaviside took a job as a telegraph operator. Electrical problems began to interest him, and in 1872 he published the first of his numerous electrical papers. He retired from this position in 1874 to live with his parents and pursue independent study and research. Among the books he studied was Maxwell's recently published (1873) *Treatise on Electricity and Magnetism.*

The above narrative may be extended by the following statement included by Heaviside in a review of Wilson's *Vector Analysis.* Heaviside began by telling the story of a boy who, enchanted by the word "Quaternion," set himself to reading Hamilton's books.

> He took those books home and tried to find out. He succeeded after some trouble, but found some of the properties of vectors professedly proved were wholly incomprehensible. How could the square of a vector be negative? And Hamilton was so positive about it. After the deepest research, the youth gave it up, and returned the books. He then died, and was never seen again. He had begun the study of Quaternions too soon.
>
> . . . . . .
>
> My own introduction to quaternionics took place in quite a different manner. Maxwell exhibited his main results in quaternionic form in his treatise. I went to Prof Tait's treatise to get information, and to learn how to work them. I had the same difficulties as the deceased youth, but by skipping them, was able to see that quaternionics could

be employed consistently in vectorial work. But on proceeding to apply quaternionics to the development of electrical theory, I found it very inconvenient. Quaternionics was in its vectorial aspects antiphysical and unnatural, and did not harmonise with common scalar mathematics. So I dropped out the quaternion altogether, and kept to pure scalars and vectors, using a very simple vectorial algebra in my papers from 1883 onward.

. . . .

Up to 1888 I imagined that I was the only one doing vectorial work on positive physical principles; but then I received a copy of Prof. Gibbs's Vector Analysis (unpublished, 1881–4). This was a sort of condensed synopsis of a treatise. Though different in appearance, it was essentially the same vectorial algebra and analysis to which I had been led.

. . . .

as time went on, and after a period during which the diffusion of pure vectorial analysis made much progress, in spite of the disparagement of the Edinburgh school of scorners (one of whom said some of my work was "a disgrace to the Royal Society," to my great delight), it was most gratifying to find that Prof. Tait softened in his harsh judgments, and came to recognise the existence of rich fields of pure vector analysis, and to tolerate the workers therein.

. . . .

I appeased Tait considerably (during a little correspondence we had) by disclaiming any idea of discovering a new system. I professedly derived my system from Hamilton and Tait by elimination and simplification, but all the same claimed to have diffused a working knowledge of vectors, and to have devised a thoroughly practical system.[29]

From this quotation a number of important facts become evident. First, it is striking that Heaviside, like Gibbs, became interested in quaternions through Maxwell's *Treatise.* Heaviside then naturally turned to Tait. Second, Heaviside, as had Gibbs, developed his system of vector analysis from the quaternion system. Heaviside explicitly stated that he did this "by elimination and simplification." Third, it is clear that until 1888 Heaviside worked in complete independence of Gibbs.[30] Fourth, Heaviside's remark concerning Tait's eventual recognition of the value of vector analysis is certainly of interest. Finally, it is significant that Heaviside made no mention of Grassmann; indeed there is no reason for believing that he had even heard of Grassmann until he received Gibbs' booklet in 1888.

VII. *Heaviside's Electrical Papers*

The first paper in which Heaviside introduced vector methods was his 1882–1883 paper "The Relations between Magnetic Force and Electric Current," published in the *Electrician.*[31] The way in which vectors were introduced by Heaviside is somewhat surprising. He began by discussing an electrical theory which involved in its mathematical treatment the Cartesian equivalent of the curl. He then gave a quasi definition of curl: "When one vector or directed quantity, *B*, is related to another vector, *C*, so that the line-integral of *B round* any closed curve equals the integral of *C through* the curve, the vector *C* is called the curl of the vector *B*." (5,I; 199) Heaviside extended this discussion of the curl by treating potentials for scalar and vector fields. This was followed by a discussion of quaternions.

After pointing out the great need for a vectorial method for representing the numerous vectorial quantities in physics, he commented that quaternions were such a method, but for some reason quaternions had not been used. He suggested the reason:

> Against the above stated great advantages of Quaternions has to be set the fact that the operations met with are much more difficult than the corresponding ones in the ordinary system, so that the saving of labour is, in a great measure, imaginary. There is much more thinking to be done, for the mind has to do what in scalar algebra is done almost mechanically. At the same time, when working with vectors by the scalar system, there is great advantage to be found in continually bearing in mind the fundamental ideas of the vector system. Make a compromise; look behind the easily-managed but complex scalar equations, and see the single vector one behind them, expressing the real thing. (5,I; 207)

He gave an illustration of the advantage of thinking in terms of vectors, and stated and proved the theorems ascribed to Gauss and Stokes. This was followed by a fuller discussion of vector potential and curl, and finally by physical applications.

It is noteworthy that Heaviside began his presentation in the middle of vector analysis. Such fundamental symbols as i, j, k, ∇, and $i\frac{\partial}{\partial x}+j\frac{\partial}{\partial y}+k\frac{\partial}{\partial z}$, never appeared. Heaviside had in fact managed to introduce vectors in this paper without having said a word about vector multiplication. Many of his proofs and discussions were thus in a qualitative or in a Cartesian form. Nothing in the paper violated quaternion usage, except the material omitted would be striking by its absence to a quaternionist. At this stage in his writings Heaviside's presentation was scarcely beyond what Maxwell had done. The interesting but probably unresolvable question arises as to whether at the time of the writing this paper Heaviside had begun to develop his nonquaternionic system.

Heaviside's next paper was entitled "Current Energy" and was published in 1883 in the *Electrician*. (5,I; 231–255) In this paper a discussion of certain electrical questions led to a quantity composed of the product of the lengths of two vectorial quantities multiplied by the cosine of the angle between them. Heaviside stated:

> Let ds' be an element of length of the current C'. The portion of M due to it alone is $-AC'\ ds' \cos (AC)$, or simply $-\overline{AC'\ ds'}$, (if we place a stroke over the product of two vector quantities to signify that the product of their magnitudes (disregarding directions) is to be multiplied by the cosine of the angle between their directions, i.e., A multiplied by component of C' along A, or C' multiplied by component of A along C'; as will be done hereafter).[32]

The remainder of the paper was devoted to further discussion of electrical questions in which Heaviside made frequent use of the scalar product. The symbols i, j, k never appeared in this paper, which as a whole was ideally suited for an introduction and illustration of the scalar product.

Heaviside's next paper, published in the *Electrician* for 1883, was entitled "Some Electrostatic and Magnetic Relations." Heaviside's discussion began with matters relating to electrostatic force and its convergence:

> Let R denote the electrostatic force, and ρ the volume density of electricity then we may say
>
> $$4\pi\rho = -\text{conv. } R, \text{ or } 4\pi\rho = \text{div. } R,$$
>
> where we use conv and div as abbreviations to be understood as follows: In terms of the components X, Y, Z of the force, we have
>
> $$4\pi\rho = \left(\frac{dX}{dx} + \frac{dY}{dy} + \frac{dZ}{dz}\right) \ldots$$
>
> The expression on the right hand side of this equation (with the $-$ sign prefixed) Maxwell called the "convergence" of the force; it is really the integral amount of force, taken algebraically, *entering* the unit volume; but since + convergence indicates negative electrification, we may as well use the term "divergence" for the same quantity with + sign prefixed. . . . (5,I; 256–257)

Heaviside here used the term "divergence" for the first time in his writings. Though Clifford had introduced this term five years earlier, it by no means necessarily follows that Heaviside had read Clifford, since Heaviside did not mention Clifford and since divergence is an obvious name for the negative of the convergence. The physical representation in terms of lines of force also suggests this name. Maxwell, it may be recalled, introduced the term "convergence" to harmonize with the first part of the quaternion equation

$$\nabla R = -\left(\frac{dX}{dx} + \frac{dY}{dy} + \frac{dZ}{dz}\right) + i\left(\frac{dZ}{dy} - \frac{dY}{dz}\right) + j\left(\frac{dX}{dz} - \frac{dZ}{dz}\right) + k\left(\frac{dY}{dx} - \frac{dX}{dy}\right)$$

Thus one significance of Heaviside's use of "divergence," instead of "convergence," is that this term harmonizes with his view of the scalar product as a positive quantity. Heaviside proceeded to apply the symbol ∇ to a scalar function; he had used this symbol before, but sparingly and with no introductory comment on it as a vectorial operation.

Later in this paper Heaviside introduced a section entitled "The Operator ∇ and Its Applications."[33] After mention of the curl, divergence, and "space-variation" (Maxwell's term for ∇P for P a scalar function of position) Heaviside wrote: "Now it is very remarkable that (as was discovered by Tait) these three operations of curl, divergence, and space-variation are really only three different forms of the same operation, the effect varying according to the nature of the function under examination." (5,I; 268) This was followed (after some explanations) by the presentation of the quaternion usage of ∇.

> For, if we denote by i, j, k three rectangular vectors of *unit* lengths parallel to x, y, z, then Xi will denote a vector of length X parallel to x, and similarly for Yj and Zk, consequently we may write $R = Xi + Yj + Zk$ with the convention that the sign of addition signifies compounding as velocities. Now the full expression of ∇ is
>
> $$\nabla = i\frac{d}{dx} + j\frac{d}{dy} + k\frac{d}{dz};$$
>
> hence
>
> $$\nabla R = \left(i\frac{d}{dx} + j\frac{d}{dy} + k\frac{d}{dz}\right)(Xi + Yj + Zk).$$
>
> Expand this expression, with the further conventions
>
> $$i^2 = j^2 = k^2 = -1, \text{ and } ij = k, jk = i, ki = j,$$
>
> and we obtain,
>
> $$\nabla R = \left(\frac{dX}{dx} + \frac{dY}{dy} + \frac{dZ}{dz}\right) + i\left(\frac{dZ}{dy} - \frac{dY}{dz}\right) + j\left(\frac{dX}{dz} - \frac{dZ}{dx}\right) + k\left(\frac{dY}{dx} - \frac{dX}{dy}\right),$$
>
> i.e., $\nabla R = \text{conv } R + \text{curl } R$. (5,I; 271)

Referring to the above usage of quaternions, Heaviside wrote: "However, this is merely parenthetical, and we shall have no more concern with quaternion expressions. . . ."[34] When this paper was republished, Heaviside added a footnote further to "emphasize the fact that the use [of quaternions] was parenthetical." (5,I; 271–272) Heaviside concluded this section with the statement:

> The operator ∇ contains the whole theory of potentials, whether of scalars or vectors. But owing to the remarkably different nature of the effects of ∇ on different functions, it conduces to clearness to distinctly separate the space-variation of a scalar, which is easily grasped, from that of a vector, and to instead speak of the curl or the divergence of the latter, as the case may be, and as we have done hitherto. (5,I; 272)

The paper was concluded by a section on fluid motion analogies. This is significant since the study of fluid motion is probably the ideal area of physics by which to illustrate the meaning of the curl and divergence of a vector function. In summary Heaviside explained the meaning of curl, divergence, and space-variation; or, in modern form, ∇P for P, a scalar function of position, and $\nabla \cdot \mathbf{R}$ and $\nabla \times \mathbf{R}$ for $\mathbf{R}$, a vector function of position. Since Heaviside had not yet defined the vector product, he could not write (as he did later) $V\nabla R$ ($\nabla \times \mathbf{R}$ in modern notation), nor did he write $\mathbf{S}\,\nabla R$ ($\nabla \cdot \mathbf{R}$ in modern notation), as he could have done, since he had defined the scalar product. He avoided these forms by writing div R and curl R.

In Heaviside's later papers of 1883 and 1884 use was made of vectors, but no new principles were introduced.[35] In the first section of his first paper of 1885, entitled "Electromagnetic Induction and its Propagation" and published in the *Electrician*, Heaviside introduced the vector product.[36] The vector product, like the scalar product, was introduced in the course of a discussion of physical questions, and the quaternion symbol V was used to designate it. Hence he wrote $C = V\epsilon E$ for certain vectors, C, ϵ, and E. His definition is equivalent to the modern and to the quaternionic definition. In the same section Heaviside gave the vector equation $C = kE$ for C (conduction current-density), E (electric force), and k (a constant). This equation, he pointed out, is true only for an isotropic medium, where C and E are parallel; if the medium is not isotropic, then k becomes a "linear vector operator." (5,I; 429–430)

Soon after publishing this section of the paper (the whole paper was published over a three-year span, and even then it was not completed), Heaviside published (1885) a paper "On the Electromagnetic Wave Surface" in the *Philosophical Magazine.* (5,II; 1–23) Herein Heaviside gave the first unified presentation of his vector system. He began the paper by stating a physical problem and led up to the statement:

> Owing to the extraordinary complexity of the investigation when written out in Cartesian form (which I began doing, but gave up aghast),

> some abbreviated method of expression becomes desirable. I may also add nearly indispensible, owing to the great difficulty in making out the meaning and mutual connections of very complex formulae. . . . I therefore adopt, with some simplification, the method of vectors, which seems the only proper method. (5,II; 3)

He pointed out that his method was similar to the quaternion method, but simplified. His criticisms of the quaternion system were at that time mild; he presented his system as a sort of half-way house between abbreviated Cartesian methods and quaternion methods.

The presentation of his system was very brief. After defining vector addition, he gave the expression for a vector in terms of i, j, and k. After defining the scalar and vector products and the rules for applying these products to combinations of i, j, and k, Heaviside defined the operator ∇ as equal to $i\frac{d}{dx}+j\frac{d}{dy}+k\frac{d}{dz}$ and showed its effects when it was applied to a scalar as well as when applied to a vector by means of the scalar and vector products. He then gave Stokes' and Gauss' theorems without proof. His presentation was confined to two pages of text, and the methods presented correspond exactly to the modern methods. (5,II; 4–5) Heaviside's treatment of the linear vector operator (presented in connection with μ, the magnetic permeability) was based on Tait's work, and a proof from Tait was included. This presentation also encompassed two pages, and from it he proceeded to his physical investigation. (5,II; 6–7)

Heaviside introduced no further innovations in vector analysis in the period from 1885 to 1888, at which time he received a copy of Gibbs' work.[37] In his papers written during this period he continued to use vector formulae, though less frequently, because of the nature of the physical subjects treated. Heaviside was impressed by, and in sympathy with, Gibbs' work (except in regard to notation); and he particularly praised Gibbs' development of the linear vector operator.

Heaviside went on to publish two other summaries of his vector method in journals, the first in the *Electrician*, beginning in 1891, and the second in the *Transactions of the Royal Society of London* for 1892. These papers require only brief mention (with the exception of some polemical parts which will be discussed later) since the important points relevant to them will be made in the discussion of his *Electromagnetic Theory* of 1893.

In a paper of 1892 Heaviside gave the following description of his system:

> It rests entirely upon a few definitions, and may be regarded (from one point of view) as a systematically abbreviated Cartesian method of investigation, and be understood and practically used by any one accustomed to Cartesians, without any study of the difficult science of Quaternions. It is simply the elements of Quaternions without the quaternions, with the notation simplified to the uttermost, and with the very inconvenient *minus* sign before scalar products done away with. (5,II; 529)

This quotation from Heaviside may be taken as a partial summary of the nature of his system and of his position in the history of vector analysis. His system was unquestionably quaternionic in origin; indeed numerous remnants from the older tradition remained in his system. But there was a second origin (using the term in a different sense) of his system; this was his ability to discern which elements of quaternion analysis should be eliminated, which altered, and which retained. Heaviside, unlike Gibbs, *did not* in general create new theorems in vector analysis, but Heaviside, unlike Gibbs, *did* associate vector analysis with a large number of new and important ideas in the ever-expanding field of electricity, particularly in the lines of development begun by Maxwell. Furthermore Heaviside should receive credit for having published the first detailed treatment of vector analysis; this appeared as a long (173 pages) chapter (compiled from a series of articles in the *Electrician*) in volume one of Heaviside's *Electromagnetic Theory* of 1893.

VIII. *Heaviside's* Electromagnetic Theory

Heaviside's *Electromagnetic Theory* was published in three volumes; the first volume appeared in 1893, the second in 1899, and the third in 1912. Of these the first volume is the most important in the history of vector analysis, since it contained the first extensive published treatment of modern vector analysis.[38] The discussion of Heaviside's chapter on vector analysis will center on (1) its contents, (2) the question of Heaviside's relation to Tait and to Gibbs, (3) the relation of Heaviside's presentation of vector analysis to his electrical writings, and (4) the forcefulness of Heaviside's polemics. Heaviside's style can be placed somewhere between brilliant and obnoxious depending on one's point of view. At least he was never dull; indeed, G. M, Minchin referred to Heaviside as "the Walt Whitman of English Physics." [39] Because of the richness of Heaviside's style and the importance of his polemics liberal use will be made of quotations.

In the preface to volume one of his *Electromagnetic Theory*[40] Heaviside stated that he would use vectors in his treatment of electricity, "for the sufficient reason that vectors are the main subject of investigation." (4; iii) Heaviside, like Maxwell before him, viewed this as a very important reason for using vectors. Heaviside has come to be known as one of the great, perhaps the greatest, developers of Maxwell's electrical theories; it is less well known that he also developed Maxwell's views on vector analysis. In his preface Heaviside described his chapter on vector analysis:

> Regarded as a treatise on vectorial algebra, this chapter has manifest shortcomings. It is only the first rudiments of the subject. Nevertheless, as the reader may see from the applications made, it is fully sufficient for ordinary use in the mathematical sciences where the Cartesian mathematics is usually employed, and we need not trouble about more advanced developments before the elements are taken up. Now, there are no treatises on vector algebra in existence yet, suitable for mathematical physics, and in harmony with the Cartesian mathematics (a matter to which I attach the greatest importance). I believe, therefore, that this chapter may be useful as a stopgap. (4; iv)

Chapter I of the book was entitled "Introduction." Herein Heaviside summarized the history of electrical theory and at one point stated: "The crowning achievement was reserved for the heaven-sent Maxwell, a man whose fame, great as it is now, has, comparatively speaking, yet to come." (4; 14)

In chapter II Heaviside began the formal development of electromagnetic theory, and by the second paragraph vectors had appeared. Later in the chapter miscellaneous vector properties were introduced to set the stage for chapter III, entitled "The Elements of Vectorial Algebra and Analysis," which extended to 173 pages.

Heaviside's polemics began with his statement that the majority of the entities dealt with in geometry and in physical science are vectors. Since ordinary algebra contains no methods for treating these quantities directly, a vector algebra is needed. (4; 132–133) He compared the Cartesian approach to the vector approach and stressed the economy of expression, the naturalness, and the susceptibility to intuitive comprehension permitted by the latter. (4; 133–134)

To this point Heaviside had written nothing that would displease even the staunchest quaternion advocate; this changed and changed rapidly when Heaviside wrote:

> But supposing, as is generally supposed, vector algebra is something "awfully difficult," involving metaphysical considerations of an abstruse nature, only to be thoroughly understood by consummately profound metaphysicomathematicians, such as Prof. Tait, for example.

> Well, if so, there would not be the slightest chance for vector algebra and analysis to ever become generally useful.
>
> . . . .
>
> There was a time, indeed, when I, although recognising the appropriateness of vector analysis in electromagnetic theory (and in mathematical physics generally), did think it was harder to understand and to work than the Cartesian analysis. But that was before I had thrown off the quaternionic old-man-of-the-sea who fastened himself on my shoulders when reading the only accessible treatise on the subject—Prof. Tait's Quaternions. But I came later to see that, so far as the vector analysis I required was concerned, the quaternion was not only not required, but was a positive evil of no inconsiderable magnitude; . . . .
>
> . . . .
>
> There is not a ghost of a quaternion in any of my papers (except in one, for a special purpose). The vector analysis I use may be described either as a convenient and systematic abbreviation of Cartesian analysis; or else, as Quaternions without the quaternions, and with a simplified notation harmonising with Cartesians. (4; 134–135)

Heaviside then mentioned that Maxwell had pointed out "the suitability of vectorial methods to the treatment of his subject . . ." (4; 135) without advocating the quaternion system. It is clear whom Heaviside wished to view as his intellectual ancestor.

Heaviside went on to argue that if one attempted to determine what processes occur most frequently in Cartesian mathematics, he would find just those processes that are systematized in vector analysis; nowhere however would the quaternion appear. Humor, history, and mathematical insight were mixed in his next statement:

> "Quaternion" was, I think, defined by an American schoolgirl to be "an ancient religious ceremony." This was, however, a complete mistake. The ancients—unlike Prof. Tait—knew not, and did not worship Quaternions. The quaternion and its laws were discovered by that extraordinary genius Sir W. Hamilton. A quaternion is neither a scalar, nor a vector, but a sort of combination of both. It has no physical representatives, but is a highly abstract mathematical concept. (4; 136)

Heaviside then turned to (or better, on) Tait by writing: "Considering the obligations I am personally under to Prof. Tait (in spite of that lamentable second chapter), it does seem ungrateful that I should complain." (4; 137) Naturally this statement was followed by a list of complaints, one of which concerned Tait's treatment of Gibbs and his "hermaphrodite monster" system. (4; 137–138) Heaviside praised Gibbs' *Elements of Vector Analysis* and expressed his agreement with Gibbs' ideas, but not with his notation. "As regards his notation, however, I do not like it. Mine is Tait's, but simplified, and made to harmonise with Cartesians."

(4; 138) In Heaviside's next remark he compared his reasons for advocating the "ex-quaternionic" system to those of Gibbs: "Prof. Gibbs would, I think, go further, and maintain that the anti- or ex-quaternionic vectorial analysis was far superior to the quaternionic, which is uniquely adapted to three-dimensions, whilst the other admits of appropriate extension to more generalized cases. I, however, find it sufficient to take my stand upon the superior simplicity and practical utility of the ex-quaternionic system." (4; 139)

Heaviside then discussed problems of notation. After explaining his objections to the practice of representing vectors by Greek letters (Hamilton, Tait, and Gibbs) or by Gothic letters (Maxwell), Heaviside stated: ". . . I found salvation in **Clarendons,** and introduced the use of th[is] kind of type . . . for vectors (*Phil. Mag.*, August, 1886), and have found it thoroughly suitable." (4; 141)

In a rather leisurely fashion for a mathematics text Heaviside then introduced vector addition, subtraction, and multiplication by using, as he had earlier, **AB** and **VAB** for the symbolic representation of the scalar and vector product. (4; 143–163) He presented vector differentiation in a manner different from that of Hamilton and Tait, and the differences were discussed in great detail. The implication of this is indirect but important; it is that Heaviside was so conscious of the fact that the form of his work was taken from the quaternion tradition that he felt obliged, in one sense, to justify any further deviations from this tradition. (4; 163–178)

Heaviside's lengthy treatment of ∇ included a large number of examples and illustrations of its use, particularly in potential theory and various part of electricity. (4; 178–256) In this section Heaviside introduced a dot or period as a separator (like the quaternionists) rather than as a symbol for the dot product (like Gibbs).[41] Heaviside then turned to the linear vector operator with a presentation based mainly on Gibbs and Tait.[42] Here as elsewhere he supplied many illustrations, chosen chiefly from electricity.

The chapter was concluded with a polemical summary of the vector analysis method and a last attack on quaternion methods. He began the summary by stating that he had limited himself to a treatment of vector analysis "in the form it assumes in ordinary physical mathematics." (4; 297) Heaviside then stated on behalf of the idea of a vector:

> And it is a noteworthy fact that ignorant men have long been in advance of the learned about vectors. Ignorant people, like Faraday, naturally think in vectors. They may know nothing of their formal manipulation, but if they think about vectors, they think of them *as* vectors, that is, directed magnitudes. No ignorant man could or would think about the

three components of a vector separately, and disconnected from one another. That is a device of learned mathematicians, to enable them to evade vectors. The device is often useful, especially for calculating purposes, but for general purposes of reasoning the manipulation of the scalar components instead of the vector itself is entirely wrong. (4; 298)

This statement represented Heaviside's position on an argument discussed in the next chapter of this history and foreshadowed in earlier sections.

Heaviside then launched into his final remarks against the quaternion system; the following quotation illustrates how the vector analysts, essentially inadvertently, blackened Hamilton's reputation:

Now, a few words regarding Quaternions. It is known that Sir W. Rowan Hamilton discovered or invented a remarkable system of mathematics, and that since his death the quaternionic mantle had adorned the shoulders of Prof. Tait, who has repeatedly advocated the claims of Quaternions. Prof. Tait in particular emphasises its great power, simplicity, and perfect naturalness, on the one hand; and on the other tells the physicist that it is exactly what he wants for his physical purposes. It is also known that physicists, with great obstinacy, have been careful (generally speaking) to have nothing to do with Quaternions; and, what is equally remarkable, writers who take up the subject of Vectors are (generally speaking) possessed of the idea that Quaternions is not exactly what they want, and so they go tinkering at it, trying to make it a little more intelligible, very much to the disgust of Prof. Tait, who would preserve the quaternionic stream pure and undefiled. (4; 301)

This mildly sarcastic statement was followed by another not so mild: "Quaternions furnishes a uniquely simple and natural way of treating *quaternions*. Observe the emphasis." (4; 301)

Having granted that the quaternion preservation of the associative law for multiplication is an advantage, Heaviside argued that quaternions (full quaternions) almost never occur in physical science and that in quaternions it is unnatural for the square (under scalar multiplication) of a vector to be negative. (4; 303) "Vectors" he stated "should be treated vectorially. When this is done, the subject is much simplified, and we are permitted to arrange our notation to suit physical requirements." (4; 303) He concluded by a brief discussion of Gibbs' and Macfarlane's notations and by the final sentence: "My system, so far from being inimical to the cartesian system of mathematics, is its very essence."[43] Throughout the remainder of the book and in the two later volumes vector methods were used extensively and fruitfully to help Heaviside to many of his important results.

In conclusion the following statements seem justified:

1. Heaviside's system was derived from the quaternionic system; this statement is amply proved by both his explicit statements and by the form and content of his presentation of vector analysis.

2. Heaviside's system was derived from the quaternionic system by means of his own mathematical originality combined with unfailing sensitivity for the needs of physical science.

3. Heaviside's system was created independently of Gibbs, whose booklet he received in 1888, and of Grassmann, whom he never mentioned and probably never read. Heaviside did however make use of some of the more advanced sections of Gibbs' presentation.

IX. *The Reception Given to Heaviside's Writings*

As previous considerations amply indicate, it is not enough to discuss a scientists' ideas; the reception accorded those ideas also demands attention. The singular aspect of Heaviside's writings on vector analysis is that they were embedded within papers and books on electrical theory. Important consequences of this fact will be evident in the following analysis of the reception of his writings.

Heaviside's first statements on vector analysis were contained in his early electrical paper published in the *Electrician.* In 1887 a change in the editorship of this journal brought Heaviside a request to discontinue his series of papers. Heaviside wrote that the new editor had informed him that "although he had made particular enquiries amongst students who would be likely to read my papers, to find if anyone did so, he had been unable to discover a single one."[44] This was not an isolated phenomenon, for as Heaviside stated, five journals had turned down papers written by him. (5,I; x) When the *Electrician* refused to print his papers, Heaviside turned to the *Philosophical Magazine* with better results; one year later, however, the editor of the *Electrician* had a change of mind and from that time regularly published Heaviside's papers.

From these facts it may be inferred that during most of the 1880's Heaviside's writings were by no means well received. This was partly due to the slowness of the acceptance of Maxwell's ideas. Perhaps the turning point for Maxwell's doctrines came when in 1887 Hertz performed his classic experiments. Partly because of the consequent increase in interest in Maxwell's ideas and partly because of confirmations of some of Heaviside's own results the tide began to turn. Lodge at this time nominated Heaviside for the

Royal Society, and he became a member in 1891. Also in that year the editors of the *Electrician* suggested that Heaviside's papers be collected and published, and hence in 1892 his two volume *Electrical Papers* appeared. The original printing of the *Electrical Papers* consisted of 750 copies; 359 were still unsold five years later when they were put on sale at reduced prices.[45]

George Francis FitzGerald published a review of Heaviside's *Electrical Papers* in the August 11, 1893, issue of the *Electrician*. FitzGerald was Professor of Natural and Experimental Philosophy at the University of Dublin and a highly respected scientist who was experienced in quaternion methods [46] and who was, like Heaviside, an early and important advocate of Maxwell's electrical ideas. FitzGerald's favorable review surely helped the reception of both Heaviside's *Electrical Papers* and *Electromagnetic Theory* of 1893. FitzGerald began by praising the publishers for bringing out these papers, which he described as "much too valuable scientifically, practically, and historically to be left buried. . . ." (6; 292) He stated that they were valuable scientifically and practically because of the numerous and important discoveries made in them, many of which FitzGerald listed, and moreover were valuable historically as a "record of the development, of methods, and scientific views of nature, in an extraordinarily acute and brilliant mind." (6; 292–293)

Concerning Heaviside's style FitzGerald wrote: "Oliver Heaviside has the faults of extreme condensation of thought, and a peculiar facility for coining technical terms and expressions that are extremely puzzling to a reader of his Papers. So much so that there seems very little hope that he will ever attain the clarity of some writers, and write a book that will be easy to read." (6; 293) FitzGerald's statement in regard to Heaviside's relation to Maxwell would be acceptable to present historians.

> Maxwell, like every other pioneer who does not live to explore the country he opened out, had not had time to investigate the most direct means of access to the country, or the most systematic way of exploring it. This has been reserved for Oliver Heaviside to do. Maxwell's treatise is cumbered with the *débris* of his brilliant lines of assault, of his entrenched camps, of his battles. Oliver Heaviside has cleared those away, has opened up a direct route, has made a broad road, and has explored a considerable tract of country. (6; 294)

Dublinite FitzGerald was somewhat critical of Heaviside's system of vector analysis (6; 295) but potential readers would hardly be put off by such criticism, especially since it was followed by a discussion of the numerous discoveries presented in these papers.

FitzGerald's concluding statement was: "Since Oliver Heaviside has written, the whole subject of electromagnetism has been remodelled by his work. No future introduction to the subject will be at all final that does not attack the problem from at least a somewhat similar standpoint to the one he puts forward." (6; 299–300) Thus the importance of Heaviside's writings for electromagnetic theory and practice was being recognized in Britain in the 1890's.

German scientists were simultaneously discovering the importance of Maxwell's work and hence turning to Heaviside. August Föppl, for example, published an exposition of Maxwell's ideas in 1894 which was widely read.[47] In this book Föppl included much from Heaviside and wrote the following concerning Heaviside in his preface: "The works of this author have in general influenced my presentation more than those of any other physicist with the obvious exception of Maxwell himself. I consider Heaviside to be the most eminent successor to Maxwell in regard to theoretical developments, just as it was Hertz—who alas was so quickly snatched from us—who was Maxwell's most eminent successor in regard to experimental developments."[48] Heaviside influenced not only Föppl's presentation of Maxwell but also his choice of mathematical methods for the book. Föppl adopted Heaviside's vector analysis and devoted the first chapter of his book to an explanation of the new methods. This is very important since it was the first presentation of modern vectorial methods in a German book. Föppl's view of Heaviside as "the most eminent successor to Maxwell in regard to theoretical developments" was shared by Felix Klein and E. T. Whittaker, both of whom pointed out that the now classic Maxwell equations appeared for the first time in modern form in Heaviside.[49] Heaviside's influence on Föppl was by no means an isolated instance; it will be shown later that Heaviside's association of vector analysis with Maxwellian electrical theory was the most influential factor in leading to the widespread acceptance of vector analysis.

After 1893 Heaviside continued to publish important scientific papers; though vectors were constantly used in these papers, no further major contributions to vector analysis came from his pen. Increasing poverty, deafness, and isolation marked his later years, and in 1925 the crisp wit and scientific creativity of this self-educated genius were stilled by death.

X. *Conclusion*

This section will be devoted to a comparison of the achievements of Gibbs and Heaviside. The origins of their vectorial investigations will be considered first.

From an interest in electricity to the reading of Maxwell, from Maxwell to the quaternionists, from the Quaternionists to modern vector analysis, this story has been twice-told in this chapter, a circumstance probably less than an accident. By the 1880's algebraic sophistication was at a high level, and the needs of physical science were sharply outlined. The times were ready for two men of mathematical ingenuity and a wealth of physical experience. And what Gibbs and Heaviside saw in the 1880's, others were ready to see in subsequent decades.[50]

The two great vectorists of the 1880's should not be and would not wish to be compared in regard to originality to the two great vectorists of the 1840's. The task of the 1880's was one of selection and alteration, not chiefly of creation. Gibbs' originality should not go unnoticed however; as Heaviside stressed, Gibbs had created some important new ideas.

Four vectorists and four men who led isolated lives; only Gibbs did extensive teaching of vectorial methods. Heaviside's influence however was the greater; he associated vector analysis with the ever-expanding field of electricity.

The mid-century saw two great vectorists working separately; the 1890's saw two great vectorists an ocean apart, united in a common cause—in which they were successful.

Notes

[1] Lynde Phelps Wheeler, *Josiah Willard Gibbs* (New Haven, 1962).

[2] Josiah Willard Gibbs, *The Scientific Papers of J. Willard Gibbs*, vol. II (New York, 1961).

[3] Peter Guthrie Tait, *Elementary Treatise on Quaternions*, 2nd ed. (Oxford, 1873).

[4] Oliver Heaviside, *Electromagnetic Theory*, vol. I (New York, 1925).

[5] Oliver Heaviside, *Electrical Papers*, 2 vols. (London, 1892).

[6] George Francis FitzGerald, "[Review of] Heaviside's Electrical Papers" in *The Scientific Writings of the Late George Francis FitzGerald*, ed. Joseph Larmor (Dublin, 1902).

[7] Peter Guthrie Tait, *Elementary Treatise on Quaternions*, 3rd ed. (Cambridge, England, 1890), vi.

[8] There are two valuable biographies of Gibbs; one (cited in note [1] above) was written by a former student of Gibbs who was also a scientist; the other (*Willard Gibbs: American Genius* [Garden City, 1942]) was written for a general audience by the poetess Muriel Rukeyser.

[9] See ch. VI of the book cited in note (1) above for an extensive discussion of the reception of this paper.

[10] As given in (1; 107–109). Through the kindness of the staff of Yale University Library I have been able to examine the majority of the preserved letters to and from Gibbs concerning vector analysis. Some of these were published in the book cited in note (1) above. All letters quoted from this book have been checked against the originals.

[11] From the unpublished Gibbs' correspondence.

[12] (1; 62). It should be noted however that Wheeler also stated that Gibbs' 1880 lectures on mechanics at Johns Hopkins University were noteworthy as "his first public use of vector methods." (1; 90) Presumably the first of these contradictory statements is the correct one.

[13] This was privately printed at New Haven in 1881 and 1884. Gibbs referred to it as "printed, not published." It was reprinted in (2; 17–90), and my references are to this printing.

[14] A detailed analysis of the precise origin of each of the elements in this and later chapters of Gibbs' booklet would be significant but beyond the scope of the present undertaking. To make such an analysis a vast knowledge of the history of calculus, electricity, mechanics, potential theory, algebra (in particular matrix theory), and of course the quaternionic developments would be needed. As almost no supporting materials exist, this would be a task of great difficulty, particularly since the size of Gibbs' booklet is misleading: when E. B. Wilson later rewrote Gibbs' system in textbook form and retained essentially the same content, Wilson was required to use 436 pages to cover what Gibbs had covered in 73. Nevertheless the major source of Gibbs' work is clear: he translated and added to the results and systematizations developed by the quaternionists.

[15] C. G. Knott, "Review of *Vector Analysis Founded upon the Lectures of J. Willard Gibbs*. By E. B. Wilson" in *Philosophical Magazine*, 6th Ser., *4* (1902), 622.

[16] See, for example, James Byrnie Shaw in his *Vector Calculus* (New York, 1922). Shaw gave (pp. 248–252) a table of comparative notations by listing the notations used by Hamilton, Tait, Gibbs, and later students of linear vector functions. This table is helpful in "translating" results from one system to another.

[17] Heaviside and others expressed this view. For Heaviside's statement see (1; 117).

[18] A bivector is a vector in which some of the scalar coefficients assume imaginary values. Hamilton and Tait both worked with biquaternions, which are directly analogous. For Tait's treatment see (3; 64–66).

[19] Edwin Bidwell Wilson, "The Contributions of Gibbs to Vector Analysis and Multiple Algebra" in *A Commentary on the Scientific Writings of J. Willard Gibbs*, vol. II (New Haven, 1936), 160. Wilson's paper rather inadequately fulfilled its title; it mainly stated what Gibbs included in his lectures with little attention to specific questions of originality. Despite Wilson's implication in the quotation that he had made no statements thirty years previously concerning Gibbs' originality, such statements may be found in writings by Wilson published in 1905 and in 1910. See Wilson, "On Products in Additive Fields" in *Verhandlungen des III. Internationalen Mathematiker Kongresses im Heidelberg, 1904* (Leipzig, 1905), 203, and Wilson, "The Unification of Vectorial Notations" in *Bulletin of the American Mathematical Society, 16* (1909–1910), 418. These statements are sufficiently indirect that Wilson may be exculpated from a charge of forgetfulness, but they are sufficiently direct that the historian can conclude that Wilson's view of Gibbs' originality was not widely different thirty years before his 1936 statement.

[20] As quoted in Rukeyser, *Gibbs*, 268. This is not footnoted, but in the bibliography Miss Rukeyser lists one work by Page: "Page, Leigh. Ms—*Gibbs' Vector Analysis*. Yale." Yale University Library has kindly written to me that they have no record of such a work. I see however no reason to question its authenticity.

[21] E. B. Wilson, "Reminiscences of Gibbs by a Student and Colleague" in *Scientific Monthly, 32* (1931), 219–220.

[22] Wilson, "Contributions of Gibbs," 159. This article summarizes the content of the course as it was taught in 1902–1903.

[23] On this see Wilson, "Products in Additive Fields," 205–215.

[24] Josiah Willard Gibbs, "On Multiple Algebra" in *Proceedings of the American Association for the Advancement of Science, 35* (1886), 37–66. My references will be to the printing of the essay in (2; 91–117). My esteem for, and indebtedness to, this paper are both large.

[25] Wilson, "Contributions of Gibbs," 129.

[26] Josiah Willard Gibbs, "On the Determination of Elliptic Orbits from Three Complete Observations" in *Memoirs of the National Academy of Sciences*, vol. IV, pt. II, 79–104. References will be to the republication in (2; 118–148).

[27] H. A. Bumstead, "Josiah Willard Gibbs" in Gibbs, *The Scientific Papers of J. Willard Gibbs*, vol. I (New York, 1961), xxi. The paper by Beebe and Phillips is "The Orbit of Swift's Comet, 1880 V, Determined by Gibbs' Vector Method" in Gould's *Astronomical Journal*, 9 (1889), 114–117, 121–124. It was also included in Klinkerfuss' third edition.

[28] Biographical details on Heaviside's life are given in many sources, including the following: (1) Rollo Appleyard, "Oliver Heaviside" in Appleyard, *Pioneers of Electrical Communication* (London, 1930), 211–260; (2) *The Heaviside Centenary Volume* (London, 1950). See especially G. F. C. Searle, "Oliver Heaviside: A Personal Sketch," 93–96, and Sir George Lee, "Oliver Heaviside—the Man," 10–17; (3) E. T. Whittaker, "Oliver Heaviside" in *Calcutta Mathematical Society Bulletin, 20* (1930), 199–220.

[29] Oliver Heaviside, *Electromagnetic Theory*, vol. III (New York, 1925), 135, 136, 137.

[30] A more exact date may be given: June, 1888. See (5,II; 529).

[31] Heaviside's papers on electricity up to 1892 were collected and published as the book cited in note (5) above. The paper referred to above may be found in (5,I; 195–231). Despite Heaviside's statement in the preface of his *Electrical Papers* (5,I; xi) to the effect that these reprinted papers are essentially identical to the original published papers, a careful check has shown that this must not be taken as completely true. Compare for example his definition of the scalar product as given in the *Electrician, 10* (Jan. 20, 1883), 224, with the definition given in (5,I; 236). I have checked all quotations against the originals and noted any important deviations.

[32] Heaviside, "Current Energy" in *Electrician, 10* (Jan. 20, 1883), 224. It should be noted that the statement given above differs substantially from the statement as given in (5,I; 236).

[33] This title does not appear in the original paper.

[34] (5,I; 271–272). The word "expressions" was followed by a qualification to the effect that one more item (which is not of importance) would be discussed.

[35] This should perhaps be qualified by the statement that ∇^2 made its first appearance in an 1884 paper. See (5,I; 338). No comment was given by Heaviside about its meaning. He seems to use it as $+\left(\frac{d^2}{dx^2}+\frac{d^2}{dy^2}+\frac{d^2}{dz^2}\right)$ rather than, as Maxwell and Tait had done, as the negative of the above.

[36] This paper is in (5,I; 429–560) and (5,II; 39–155). The paper was published in sections, the last of which appeared in 1887. The vector product is introduced in (5,I; 431).

[37] The only exception to this is that in 1886 he introduced the practice of indicating vectors by boldface type. See (5,II; 172).

[38] Since Heaviside's treatment of vector analysis in this book was highly polemical (whereas Gibbs' earlier presentation was not), there would be good reasons for postponing the discussion of it to the next chapter. There are better reasons for its inclusion at this point; however, that it has relevancies for the following chapter should now be made clear.

[39] George M. Minchin, "[Review of] *Electromagnetic Theory.* By Oliver Heaviside" in *Philosophical Magazine*, 5th Ser., *38* (1894), 146.

[40] Oliver Heaviside, *Electromagnetic Theory*, vol. I (London, 1893). I have used the 1925 reprint cited in note (4) above, and all references are given in terms of it.

[41] (4; 179). For an example of Tait's use of the dot see (3; 27).

[42] For Heaviside's treatment of the linear vector operator see (4; 256–297). For statements by Heaviside indicating his use of Gibbs' treatment see for example (4; 263, 295, 300).

[43] (4; 305). By "cartesian system" Heaviside meant the traditional nonvectorial method for treating problems in space by means of Cartesian co-ordinates.

[44] E. T. Whittaker, "Oliver Heaviside" in *Calcutta Mathematical Society Bulletin, 20* (1930), 209.

[45] *Ibid.*, 211.

[46] See for example papers 8 and 22 in *The Scientific Writings of the Late George Francis FitzGerald*, ed. Joseph Larmor (Dublin, 1902).

[47] August Föppl, *Einführung in die Maxwell'sche Theorie der Elektricität* (Leipzig, 1894). Föppl's book went through four editions in eighteen years.

[48] *Ibid.*, VII.

[49] Felix Klein, *Vorlesungen über die Entwicklung der Mathematik im 19 Jahr-*

hundert, pt. II, ed. R. Courant and St. Cohn-Vossen (New York, 1956), 47, and Whittaker, "Heaviside," 202–203.

[50] Mention should be made of a work by Alphonse Demoulin published in 1894 and entitled *Mémoire sur l'Application d'une Méthode Vectorielle à l'Etude de Divers Systèmes de Droites* (Bruxelles, Paris), vi and 118 pp. In this work the author used the modern scalar and vector products and cited as his sources M. Resal, *Traité de Cinématique Pure* (Paris, 1862) for the scalar product and J. Massau, *Cours de Mécanique de l'Université de Gand* (1st ed., 1879; 3rd ed., Paris and Gand, 1891) for the vector product. Demoulin knew both Grassmann's and Hamilton's works. This information has been derived from Demoulin's book, pp. V–VI of an untitled section, essentially a preface. No further mention of this work will be made because of its brevity, because no further publications seem to have come as a result of it, and because of the complexity involved in making any brief interpretations as to its ancestry. I have seen only the 1891 edition of Massau's book, which must have had a very limited circulation. This is indicated by its title and by the fact that the text was published in handwritten form. Massau's 1891 edition mentioned both Grassmann and Hamilton, and, like Demoulin's book, had little or no influence. Massau's and Demoulin's books are significant as reflective of the interests of that time and are notable in respect to the small interest in any form of vectorial analysis in France at that time.

CHAPTER SIX

A Struggle for Existence in the 1890's

I. *Introduction*

The following prediction was made by Gibbs in an 1888 letter to Thomas Craig: "I believe that a Kampf ums Dasein [struggle for existence] is just commencing between the different methods and notations of multiple algebra, especially between the ideas of Grassmann & of Hamilton." [37] Gibbs' prediction was fulfilled, for in the early 1890's a widespread and vigorous debate on vectorial methods took place. No less than eight journals, twelve scientists, and thirty-six publications were involved. The spirit of the debate is aptly characterized by Lord Rayleigh's paraphrase of Tertullian: "Behold how these vectorists love one another." [38]

This debate played an important role in the history of vector analysis. That interest in the questions debated was at a high level is indicated by the fact that eight leading scientific journals permitted publication of these articles. That interest in vectorial methods in general and the Gibbs-Heaviside system in particular greatly increased during the 1890's may in large measure be attributed to the forceful, timely, and stimulating presentations in these articles. From the arguments advanced in the debate much can be learned about how these early vectorists viewed their systems and the systems of their opponents.

All publications from the period 1890 to 1894 which contain arguments relevant to vectorial methods will be discussed in this chapter.[39] This time limitation is by no means arbitrary; the number of polemical articles published in the years immediately before and after this period is very small. Moreover the discussion is unified by the fact that nearly every publication from this group refers to one or more other publications from the group. Though slightly more than half of the writings appeared in the important British scientific weekly *Nature,* the discussion certainly went beyond its pages and touched on important questions not dealt with in its pages.[40]

II. *The "Struggle for Existence"*

The beginning of this "struggle for existence" may be dated as 1890, a year in which Peter Guthrie Tait wrote two strongly worded pleas for recognition of the fitness of the quaternion system. The occasion was the publication in that year of the "Third Edition, Much Enlarged" of his *Elementary Treatise on Quaternions.* The January, 1890, issue of the *Philosophical Magazine* carried an article entitled "On the Importance of Quaternions in Physics," which was an abstract of Tait's address to the Physical Society of the University of Edinburgh, given November 14, 1889.

Tait began indirectly by posing the question "Whether is experiment or mathematics the more important to the progress of physics?" (1; 84) He concluded that the question was absurd because "to their combined or alternate assaults everything penetrable must, some day, give up its secrets." (1; 84) Their "inseparable connexion," he stated, leads to the important conclusion "that every formula we employ should as openly as possible proclaim its physical meaning." (1; 85) For this reason the primary characteristic to be sought in selecting mathematical methods for physical problems is neither compactness nor elegance, but "expressiveness." In this way Tait led the discussion to quaternions, which he described as "transcendently expressive," compact, and elegant. Tait here as elsewhere invoked the name and reputation of Hamilton on behalf of the quaternion cause. This was not without effect, especially since Graves' immense *Life of Hamilton* had just been published, and this two-thousand-page tribute to the originator of quaternions must have brought the memory of Hamilton before the minds of many men of the time.

Tait proceeded to an attack on the artificiality of Cartesian coordinates, "one of the wholly avoidable encumbrances which now retard the progress of mathematical physics." (1; 86) This encumbrance, Tait argued, could be avoided if quaternions were widely adopted, and he cited as one reason for this that quaternions are "uniquely adapted to Euclidian space. . . ." (1; 87) Tait then gave a number of illustrations of the simplicity and brevity of the quaternion expressions as compared with equivalent Cartesian expressions. He stressed these advantages above all in relation to the differential calculus of quaternions and particularly in applications of ∇. At one point Tait made a statement that he would often repeat in similar forms. Concerning ∇ he wrote: "No doubt, it was originally defined in the cumbrous and unnatural form $i\frac{d}{dx} + j\frac{d}{dy} + k\frac{d}{dz}$. But that was in the very infancy of the new calculus, before its inventor had succeeded in completely removing from its formulae the fragments of the Cartesian shell, which were still persistently clinging about them." (1; 91) Essentially Tait was stating that Hamilton's frequent recourse to Cartesian equivalents was a blot on the system and that this had greatly hindered recognition of its many advantages. Then followed a number of illustrations of the power and simplicity of ∇.

The final point in the paper centered around a paraphrase of Horace; Tait wrote: "*The highest art is the absence* (not, as Horace would have it, the *concealment*) *of artifice*." (1; 86) Thus Cartesian methods were artificial, quaternions natural. This led Tait to an interesting characterization of nineteenth-century physical science:

> The magnificent artificers of the earlier part of the century were, in many cases, blinded by the exquisite products of their own art. To Fourier, and more especially to Poinsot, we are indebted for the practical teaching that a mathematical formula, however brief and elegant, is merely a step towards knowledge, and an all but useless one, until we can thoroughly read its meaning. It may in fact be said with truth that we are already in possession of mathematical methods, of the artificial kind, fully sufficient for all our present, and at least our immediately prospective, wants. What is required for physics is that we should be enabled at every step to feel intuitively what we are doing. Till we have banished artifice we are not entitled to hope for full success in such an undertaking. (1; 96–97)

In conclusion it may be noted that Tait's arguments were mainly directed to the survival struggle between vectorial and Cartesian methods and that hence the majority of the arguments could be applied equally well on behalf of the Gibbs-Heaviside system.

The second writing of Tait from the year 1890 that was seminal for the debate was the preface[2] to the third edition of his *Treatise on Quaternions*. Tait's book was directed mainly toward the physical applications of quaternions; to rectify this "bias," as he called it, he added for this edition "an entire Chapter, on the Analytical Aspect of Quaternions" through "the unsolicited kindness of Prof. Cayley." (2; v) Since Cayley four years later made an attack on quaternions, Tait may have come to regret his decision to include this chapter.

Tait stated that "little progress has recently been made with the development of Quaternions." (2; vi) In a paragraph rich with literary quotations he ascribed this partly to excessive efforts to modify notations, especially in France, and concluded with the statement: "Even Prof. Willard Gibbs must be ranked as one of the retarders of Quaternion progress, in virtue of his pamphlet on *Vector Analysis;* a sort of hermaphrodite monster, compounded of the notations of Hamilton and of Grassmann." [41] Turning to priority questions, Tait argued that Grassmann certainly did not have quaternions before Hamilton (correct), that Hamilton published his system first (correct, but misleading), and that in the 1830's Hamilton had published Grassmann's internal and external products (incorrect and refuted by Gibbs [7]).

Tait proceeded to argue for the superiority of quaternion methods over Cartesian methods and to suggest that in quaternion work even to have "recourse to quasi-Cartesian processes is fatal to progress." (2; vii) He concluded by a quotation from a letter he had received long ago from Hamilton: "*Could* anything be simpler or more satisfactory? Don't you *feel,* as well as think, that we are on a *right track,* and shall be *thanked* hereafter? Never mind when." (2; viii) This was of course to draw upon Hamilton's great reputation to advance his cause. As time passed, however, statements such as this led to a diminishing of Hamilton's stature.

In the April 2, 1891, issue of *Nature* there appeared an article [3] by Gibbs written in response to Tait's references to Gibbs as one of the "retarders of quaternion progress" and to the Gibbs' system as a "hermaphrodite monster." Gibbs began by quoting Tait's statement and followed this by a paragraph which well illustrates Gibbs' tactfulness and his staid but forceful style.

> The merits or demerits of a pamphlet printed for private distribution a good many years ago do not constitute a subject of any great importance, but the assumptions implied in the sentence quoted are suggestive of certain reflections and inquiries which are of broader interest, and seem not untimely at a period when the methods and results of the various forms of multiple algebra are attracting so much attention. It seems to be assumed that a departure from quaternionic usage in the treatment of vectors is an enormity. If this assumption is true, it is an important truth; if not, it would be unfortunate if it should remain unchallenged, especially when supported by so high an authority. The criticism relates particularly to notations, but I believe that there is a deeper question of notions underlying that of notations. Indeed, if my offence had been solely in the matter of notation, it would have been less accurate to describe my production as a monstrosity, than to characterize its dress as uncouth. (3; 511)

The first "notions" with which Gibbs dealt were the scalar and vector products. He argued that these products are fundamental since they represent the most important relations in physics and geometry, whereas few, if any, correlates are found for the quaternion product or for the quaternion itself. He suggested that this conclusion was evident even from an examination of the practices of the quaternionists in dealing with spatial relations. And he added that vector analysis, unlike quaternion analysis, could be extended to apply to space of four or more dimensions. (3; 511–512) Gibbs introduced his next point by admitting that the "quaternion affords a convenient notation for rotations" (3; 512), but he added that the representation of rotations by the linear vector function in the Gibbsian system "seems to leave nothing to be desired. . . ." (3; 512) After a comparison of the use of ∇ in the two systems Gibbs wrote in summary: "These considerations are sufficient, I think, to show that the position of the quaternionist is not the only one from which the subject of vector analysis may be viewed, and that a method which would be monstrous from one point of view, may be normal and inevitable from another." (3; 512)

Gibbs turned to the question of notation by correcting Tait's statement that his system made use of Grassmann's notation. He then suggested that his notation was simpler, clearer, and more expressive than the quaternionic. In his statements in this regard he exhibited a noteworthy openness of mind and in general placed notation questions in a subordinate position. (3; 512–513)

Tait's reply[4] to Gibbs' article was published within the month. It is perhaps ironic that one of the aspects of Gibbs' system attacked strongly in Tait's reply was Gibbs' dyad, or what Gibbs called the indeterminate product of two vectors. Tait said this was confusing and "undoubtedly artificial in the highest degree. . . ." (4; 608) The irony is that nearly half a century prior to this time mathematicians attacked the quaternion (noncommutative) product on nearly the same grounds; hence it is in one way surprising that a quaternionist would balk at a further extension in the meaning of "product."

Tait wrote in response to Gibbs: "It is singular that one of Prof. Gibbs' objections to Quaternions should be precisely what I have always considered (after perfect inartificiality) their chief merit:—viz. that they are '*uniquely adapted to Euclidian space*, and therefore specially useful in some of the most important branches of physical science.' What have students of physics, as such, to do with space of more than three dimensions?" (4; 608) Fate seems to have been against Tait, at least in regard to the last point. Tait concluded by a discussion of the comparative compactness of expressions in the two systems; in this the quaternionists had a slight advantage, for quaternion products retain associativity, whereas the Gibbsian vector product does not. (Thus in the Gibbsian system $i \times (j \times j) = 0 \neq (i \times j) \times j = -i$.)

Gibbs' second article [5] was published in *Nature* four weeks after Tait's article. Entitled "Quaternions and the 'Ausdehnungslehre,'" it was written in response to historical statements made by Tait in his *Encyclopaedia Britannica* article "Quaternions" (of 1886) and in the preface to his quaternion *Treatise* of 1890. Gibbs' motivation in writing this article may be inferred from a letter he wrote in 1888 to Thomas Craig to request the latter to publish Grassmann's 1840 *Theorie der Ebbe und Flut.* It was in this letter that Gibbs predicted that a "struggle for existence" among the vectorial systems was about to begin, and to this statement he added: "The most important question is of course that of merit, but with this questions of priority are inextricably entangled, & will be certain to be the more discussed, since there are so many persons who can judge of priority to one who can judge of merit." [37] Thus though Gibbs dealt primarily with priority questions, he was well aware that much more was at stake: by correcting Tait's excessive priority statements he added prestige to Grassmann, and in comparing Grassmann's ideas to Hamilton's (overtly in regard to historical questions) he put forth arguments for the superiority of Grassmann's methods. And to praise and to recommend Grassmann's system was of course to praise and to recommend his own system, for he discussed primarily those aspects of Grassmann's system that were also to be found in the Gibbs-Heaviside system.

Gibbs began by admitting that Hamilton was the first to announce his discovery, whereas Grassmann "seems to have been in no haste to place himself on record, and published nothing until he was able to give the world the most characteristic and fundamental part of his system with considerable development in a treatise of more than 300 pages, which appeared in August 1844." (5; 79) Both were, wrote Gibbs, memorable discoveries, and hence "Historical justice, and the interests of mathematical science," require that Tait's historical statements concerning the two systems "should not be allowed to pass without protest." (5; 79) Gibbs' approach to the priority question was to state that the systems should first be compared in terms of what they have in common and then in terms of what is peculiar to each.

Both systems have vector addition and the scalar and vector products, Gibbs stated, but the quaternion product is found only in the quaternion system, while Grassmann had the linear vector function first. Gibbs then added:

> To what extent are the geometrical methods which are usually called quaternionic peculiar to Hamilton, and to what extent are they common to Grassmann? This is a question which anyone can easily decide for himself. It is only necessary to run one's eye over the equations used by quaternionic writers in the discussion of geometrical or physical subjects, and see how far they necessarily involve the idea of the quaternion, and how far they would be intelligible to one understanding the functions $S\alpha\beta$ and $V\alpha\beta$, but having no conception of the quaternion $\alpha\beta$, or at least could be made so by trifling changes of notation, as by writing S or V in places where they would not affect the value of the expressions. For such a test the examples and illustrations in treatises on quaternions would be manifestly inappropriate, so far as they are chosen to illustrate quaternionic principles, since the object may influence the form of presentation. But we may use any discussion of geometrical or physical subjects, where the writer is free to choose the form most suitable to the subject. (5; 80)

Gibbs wrote that if for example pages 160–371 of Tait's *Treatise* were to be examined, it would be evident that "for the most part the methods of representing spatial relations used by quaternionic writers are common to the systems of Hamilton and Grassmann." (5; 80) After posing the question of the importance of the remaining cases where the quaternion played a fundamental part, Gibbs suggested that these were "very exceptional." Thus Gibbs restated in a very effective way what he had argued earlier: scalar and vector products are more useful and more fundamental than the quaternion product. And the evidence he cited was the practice of the quaternionist Tait!

Gibbs went on to discuss Grassmann's point analysis, his "wealth of multiplicative relations," and in general the vast scope of his system as compared to Hamilton's. He attempted to show that the discovery of matrices came in Grassmann's 1844 work, a full fourteen years before Cayley's famous publication of 1858. Gibbs proceeded to give two quotations from Tait which stated in part that Hamilton had discovered the scalar and vector products in the 1830's in his "Theory of Conjugate Functions or Algebraic Couples." This statement Gibbs simply rejected, and with good reason. Tait had gone too far on too shaky a foundation, for his knowledge of Grassmann's system was very limited. In this regard Gibbs had the distinct advantage that he was well acquainted with Grassmann's *and* Hamilton's systems.

Gibbs' carefully reasoned and well-documented paper exhibited an openness and flexibility which must have made it appealing to sympathetic readers without being antagonistic to readers of another persuasion. He argued forcefully that what was important in Hamilton's system was also in Grassmann's and that moreover

Grassmann's system contained a wealth of applications unique to it.

Tait's very brief reply[6] to Gibbs' article was published just one week later in *Nature.* In it Tait reasserted some of the views Gibbs had attacked but essentially gave no new reasons for, or clarifications of, those views. Tait admitted his lack of familiarity with Grassmann's writings and in general wrote as one might write who was surprised that anyone would attack quaternions on such grounds. The quaternion system was at that time far better known, and Tait must have felt little motivation for giving a detailed, tactful, and comprehensive rejoinder. Tait had long experience in defending quaternions, but never on such grounds as these.

This is the last article of 1891 that will be discussed; it should be pointed out however that in the November 13, 1891, issue of the *Electrician* Heaviside began a series of papers on vector analysis which later formed the polemical third chapter of volume one of his *Electromagnetic Theory* (1893). The arguments given by Heaviside in this chapter (and hence in those papers) were discussed in the previous chapter.

In June, 1892, the debate resumed. The article[7] that appeared in that month's issue of the *Philosophical Magazine* was by Alexander McAulay and was entitled "Quaternions as a practical Instrument of Physical Research." Alexander McAulay (1863–1931) was an 1886 graduate of Cambridge (forty-ninth Wrangler), who in 1892 was tutor and lecturer in mathematics and physics at Ormond College in Melbourne, Australia. The paper had been given at the meeting of the Australasian Association for the Advancement of Science in January, 1892. The article is interesting above all because of the historical judgments made by McAulay.

McAulay began by stating that although Hamilton was a great mathematician and although quaternions were fairly well known, Hamilton "has left scarcely a successor." (7; 477) McAulay then asked: "Can any cause be assigned for this extraordinary case of arrested development?" (7; 477) McAulay suggested that if physicists were asked this question, they would reply that quaternions are not used because they have not been fruitful in leading to scientific discoveries. He then stated: "The chief object of the present paper is to *shake* the belief of mathematical physicists. It is too much to hope to *overturn* that belief." (7; 478) McAulay commented further on the apathy in regard to quaternions: "I confess that the more I think of this apathy the more extraordinary does it appear, and, as already hinted, it will probably prove an insoluble problem to the future historian of Mathematics." (7; 478) It was argued that most physicists who had rejected quaternions had studied them at a time after "their mathematical ideas and methods" had "nearly or completely crystallized." (7; 479) After a limited period of study and before they have seen the powerfulness of the methods, they cease their study. A major factor in this too early cessation of their labors has been Maxwell, who "is responsible to a large extent for the discredit into which quaternions have fallen among physicists." (7; 478–479) Young physicists abandon their study of quaternions too early, "consoling themselves that Maxwell . . . had had more experience than themselves" and yet had found quaternions very limited in their usefulness. (7; 479)

McAulay stated that physics would advance with great rapidity if quaternions were "introduced to serious study to the almost complete exclusion of Cartesian Geometry, except in an insignificant way, as a particular case of the former." (7; 480) McAulay then gave an elaborate series of quaternion applications in an attempt to illustrate his point.

The next paper[8] for discussion is Alexander Macfarlane's "Principles of Algebra of Physics," which was delivered at the August, 1891, meeting of the American Association for the Advancement of Science and published in the *Proceedings* of that society in July, 1892. Macfarlane (1851–1913) had studied under Tait at Edinburgh and in 1891 was teaching at the University of Texas. He was in that year secretary of the physics section of the American Association and was until his death one of the most active participants in the debate on systems of vector analysis.

Macfarlane began his paper by two quotations from the preface of Tait's *Treatise on Quaternions* and then stated in effect that progress in quaternion development has been slow, but there is reason for hope. Macfarlane remarked that in his opinion the quaternion system was on the right track, while he proceeded immediately to qualify his statement rather drastically. He wrote:

> But at the same time I am convinced that the notation can be improved; that the principles require to be corrected and extended; that there is a more complete algebra which unifies Quaternions, Grassmann's method and Determinants, and applies to physical quantities in space. The guiding idea of this paper is generalization. What is sought for is an algebra which will apply directly to physical quantities, will include and unify the several branches of analysis, and when specialized will become ordinary algebra. (8; 65)

After mentioning the Tait-Gibbs debate in *Nature,* Macfarlane gave a series of criticisms of the quaternion system. A key point in

Macfarlane's discussion is the following. He insisted that the quaternion symbols i, j, and k have not one, but two meanings. For example, he argued that i may be viewed either as a certain unit vector or as a "versor," or turner. Hence if i and j are viewed as "versors," then $ij = k$ means a right-handed rotation through one quadrant in the plane perpendicular to j compounded with a right-handed rotation through one quadrant in the plane perpendicular to i, which is equivalent in result to a right-handed quadrant rotation about k. But, said Macfarlane, Tait was too hasty in allowing such an interpretation since i, j, and k symbolize vectors, not versors. The crucial point is the meaning of ii. If the i's in this product are viewed as versors, then ii means a rotation of 180° around the i axis, which when applied to j would produce $-j$. But if the i's be viewed as vectors, then Macfarlane stated that it was better to define the product as positive, for example, $(ai)(bi) = +ab$, so that the the product would be in harmony with such things as $\frac{1}{2}mv^2$ (kinetic energy).

As a result of such considerations Macfarlane constructed a new system of vector analysis more in harmony with the Gibbs-Heaviside system than with the quaternion system. As part of this project he introduced new notations and defined a full product of two vectors which was comparable to the full quaternion product except that the scalar part was positive, not negative as in the older system. This system apparently never became popular or widely employed, though expositions of it were published rather frequently. Macfarlane explained his system in a chapter in Mansfield Merriam and Robert S. Woodward (editors), *Higher Mathematics,* of which editions appeared in 1896, 1898, and 1900, and in addition this chapter was published as a book in 1906.[42]

Macfarlane's article was intelligently written and in many ways was impressive; nonetheless it further complicated an already complex situation. The introduction in 1892 of another system of vector analysis, even a sort of compromise system such as Macfarlane's, could scarcely be well received by the advocates of the already existing systems and moreover probably acted to broaden the question beyond the comprehension of the as-yet uninitiated reader.[43]

In November, 1892, the debate returned to the pages of *Nature* with a review[9] probably written by Alfred Lodge[44] of Macfarlane's "Principles of the Algebra of Physics."

Lodge began by writing: "This is a very suggestive contribution to the foundations of the Algebra of Vectors as recently so strongly advocated in America by Prof. Willard Gibbs, and in this country by

Mr. Oliver Heaviside." (9; 3) Lodge then summarized Macfarlane's work and made numerous comparisons with the Gibbs-Heaviside and the quaternion systems. The review was more descriptive than critical; it was in general favorable to Macfarlane's system as worthy of consideration. It concluded: "A text book of vector algebra . . . is much needed, as many physicists are becoming interested in the new algebra, owing in great measure to Mr. O. Heaviside's able exposition of its principles and applications in the *Electrician* and elsewhere." (9; 5)

Oliver Heaviside's paper[10] "On the Forces, Stresses, and Fluxes of Energy in the Electromagnetic Field" may be considered next. It was read to the London Royal Society in 1891 and published in the Transactions of that society in 1893, although it had been published in 1892 in Heaviside's *Electrical Papers.*

Early in the paper Heaviside presented an exposition of his system of vector analysis which was prefaced by an attack on quaternions. After pointing out that vectors would be used in his paper because the physical quantities were vectors, Heaviside (in mockery of Tait's preface) discussed the "retardation" of vector analysis because of the lack of an adequate treatise on the subject, "Professor TAIT'S well-known profound treatise being, as its name indicates, a treatise on Quaternions." (10; 427) He referred to the antiquaternionic arguments recently given by Gibbs in *Nature* and described this dispute as "rather one-sided." Gibbs was called "anything but a retarder of progress in vector analysis and its application to physics." (10; 428) In this instance, as in others, Heaviside spoke very favorably of Gibbs' booklet. The fact that Heaviside and Gibbs were in agreement on all aspects of vector analysis except notation must have strengthened their position.

Heaviside then briefly presented his main arguments for his system as against the quaternion system. Statements such as the one below (which is in one sense very true and in another very false) illustrate the directness, humor, and humility of Heaviside's style. He wrote: ". . . I ought to also add that the invention of quaternions must be regarded as a most remarkable feat of human ingenuity. Vector analysis, without quaternions, could have been found by any mathematician by carefully examining the mechanics of the Cartesian mathematics; but to find out quaternions required a genius." (10; 461)

In summary, this paper is most appropriately viewed as another publication by Heaviside in which polemics for vector analysis were found associated with important new physical results. One might wonder what better rhetoric could be used than advocacy of a method followed by the testimony of actual success with that method.

The next paper,[11] published in *Nature*, was occasioned by Heaviside's remarks in his *Philosophical Transactions* paper. Alexander McAulay, the author, began by giving his opinion of the present status of the vector question: "There are two widely-known systems of vector analysis before the public—Quaternions and the Ausdehnungslehre—and quite a multitude of less known ones, of which Prof. Gibbs's seems to be one of the least open to objection, and of which, in my opinion, Mr. Heaviside's is by no means so." (11; 151) McAulay proceeded to make what must be viewed as a hopelessly idealistic appeal to Gibbs and Heaviside "on grounds independent of the merits or demerits of their particular systems." (11; 151) He suggested that the "woefully small" band of vector analysists should concentrate on making vectorial methods better known, and to do this they should limit the debate to quaternions versus Grassmann's system. "The day for Prof. Gibbs's improvements is not yet. Prof. Gibbs and Mr. Heaviside have not yet convinced the rest of the small band. . . . Let me implore them to sink the individual in the common cause, and content themselves with the faith that posterity will do them justice." (11; 151)

McAulay concluded his paper with the following somewhat ambiguous, certainly inflammatory, statement:

> To vary the metaphor, Maxwell, Clifford, Gibbs, Fitzgerald, Heaviside prescribe a course of spoon-feeding the physical public. Hamilton and Tait recommend and provide strong meat. I do not think that harm, but rather good, will come from this double treatment, as one course will suit some patients and the other others. *But* let the spoon-feeders provide spoon-meat of the same *kind* as the other physicians. Is not Maxwell, Clifford, and Fitzgerald's food as digestible as Prof. Gibbs's and Mr. Heaviside's? (11; 151)

Needless to say, McAulay's attempt to pour oil on the waters failed. Both his statements and his brash attitude encouraged the continuance of the debate.

It was remarked previously that one major significance of Heaviside's 1892 *Philosophical Transactions* paper was that the polemical sections were embedded in a paper containing important physical results. In the same volume of this journal a paper[12] by Alexander McAulay was published under the title "On the Mathematical Theory of Electromagnetism." This paper, nearly twice as long as Heaviside's, was in ways similar to it. It was directed at extending electrical theory within the Maxwell tradition, and it contained polemics for a system of vectorial analysis—for quaternions. McAulay in his paper made extensive use of quaternions, and although his paper seems to have been scientifically less important than Heaviside's, it was an impressive display. Early in the paper McAulay stated:

> As might be expected, the mathematical machinery that appears to be most convenient for investigating as fully as possible the consequences of these assumptions, and others intimately connected with them, is novel. And I may remark in passing that what Professor TAIT persistently and with complete justice emphasizes as one of the greatest boons that Quaternions grant to ungrateful physicists, viz., their *perfect naturalness*, seems to me to receive illustration in the methods about to be described. (12; 686–687)

McAulay's presentation must have helped the quaternion cause, though he deviated somewhat from the pure quaternion tradition by introducing much new notation. Concerning this paper it may be noted that though it perhaps balanced out the effect of Heaviside's paper, it with Heaviside's paper must have sharpened the question (vectors vs. quaternions) for many readers of the *Philosophical Transactions.* Moreover it is noteworthy that more than 150 pages of the 1892 issue of the leading British scientific journal of that time was taken up by these two papers written in vectorial language.

In 1893 McAulay published a short book [13] with a highly polemical preface that played a part in this debate. McAulay's style in his preface is indicated by a comment made by Tait in a review [15] of the book. Tait referred to the preface as "extremely interesting as the perfervid outburst of an enthusiast." (15; 193)

McAulay (who by 1893 was Lecturer in Mathematics and Physics at the University of Tasmania) stated at the beginning of his preface that the book was originally an essay submitted (in 1887) for the Smith's Prize Competition at Cambridge.[45] McAulay bemoaned the "mournful thing" that Cambridge mathematics did not in general include quaternions. As to why Cambridge had turned a deaf ear, McAulay stated that he could not "believe that she is in her dotage and has lost her hearing." (13; v) Then followed some passages enlightening for the picture they give of interest in quaternions at Cambridge, the center of British mathematics.

> When I sent in the essay I had a faint misgiving that perchance there was not a single man in Cambridge who could understand it without much labor. . . .
>
> . . . .

> There is no lack in Cambridge of the cultivation of Quaternions *as an algebra,* but this cultivation is not Hamiltonian . . . Hamilton looked upon Quaternions as a *geometrical* method, and it is in this respect that he has as yet failed to find worthy followers resident in Cambridge. (13; vi)
>
> . . . .
>
> The only way to convince the nurses [the Cambridge tutors] that Quaternions form a healthy diet for the *young* mathematician is to prove to them that they will "pay" in the first part of the Tripos. Of course this is an impossible task while the only questions . . . are in the second part and average one in two years. (13; vii)

McAulay then addressed himself to the Cambridge student by pleading that he "steep" himself in the "delirious pleasures" of quaternions and promising: "When you wake you will have forgotten the Tripos and in the fulness of time will develop into a financial wreck, but in possession of the memory of that heaven-sent dream you will be a far happier and richer man than the millionest millionaire." (13; vii–viii) Other passages no less strongly worded could be cited from this preface, as well as mixed references to Tait. The content of the concluding passage is probably obvious to the reader; nevertheless it may be quoted: "Let me in conclusion say that even now I scarcely dare state what I believe to be the proper place of Quaternions in a Physical education, for fear my statements be regarded as the uninspired babblings of a misdirected enthusiast, but I cannot refrain from saying. . . ." (13; xi) Then followed the essay itself, which was also rich in polemical statements, surrounded by numerous highly technical applications of quaternions to elastic solid theory, electrical theory, hydrodynamics, and the vortex-atom theory. The majority of the polemical statements from the essay itself have already been considered in the discussion of McAulay's papers in the *Philosophical Magazine* and the *Royal Society Transactions.*

In conclusion it may be stated that McAulay's preface qualifies him for consideration as one of the most vociferous mathematicians of the century. However, since success in debate is not always a function of the expressed enthusiasm of the participants, it may be wondered whether McAulay's writing produced much more than controversy.

Two reviews of McAulay's book merit discussion,[46] one by Macfarlane, the other by Tait.

Alexander Macfarlane's review [14] appeared in the first volume of the *Physical Review* (1893), and is above all interesting as a comparative study. Macfarlane, like Gibbs and Heaviside, was an advocate of a particular vector system which had been derived from the quaternion system. He, unlike Gibbs and Heaviside, had *long* been a quaternion advocate before his departure from that system and had been a student of Tait. Macfarlane's remarks in this review are especially interesting when they are compared with the statements of Gibbs and Heaviside on Hamilton and on quaternions.

In the earlier parts of his review Macfarlane showed sympathy with McAulay's enthusiastic championing of the quaternion cause, though he maintained that a large part of the essay was "a translation into quaternion notation of known results. . . ." (14; 388) The transitional passage is the following:

> I agree with the author in his estimate of the value of Hamilton's quaternion researches: they constitute, in my opinion, the greatest mathematical work of the century. They contain what was long sought after—*a* veritable extension of algebra to space: I do not say *the*, for I believe that there is more than one. The Cartesian analysis is also an extension of algebra to space, but it is fragmentary and incomplete; whereas the quaternion analysis is the true spherical trigonometry in which the axis of an angle as well as its magnitude is considered. (14; 389)

Macfarlane then disagreed with McAulay's explanation of the neglect of quaternions and argued that the neglect was mainly due to a small group of defects in the quaternion system (which were remedied in Macfarlane's own system). Neither Gibbs nor Heaviside ever made such laudatory statements about Hamilton's work as those given in the quotation from Macfarlane. Gibbs did not do this, partly because he was convinced of the superiority of Grassmann's system; Heaviside did not, partly for reasons of temperament and partly because his strategy, like that of Gibbs, was aimed at dissociating their system from that of Hamilton. Macfarlane could make and did make such a statement, it would seem, because his strategy was different; he wished to present his system as within the quaternion tradition but with the few quaternion defects removed. Thus he felt no compulsion toward making a vigorous attack on either the quaternion system or on Hamilton and Tait. Thus the review may be described as favorable to McAulay's book but more favorable to his own system.

Tait's review [15] of McAulay's book was the lead article in the December 28, 1893, issue of *Nature*. Tait began his review by praising McAulay's book, particularly as compared to another work by an unnamed author (Macfarlane) and as compared to the works mentioned in the following passage: "It is positively exhilarating to dip into the pages of a book like this after toiling through the arid wastes presented as wholesome pasture in the writings of Prof.

Willard Gibbs, Dr. Oliver Heaviside, and others of a similar complexion."[47] Tait's remarks on McAulay's preface were not so favorable; in fact, he passionately attacked McAulay's tendency to write passionately.

> It is much to be regretted that Mr. McAulay has not determined simply to let his Essay speak for itself. His Preface, though extremely interesting as the perfervid outburst of an enthusiast, assumes here and there a character of undignified querulousness or of dark insinuation, which is not calculated to win sympathy. It has too much of the "Rends-toi, coquin" to make willing converts; and in some passages it runs a-muck at Institutions, Customs and Dignities. Nothing seems safe. It is a study in monochrome:—the lights dazzingly vivid, and the shades dark as Erebus! (15; 193)

Tait described McAulay as a man of "genuine power and originality," whose reaction to the Cambridge restrictions "must have been gall and bitterness." (15; 193)

> Intuitively recognising its power, he snatches up the magnificent weapon which Hamilton tenders to all, and at once dashes off to the jungle on the quest of big game. Others, more cautious or perhaps more captious, meanwhile sit pondering gravely on the fancied imperfections of the arm; and endeavour to convince a bewildered public (if they cannot convince themselves) that, like the Highlander's musket, it requires to be treated to a brand-new stock, lock and barrel, *of their own devising*, before it can be safely regarded as fit for service. (15; 193)

Tait then commented, more benevolently than favorably, on some of McAulay's innovations.

Such was Tait's manner of welcoming a new zealot into the quaternion fold. McAulay, like Tait before him, had discovered the quaternion system at Cambridge despite Cambridge. All in all Tait must have been quite elated about this new convert. Finally a remark of Heaviside on McAulay, written in 1894 in a letter to Gibbs, may be given: "He seems to be a very clever fellow, and he knows it and shows that he knows it a little too much sometimes."[48]

Readers of the January 5, 1893, issue of *Nature* found therein a brief article[16] by Tait. Tait began by stating that he had assumed that his 1891 replies to Gibbs had been sufficient to show the "necessary impotence" and "inevitable unwieldiness" of every vectorial system that lacks the quaternionic product. But, wrote Tait, this illusion was dispelled by Heaviside's Royal Society paper.[10] Of this fifty-seven-page paper Tait read four pages—then: ". . . I met the check-taker as it were:—and found that I must pay before I could go further. I found that I should not only have to unlearn Quaternions (in whose disfavour much is said) but also to learn a new and most uncouth parody of notations long familiar to me; so I had to relinguish the attempt." (16, 225)

Tait then mentioned the criticisms of quaternions published by Heaviside in the *Electrician.* To these Tait responded only by quoting two short passages from his 1890 *Philosophical Magazine* article. Here as elsewhere Tait either underestimated his opponents or followed the questionable strategy of treating his opponents' statements as though they did not merit a detailed reply. Tait then declared that the main object of his present note was to call attention to a paper by Knott recently given before the Edinburgh Royal Society (the discussion of this paper is best delayed until later). Tait stated that the paper "is a complete exposure of the pretensions and defects of the (so-called) Vector Systems." (16; 226) From reading this paper Tait claimed that he had come to understand the vectorial ideas of his opponents; his reaction was expressed in his final sentence: "I find it difficult to decide whether the impression its revelations have left on me is that of mere amused disappointment, or of mingled astonishment and pity." (16; 226)

In the March 16, 1893, issue of *Nature* Gibbs re-entered the debate with an article [17] entitled "Quaternions and the Algebra of Vectors." It was written directly in response to McAulay's paper [11] in the December 15, 1892, issue of *Nature* (and indirectly in response to other of McAulay's writings). Gibbs began by discussing the slowness of the acceptance of quaternions. McAulay (and Tait earlier) had placed much blame on the lack of uniformity in notation; Gibbs wisely commented that this cause could not be accepted, since almost no comparable mathematical system had preserved uniformity of notation for a longer time than the quaternion system. Having rejected this explanation, Gibbs suggested another: it was that Hamilton's method of presentation had obscured the "simplicity, perspicuity, and brevity" of the *vectorial* approach and had moreover put the "geometrical relations of vectors" in a secondary position. (17; 463) After making the most of one of McAulay's poorly chosen metaphors, Gibbs suggested a law of evolution for mathematical systems.

> Whatever is special, accidental, and individual, will die, as it should; but that which is universal and essential should remain as an organic part of the whole intellectual acquisition. If that which is essential dies with the accidental, it must be because the accidental has been given the prominence which belongs to the essential.
>
> . . . .
>
> In mechanics, kinematics, astronomy, physics, all study leads to the consideration of certain relations and operations. These are the capital no-

tions; these should have the leading parts in any analysis suited to the subject. (17; 464)

Gibbs' "capital notions" (in the second quotation) are clearly those which are "essential and universal" (first quotation). In this way he led up to an argument central to this and to the preceding papers.

If I wished to attract the student of any of these sciences to an algebra for vectors, I should tell him that the fundamental notions of this algebra were exactly those with which he was daily conversant. I should tell him that a vector algebra is so far from being any one man's production that half a century ago several were already working toward an algebra which should be primarily geometrical and not arithmetical, and that there is a remarkable similarity in the results to which these efforts led. . . . I should call his attention to the fact that Lagrange and Gauss used the notation $(\alpha\beta\gamma)$ to denote precisely the same as Hamilton by his $S(\alpha\beta\gamma)$, except that Lagrange limited the expression to unit vectors, and Gauss to vectors of which the length is the secant of the latitude, and I should show him that we have only to give up these limitations, and the expression (in connection with the notion of geometrical addition) is endowed with an immense wealth of transformations. I should call his attention to the fact that the notation $[r_1r_2]$, universal in the theory of orbits, is identical with Hamilton's $V(\rho_1\rho_2)$, except that Hamilton takes the area as a vector, i.e. includes the notion of the direction of the normal to the plane of the triangle, and that with this simple modification (and with the notion of geometrical addition of surfaces as well as of lines) this expression becomes closely connected with the first-mentioned, and is not only endowed with a similar capability for transformation, but enriches the first with new capabilities. In fact, I should tell him that the notions which we use in vector analysis are those which he who reads between the lines will meet on every page of the great masters of analysis, or of those who have probed deepest the secrets of nature. . . . (17; 464)

The above quoted passage is typical of Gibbs' ability to present powerful and sensible arguments. The remark following this passage referred to McAulay's lamentations at the poor acceptance of the quaternion system; the remark is both appealing and historically true. "There are two ways in which we may measure the progress of any reform. The one consists in counting those who have adopted the *shibboleth* of the reformers; the other measure is the degree in which the community is imbued with the essential principles of the reform. I should apply the broader measure to the present case, and do not find it quite so bad as Mr. McAulay does." (17; 464)

In summary, Gibbs maintained that there were two discernible traditions leading up to the present situation. The first tradition manifested itself in the great physical treatises of the past wherein stress gradually came to be placed on certain fundamental notions and operations. The second tradition, running parallel to the first and in part stemming from it, consisted in the creation of formal vectorial systems. Gibbs argued that these two traditions were at that time (the 1890's) converging and that the vectorial approach was becoming common, even though dispute was rampant as to which vectorial system was preferable. To decide between the various systems emerging from the second tradition, one need only analyze the content of the first tradition.

Gibbs concluded with a tactful and true remark which, though it criticized Hamilton, praised Tait and other second-generation quaternionists for presenting the quaternion system in a more acceptable form.

> Now I appreciate and admire the generous loyalty toward one whom he regards as his master, which has always led Prof. Tait to minimise the originality of his own work in regard to quaternions, and write as if everything was contained in the ideas which flashed into the mind of Hamilton at the classic Brougham Bridge. But not to speak of other claims of historical justice, we owe duties to our scholars as well as to our teachers, and the world is too large, and the current of modern thought is too broad, to be confined by the *ipse dixit* even of a Hamilton. (17; 464)

Gibbs' paper was soon followed by another which, though fully in agreement with Gibbs' arguments, differed markedly in style. It was written by Oliver Heaviside, whose humor and brashness complemented Gibbs' seriousness and tact. Heaviside wrote in response to McAulay[11,12] and Tait[16] and in support of Gibbs. Concerning McAulay's Royal Society paper Heaviside wrote: "As the heart knoweth its own wickedness, he will not be surprised when I say that I seem to see in his mathematical powers the 'promise and potency' of much future valuable work of a hard-headed kind." (18; 534) Heaviside then suggested that McAulay simply give up quaternions! "A difficulty in the way," wrote Heaviside "is that he has got used to quaternions. I know what it is, as I was in the quaternionic slough myself once." (18; 534)

After making criticisms of some of McAulay's arguments, Heaviside turned to Tait and indirectly to Knott. He chided them about the irrelevance of their arguments on the superiority of quaternionic notations, and then he attempted to show that they were wrong anyway and that his system was superior in brevity and clearness of expression. Heaviside went on to state what he believed was the significance of Tait's paper. "The quaternionic calm and peace have been disturbed. There is confusion in the quaternionic citadel; alarms and excursions, and hurling of stones and pouring of boiling water upon the invading host. What else is the meaning of his letter, and more especially of the concluding paragraph? But the worm may turn; and turn the tables." (18; 534) Heaviside concluded the article with a discussion of some specific considerations on the quaternion notation and methods of treating rotations. The major significance of the article is that through its humor, frankness, and partial refutation of arguments of the quaternionists it gave support to Gibbs' more closely reasoned article. Heaviside's brashness must have antagonized not a few people; nevertheless his brashness was not that of the young and little known Alexander McAulay.

One of the most important papers published during this debate was Cargill Gilston Knott's lengthy paper[19] "Recent Innovations in Vector Theory" published in the *Proceedings of the Royal Society of Edinburgh* for 1892. It had been read December 19, 1892, was published in 1893, and was twice abstracted in *Nature* in 1893. The first abstract was very brief;[49] the second[20] was longer than any other paper in this debate in *Nature* with the exception of Gibbs' response to it, which was only slightly longer.

Knott had been Tait's assistant at Edinburgh from 1879 to 1883, then Professor of Physics at the Imperial University of Japan, and from 1892, Lecturer on Applied Mathematics in Edinburgh University. He was to become Tait's biographer and remained throughout his life a staunch advocate of quaternions.[50] The following discussion will concentrate on the full-length paper as published in the *Edinburgh Royal Society Proceedings.*

This paper, as nearly all others in the debate, was written, as Knott stated, "wholly from the point of view of mathematical physics. . . ." (19; 212) The paper was aimed at criticizing the work of Macfarlane, Heaviside, and, above all, Gibbs, whom Knott called the "high-priest" of the "clique of vector analysis." (19; 212) After mentioning some of the earlier papers in the debate, Knott briefly discussed the work of Rev. Matthew O'Brien and concluded that "the anti-quaternionic vector analysts of today have barely advanced beyond the stage reached by O'Brien . . ." (19; 212) in 1852. Knott's aim in making this statement (or better, overstatement) was to portray the Gibbs-Heaviside system as nothing new, rather as an old system that had not survived.

Knott then attacked Gibbs' statement that the scalar and vector products were fundamental, while the quaternion product was not Knott argued that without the quaternion the division of one vec by another is not possible, that this operation is fundamentai and that hence the quaternion is fundamental. To Gibbs' argument that his vectorial system could be extended to higher dimensions, while quaternions could not, Knott responded that this was like trying "to solve the Irish question by a discussion of the social life in Mars. . . ." (19; 217) After a discussion of notation Knott turned to the question of why the square of a unit vector should be equal to −1. He showed that if we assume the associative law and that $ij = k = -ji$, $jk = i = -kj$, and $ki = j = -ik$, then we can write: "$i(i + j)j = (i^2 + ij)j = i^2j + kj = +i^2j - i$" and "$i(i + j)j = i(ij + j^2) = ik + ij^2 = -j + ij^2$." But since $+i^2j - i$ must equal $-j + ij^2$, i^2 and j^2 must equal −1. (19; 221–222) Thus a necessary condition for the associative law is that the square of a unit vector be equal to minus one. Knott made an issue of this, and it is not a small point.

Repeatedly in the paper Knott took an algebraically narrow point of view. He was unable to see the legitimacy of defining two distinct products of two vectors; there was only one product, and that was the heaven-sent and Hamilton-discovered quaternion product. Thus Knott criticized the Gibbs-Heaviside system because it lacked anything corresponding to the quaternion $\nabla\omega$ (for ω, a vector) (19; 223–224) and because $\nabla \cdot \omega$ and $\nabla \times \omega$ (expressed in Gibbs' notation) had to be defined separately. (19; 223–224)

Gibbs had introduced the expressions "Pot," "New," "Lap," and "Max" as abbreviations in certain theorems in potential theory.[51] These had been introduced primarily to avoid writing a series of ∇'s. Knott dealt with this jokingly by suggesting the introduction of a "Tai" and a "Ham." (19; 225) This evoked one of Gibbs' few humorous responses.

Knott then turned to Gibbs' treatment of the linear vector function. Knott's position in this regard is well summarized in the following quotation:

> In the course of the development of the theory of the dyadic, Gibbs, with his usual proneness to lexicon products, invents a few names (or new meanings to old ones), such as Idemfactor, Right Tensor, Tonic, Cyclotonic, Shearer, and so on. These are all special forms of the linear and vector function; and, excepting possibly the names, Professor Gibbs does not seem to have contributed anything of value to Hamilton's beautiful theory. In no case, so far as I have been able to see, do his methods compare, for conciseness and clearness, at all favourably with Hamilton's and Tait's. (19; 229)

One line of attack used by Knott was to attempt to show that Gibbs was forced to bring in the quaternion. (19; 235) Knott concluded with a summary and by an invocation of the name Hamilton.

Knott's paper is important in that it was the first detailed criticism of the Gibbs-Heaviside system written by a quaternionist. It seems highly probable that it made impressive reading for the interested reader. Though Knott had read Gibbs' booklet with care and written with a certain Stoic control, nonetheless the bitterness that had characterized the earlier papers by Tait and McAulay was not absent.

Alexander Macfarlane was the first to reply to Knott's attack on "Recent Innovations." His paper[21] appeared in late May of 1893 in *Nature,* slightly more than a month after the longer abstract of Knott's paper had appeared. It was also written partly in reply to Lodge's review.[9]

Much of Macfarlane's paper was simply a restatement of his views and hence need not be discussed. The major point of interest is Macfarlane's discussion of Knott's views concerning whether the square of a vector should be positive or negative. Knott had at the end of his paper made the statement: "The *assumption* that the square of a unit vector is positive unity leads to an algebra whose characteristic quantities are non-associative, but in no sense more general than the corresponding but associative quaternion quantities, and whose ∇ is not the real efficient *Nabla* of quaternions." (19; 236–237) Knott's disagreement with the Heaviside-Gibbs-Macfarlane contention that the square of a vector should be positive was evident not only in his writing of "assumption" in italics but also ran through his whole paper. Macfarlane responded by arguing (correctly) that both views were assumptions (or better definitions) and were hence completely arbitrary algebraically. He cited as evidence a passage from Kelland and Tait's *Introduction to Quaternions.* Again history had come full cycle, for Hamilton in the 1840's had argued that there was nothing algebraically wrong in defining a product that violates the commutative law, and in the 1890's Knott had argued in defense of the quaternion cause that there was something unnatural in a system that violated the associative law.

Knott (in close relation with the previous question) had argued that $\nabla^2 u$ should equal $-\left(\dfrac{d^2u}{dx^2}+\dfrac{d^2u}{dy^2}+\dfrac{d^2u}{dz^2}\right)$, since the quaternion (the "real") Nabla gave this result. To this Macfarlane replied by pointing out that Tait used "the unreal $\nabla^2 u$ in his 'Treatise on Natural Philosophy,'" (21; 76) and by suggesting that "The *onus probandi* lies on the minus men." (21; 76)

In conclusion it may be noted that the controversy on the important plus or minus question was here as elsewhere being argued by the quaternionists on grounds of *algebraic* elegance, simplicity, and naturalness, whereas the vector analysts argued on pragmatic grounds—the test for them was what most *conveniently* corresponded to the most frequent relations found in physics and geometry. It is important to note that Macfarlane's arguments (unintentionally of course) were helpful to the Gibbs-Heaviside cause, for what Macfarlane had written on behalf of his system applied equally to the Gibbs-Heaviside system.

Knott's reply[22] to Macfarlane's paper was published three weeks later in *Nature.* It said little that was new. Since the minus or plus question is mathematically unanswerable, either being legitimate, Knott could do little more than restate old views couched in new rhetoric. Typical of this, and interesting because it contained Knott's admission that the answer was arbitrary, was Knott's concluding paragraph:

> In conclusion, let me say that no reasonable man can possibly object to investigators using any innovations in analysis they may find useful. But in the present case there is a very serious objection to the innovators condemning the system, for which they have one and all drawn inspiration, as "unnatural" and "weak," without in any way showing it so to be. That they can re-cast many quaternion investigations into their own mould does not prove their mould to be superior or even comparable to the original. Yet, in so far as they possess much in common with quaternions, the modified systems used by Gibbs, Heaviside, and Macfarlane cannot fail to have many virtues.
>
> "His form had not yet lost
> All her original brightness, nor appeared
> Less than Archangel ruined."[52]

Two weeks after Knott's response to Macfarlane, Alfred Lodge published an unpolemical paper[23] suggesting a solution to the plus or minus question. Lodge's suggestion was not accepted; at most it provoked Knott to give a bevy of reasons why it should not be accepted. Its sole importance is that it was another article within the debate.

Gibbs' reply[24] to Knott was published in *Nature* on August 17, 1893, nearly four months after the longer abstract of Knott's paper had been published. In it Gibbs replied in detail to the detailed criticisms of Knott; since such criticisms and replies are of interest only when they deal with major aspects of the system, the emphasis in the following discussion will be on illustration rather than summarization.

Gibbs began by stating that he felt bound to reply to Knott's paper because Knott had attacked the whole system of vector analysis primarily on the basis of faults found in Gibbs' *Elements of Vector Analysis.* Gibbs replied first to Knott's assertion that he had been forced to introduce the quaternion into his booklet. This charge was in one case due to an inadvertence (a tactful way of describing Knott's mathematical error) and in two other cases to overstatement—"My critic is so anxious to prove that I use quaternions that he uses arguments which would prove that quaternions were in common use before Hamilton was born." (24; 365)

To Knott's criticism of Gibbs' use of ∇, Gibbs replied by asking the reader to consider the following statement made by Kelvin:

> Helmholtz first solved the problem—Given the spin in any case of liquid motion, to find the motion. His solution consists in finding the potentials of three ideal distributions of gravitational matter having densities respectively equal to $1/\pi$ of the rectangular components of the given spin; and, regarding for a moment these potentials as rectangular components of velocity in a case of liquid motion, taking the spin in this motion as the velocity in the required motion. (27; 365)

Gibbs then translated Kelvin's statement as he thought Knott might prefer it: "Helmholtz first solved the problem—Given the Nabla of the velocity in any case of liquid motion, to find the velocity. His solution was that the velocity was the Nabla of the inverse square of Nabla of the Nabla of the velocity. Or, that the velocity was the inverse Nable of the Nabla of the velocity." (24; 365) Gibbs then argued that his introduction of the symbolic abbreviations "Pot," "New $= \nabla$Pot," "Lap $= \nabla \times$ Pot," and "Max $= \nabla .$ Pot" were as easy to remember, as expressive, and more conducive to rigor than the quaternion equivalents. Gibbs proceeded to answer a charge by Knott that there was a "tangle" and "jangle" in sections 91 to 104 of his *Elements of Vector Analysis* by thoroughly agreeing and explaining that this was due to a time lapse of a "couple of years" between writing parts of the sections. Section 101 was the last section of the 1881 part of Gibbs' booklet. Gibbs then capitalized on Knott's reference to this "tangle" and "jangle" in the terms: "No finer argument [against Gibbs' system] . . . can be found." (24; 366) Gibbs gave a detailed discussion of Knott's attack on his treatment of the linear vector function. From this discussion the reader could gain some knowledge of the superiority of Gibbs' treatment.

This paper alone, because of its many details and because it was written as a defense, could not have persuaded the uninitiated. It is best viewed as a capable and thorough response and hence as serving to remove the doubts of someone acquainted with, but undecided between, the rival systems.

The issue of *Nature* that appeared the week following Gibbs' paper contained a short article[25] by Robert S. Ball, who had written a work entitled *The Theory of Screws* (Dublin, 1876) that bore a relation to vector analysis. Ball (1840–1913) had been the astronomer royal of Ireland and was in 1893 Lowndean Professor of Astronomy and Geometry at Cambridge.

Ball's major point is summarized in the following quotation:

> It has always appeared to me that the student of physical science would better employ his time by studying the "Ausdehnungslehre" to which some of your correspondents have referred than by studying quaternions.
>
> The wonderful work of Grassmann is contained in a moderate-sized book in remarkable contrast to the two terrific volumes of Hamilton, which even Prof. Tait admits that he has not read entirely. The fact that the ausdehnungslehre could be mastered in a mere fraction of the time that would have to be devoted to the mastery of quaternions, is not however the important point. (25; 391)

Ball concluded by giving an illustration of Grassmann's methods and by a plea for the translation of Grassmann's 1862 *Ausdehnungslehre.*

The September 28, 1893, issue of *Nature* contained two articles, a short one by Knott and a shorter one by R. W. Genese. Genese's paper [26] will be considered first.

Robert William Genese (1848–1928) had received his B.A. (1871, eighth Wrangler) and M.A. (1874) from Cambridge and was in 1893 Professor of Mathematics at University College, Aberystwyth. Genese's paper constituted a seconding of Ball's motion that Grassmann's book be translated; he even pledged to subscribe ten pounds toward the accomplishment of that task. Lamented was the fact that Grassmann's ideas were not as yet taught in England, and Clifford was cited on behalf of the value of Grassmann's work.

Knott's paper [27] in the September 28 issue of *Nature* was written in reply to Gibbs and to Lodge. Herein Knott briefly commented on a number of Gibbs' replies but gave no new arguments. This part of the paper is best viewed as no more than an attempt to prevent the belief that he had acquiesced to Gibbs' arguments. Knott's last paragraph was his answer to Lodge's suggestion for an innovation, against which Knott gave five arguments.

Knott's reply [22] to Macfarlane's reply [21] to Knott's original paper was discussed by Macfarlane in an article [28] in the October 5, 1893, issue of *Nature.*

Macfarlane began by stating that he wished to aid the discussion by restating his fundamental position. This was in brief that quaternion notations as well as some of the fundamental quaternion principles needed revision, including the principles that the square of a vector is negative and that $\nabla^2 = -\left(\frac{d^2}{dx^2} + \frac{d^2}{dy^2} + \frac{d^2}{dz^2}\right)$. He concluded first by calling attention to two papers he had recently delivered at the International Mathematical Congress at Chicago in which he had extended his system and then by presenting the following statement: "As regards Prof. Knott's closing quotation from 'Paradise Lost,' I feel like the Senior Wrangler who, having read through the poem, remarked that it was all very pretty, but he didn't quite see what it proved. I close with a quotation which is from as good a book, and possesses more logical force: 'Ye shall know them by their fruits. Do men gather grapes of thorns, or figs of thistles?' " (28; 541)

In conclusion it may be noted that the important question to be asked concerning this paper is not as to how much it advanced Macfarlane's system, but whether it helped the quaternion or the Gibbs-Heaviside vector analysis cause. The answer is clear; it indirectly but indisputably helped the latter, for the majority of Macfarlane's conclusions were in harmony with the Gibbs-Heaviside vector system.

The last word, literally but not symbolically, in the debate in *Nature* went to Heaviside. This paper[29] was published in the January 11, 1894, issue and was written mainly in response to Tait's review[15] of McAulay's book.[13]

Heaviside began by arguing that Tait failed to recognize the values of the systems put forward by the opponents of the quaternion system. "He does not know their ways, either of thinking or of working, as is abundantly evident in all that he has written adversely to Prof. Willard Gibbs and others." (29; 246) Heaviside went on to state that he was surprised at Tait's warm welcome to McAulay's innovations. Then Heaviside argued that the illustration of the value of McAulay's work selected by Tait for discussion could be and essentially had been dealt with in a better manner in the vector analysis system. Heaviside concluded by a second invitation to McAulay to convert to vector analysis.

The final three papers from 1893 to be discussed were read before the Edinburgh Mathematical Society and published in the *Proceedings* of the Society sometime after June 9, 1893 (the date of the last paper in volume eleven). These papers were never referred to in the course of the *Nature* debate, and hence their discussion has been postponed until the completion of the discussion of the *Nature* articles.

The first paper[30] was by Knott (who was to assume the presidency of the Society in November, 1893) and was entitled "The Quaternion and its Depreciators." Most of Knott's paper was either repetition (at times verbatim) or elaboration of his statements in his slightly earlier "Recent Innovations in Vector Theory." Hence this long paper may be treated rather briefly.

Knott began by classifying the statements of the quaternion depreciators (Gibbs, Heaviside, and Macfarlane) under three heads: first, the question of "the value of the quaternion as a fundamental geometrical conception"; second, "the question of notation"; and third, "the question of the sign of the square of a vector. . . ." (30; 62) After repeating a number of his earlier arguments on the first question, Knott concluded by suggesting an analogy. He stated that although sin θ and cos θ occur more frequently than θ itself, we should not conclude that θ plays no fundamental role. Similarly we should not infer that $\alpha\beta$ is not fundamental simply because $V\alpha\beta$ and $S\alpha\beta$ occur more frequently.

Knott then turned to the question of notation and cited such arguments as were available on behalf of the quaternion system. From this he proceeded to the question concerning the sign appropriate to the square of a vector. His main positive argument was, as before, that the quaternion minus preserved the associative law for multiplication, and his main negative arguments were in response to Macfarlane's charge that essentially unit vectors and versors should not be identified, since the square of the first should be $+1$ and of the later -1.

Knott concluded by referring the reader to his paper in the *Proceedings of the Royal Society of Edinburgh* which discussed these questions and others.

This paper, like Knott's other long paper, was carefully written and contained no special excess of scorn. Neither Gibbs nor Heaviside replied to it, perhaps because they felt it was unnecessary or possibly because they felt it was useless (considering the audience).

Knott's paper had been delivered at the January 13, 1893, meeting of the Edinburgh Mathematical Society. At the next meeting (February 10, 1893) Dr. William Peddie, "Assistant to the Professor of Natural Philosophy" at Edinburgh University (hence Tait's assistant), delivered a paper[31] on behalf of the quaternion system.

After commenting that he believed that new and very general mathematical methods (like quaternions) were difficult for students to assimilate, Peddie stated: "And it seems as if it were for this reason that, in recent years, attempts have been made, by men of known mathematical ability, to smooth the paths." (31; 85) He was of course referring to Gibbs, Heaviside, and Macfarlane. Peddie's view as to their success was evident from his next statement: "Practically, all these attempts consist in using, instead of Hamilton's, another system of quaternions, cut up into parts; the parts of that system being used because they are imagined to be superior to the corresponding parts of Hamilton's system in respect of naturalness." (31; 85) Peddie then discussed three points in Tait's *Treatise* which Heaviside had called "sticking points." The discussion of the third point turned on the fact that the quaternionists began with fewer definitions, but they consequently were forced to go through more elaborate developments to get some of the fundamental results. In response to the charge that the quaternion is not a fundamental entity, Peddie, presumably on the mistaken assumption that the vector system had no method of treating rotations, argued that the quaternion was essential as a method of dealing with rotations. Peddie concluded the article by an attempt to show that the "notions of quaternions *are* applicable to space of four or any number of dimensions." (31; 90)

In conclusion, Peddie's paper was competent but no more than that. He had not however read Gibbs' work, and thus a number of his arguments were vague or missed the point.

The papers by Knott and Peddie were given at the first two 1893 meetings (January and February) of the Edinburgh Mathematical Society. This sequence was continued in the March and April meetings by Peddie, who presented at these meetings two parts of a paper entitled "Elements of Quaternions." From the published one-sentence summary of the first paper [53] it may be inferred that Peddie discussed therein vector addition and subtraction and the usefulness of these operations. A six-page abstract [32] of the second part of his paper (April meeting) was published. From this abstract it is clear that Peddie's intention in the paper was to intersperse a review of the fundamental principles of quaternion analysis with periodic forays against the vector analysts. Since these asides were in part repetitive of earlier arguments used by Peddie and since reasons for his statements were not in general published because of the fact that only an abstract was printed, nothing need be said of this paper beyond mention of its existence and spirit.

Two 1894 reviews of the first volume of Heaviside's *Electromagnetic Theory* merit discussion, since both commented on

Heaviside's vectorial ideas. The shorter of these reviews was written by Alexander Macfarlane and published in the *Physical Review*.[33] Macfarlane described the book and praised all parts except the third chapter, which was on vector analysis; he wrote:

> The third chapter expounds the elements of vector analysis. The exposition, while clear, and suitable for the class of readers addressed, contains principles which appear of doubtful validity to those who have made a special study of the matter. For instance, the scalar product of two vectors is not distinguished by any prefix, though the vector product is, apparently on the same principle that the eldest daughter is sufficiently distinguished by Miss, provided all the younger sisters have their Christian names appended. But the scalar product and the vector product are only partial products; the simpler name belongs properly to the complete product, which is their sum. Here we have an indication of one of the principal defects of the analysis; it is fragmentary, not a method *totus teres atque rotundus.* (33; 153–154)

This statement is illustrative of two practices common throughout the whole debate. The first is the penchant of vectorists for metaphorical expressions; the second and the more important is the tendency of many of the writers, particularly the quaternionists, to treat the question of products as something more than a question of arbitrary definition. In Macfarlane's case his statement that the scalar and vector products are only partial products was presumably not made through his own ignorance; rather it is best viewed as an argument against Heaviside's ideas based on the assumed algebraical "naiveté" of Macfarlane's readers.

The second review[34] of Heaviside's *Electromagnetic Theory* was written by George M. Minchin and published in the *Philosophical Magazine.* Minchin's rather lengthy review was in general very favorable and concentrated on Heaviside's innovations. Minchin commented upon Heaviside's style in the following way:

> A reader of Mr. Heaviside's writings is at once struck by the extraordinary style which distinguishes him from every other English writer on Mathematics or Physics; and the impression which is produced by this style is often the reverse of pleasing. There is a complete absence of the conventionalities which are generally recognized as proper to the writing of a scientific treatise. Mr. Heaviside is the Walt Whitman of English Physics; and, like the so-called "poet," he is certain to raise aversion to his peculiarities. (34; 146)

Minchin however was not in all ways averse to Heaviside's style; he went on to praise him for his clearness in exposition.

After an extended discussion of some of the electrical terms introduced by Heaviside, Minchin discussed Heaviside's chapter on vector analysis. He criticized first of all Heaviside's notation, for example, his use of Clarendon type for the representation of vectors. Minchin argued, as others had argued, that this convention was very inconvenient when one wished to write vectors in a manuscript, but it was at least better than Maxwell's use of Gothic capitals. Turning to the ideas in what he called Heaviside's "*Heretical Vector Analysis*," Minchin pointed out that he used this term "without necessarily implying any censure. . . ." (34; 154) Minchin presented no real arguments against Heaviside's vector system; his tone however revealed that he was not favorably inclined toward this innovation, this "Heretical" system. In summary, Minchin praised most of Heaviside's electrical ideas, but he was not so favorably inclined to his vector system. His review may be compared with FitzGerald's review of Heaviside's *Electrical Papers* which was discussed in the last chapter. Both reviews must have helped Heaviside to become better known, and their discussions of Heaviside's vectorial ideas (which had helped Heaviside to many important results) must have at least served to make that system (which was little known in 1890) much better known.

The final two papers of the debate are among the most interesting. The disputants were Cayley and Tait, and the question was one that had nowhere else been discussed.

Arthur Cayley (1821–1895), whose paper appeared first, was an extremely prolific and very well-known mathematician. He held the Sadlerian Chair for Mathematics at Cambridge University; the remarks in his paper are thus especially interesting, since they probably represent the point of view that he passed on to many Cambridge students. It is probable that McAulay's statements about quaternions at Cambridge were directed against Cayley and his view of quaternions. Cayley had long been interested in certain aspects of quaternions and was in fact the first person (after Hamilton) to publish a paper on quaternions. Moreover Cayley had, as mentioned previously, written a chapter entitled "A Sketch of the Analytical Theory of Quaternions" for the 1890 edition of Tait's *Treatise on Quaternions.* Tait described this as being included through Cayley's "unsolicited kindness." Cayley and Tait had frequently corresponded, and this correspondence more or less directly led to the two papers to be discussed.[54] Most of the arguments that were eventually presented in these papers may be found in these letters. As early as 1888 Cayley had commented concerning their different views of quaternions: "we are irreconcileable and shall remain so. . . ."[55] Nevertheless they continued the discussion in later letters, and in June, 1894, Cayley sent Tait a copy of a paper entitled "Coordinates versus Quaternions" which Cayley planned to publish in the *Messenger of Mathematics*. Tait suggested that Cayley allow his paper to be read before and published by the Royal Society of Edinburgh.[56] After Cayley agreed, his paper[35] and Tait's reply were read at the July 2, 1894, meeting and subsequently published in the *Proceedings of the Royal Society of Edinburgh*.

Cayley's paper began with a quotation from the preface to the first edition of Tait's *Treatise*, a preface which was reprinted in the later editions. Tait had written:

> It must always be remembered that Cartesian methods are mere particular cases of quaternions where most of the distinctive features have disappeared; and that when, in the treatment of any particular question, scalars have to be adopted, the quaternion solution becomes identical with the Cartesian one. Nothing, therefore, is ever lost, though much is generally gained, by employing quaternions in place of ordinary methods. In fact, even when quaternions degrade to scalars, they give the solution of the most general statement of the problem they are applied to, quite independent of any limitations as to choice of particular coordinate axes. (35; 271)

Cayley summarized his own views in the following paragraph:

> My own view is that quaternions are merely a particular method, or say a theory, in coordinates. I have the highest admiration for the notion of a quaternion; but . . . as I consider the full moon far more beautiful than any moonlit view, so I regard the notion of a quaternion as far more beautiful than any of its applications. As another illustration . . . I compare a quaternion formula to a pocket-map—a capital thing to put in one's pocket, but which for use must be unfolded: the formula, to be understood, must be translated into coordinates. (35; 271–272)

In short, Cayley argued that[35] the quaternion as an entity for the pure mathematician is interesting and important, but the quaternion method as a method for the applied mathematician (considering the geometer in this case as an applied mathematician) is nothing more than a shorthand or a method of abbreviation, and even under this aspect it is not a useful method.

Cayley gave three elementary illustrations of his point of view. The second illustration consisted in a comparison of the solutions, worked out by quaternions and by Cartesian methods, of the problem: Given two lines OA and OB, find the line OC perpendicular to the plane containing OA and OB. He stated that the quaternion solution would be $m\gamma = V\alpha\beta$ where $\gamma = OC$, $\alpha = OA$, $\beta = OB$, and m is a scalar. He commented that this was much briefer than the Cartesian solution, but was unintelligible and useless until translated into the Cartesian equivalent.

Cayley's final statement was the following: "In conclusion, I would say that while coordinates are applicable to the whole science of geometry, and are the natural and appropriate basis and method in the science, quaternions seem to me a particular and very artificial method for treating such parts of the science of three-dimensional geometry as are most naturally discussed by means of the rectangular coordinates *x*, *y*, *z*." (35; 275) Cayley's arguments could of course be translated directly into arguments against almost any vectorial system.

Tait's reply[36] to Cayley was entitled "On the Intrinsic Nature of the Quaternion Method." It is especially interesting for the historical views presented therein by Tait.

Tait began the paper by admitting that quaternions were not applicable to spaces of higher dimension than three; then he argued that Hamilton was of all people most able to dispense with abbreviations, and yet he had devoted the last twenty years of his life to quaternions. In the following quotation Tait stated his primary argument:

> It will be gathered from what precedes that, in my opinion, the term Quaternions means one thing to Prof. Cayley and quite another thing to myself: thus
>
> To Prof. Cayley Quaternions are mainly a Calculus, a species of Analytical Geometry; and, as such, *essentially* made up of those co-ordinates which he regards as "the natural and appropriate basis of the science." They artfully conceal their humble origin, by an admirable species of packing or folding:—but, to be of any use, they . . .
>
> doubly dying, must go down
To the vile dust from which they sprung!
>
> To me Quaternions are primarily a Mode of Representation:—immensely superior to, but essentially of the same kind of usefulness as, a diagram or a model. They *are*, virtually, the thing represented: and are thus antecedent to, and independent of, co-ordinates: giving, in general, all the main relations, in the problem to which they are applied, without the necessity of appealing to co-ordinates *at all*. Co-ordinates may, however, easily be *read into* them:—when anything (such as metrical or numerical detail) is to be gained thereby. Quaternions, in a word, *exist* in space, and we have only to recognize them:—but we have to *invent* or *imagine* co-ordinates of all kinds. (36; 277–278)

Tait then proceeded to discuss the history of quaternions. He first stated that the quaternion of the latter half of the century must be viewed "as having, from at least one point of view, but little relation to that of the seven last years of the earlier half." (36; 278) Tait argued that Cayley's view of the quaternion was the earlier view, and he suggested that Hamilton's greatest contribution was that through him "*From the most intensely artificial of systems arose, as if by magic, an absolutely natural one!*" (36; 279) Tait's meaning is clarified by the following quotation:

> Most unfortunately . . . Hamilton's nerve failed him in the composition of his first great Volume. Had he then renounced, for ever, all dealings with *i*, *j*, *k*, his triumph would have been complete. He spared Agag, and the best of the sheep, and did not utterly destroy them! He had a paternal fondness for *i*, *j*, *k*. . . . He had a fully recognized, and proved to others, that his *i*, *j*, *k* were mere excrescences and blots on his improved method:—but he unfortunately considered that their continued (if only partial) recognition was indispensable to the reception of his method by a world steeped in Cartesianism! Through the whole compass of each of his tremendous volumes one can find traces of his desire to avoid even an allusion to *i*, *j*, *k*; and, along with them, his sorrowful conviction that, should he do so, he should be left without a single reader. . . . And I further believe that, *to this cause alone*, Quaternions owe the scant favour with which they have hitherto been regarded. (36; 279–280)

Tait then admitted that the same defect could be found in his own book, a defect he planned to remedy if a fourth edition was called for. Tait stated that from his abundant correspondence with Hamilton he had learned of Hamilton's changed view of quaternion analysis. Thus, in summary, Tait's argument was that Cayley's criticisms simply did not apply to the modern quaternion methods, which avoid use of *i*, *j*, *k*, and hence of coordinates, of which the quaternion is "altogether independent" and to which it is "antecedent."

Among Tait's concluding statements is the suggestion of an analogy to replace Cayley's "pocket-map" analogy.

> A much more natural and adequate comparison would, it seems to me, liken Co-ordinate Geometry . . . to a steam-hammer, which an expert may employ on any destructive or constructive work *of one general kind*, say the cracking of an egg-shell, or the welding of an anchor. But you must have your expert to manage it, for without him it is useless. He has to toil amid the heat, smoke, grime, grease, and perpetual din of the suffocating engine-room. The work has to be brought to the hammer, for it cannot usually be taken to its work . . . Quaternions, on the other hand, are like the elephant's trunk, ready at *any* moment for *anything*, be it to pick up a crumb or a field-gun, to strangle a tiger, or to uproot a tree. Portable in the extreme, applicable anywhere . . . directed by a little native who requires no special skill or training, and who can be transferred from one elephant to another without much hesitation. Surely this, which adapts itself to its work, is the grander instrument! But then, *it* is the natural, the other the artificial, one. (36; 283)

If Heaviside was the "Walt Whitman of English Physics," the above quotation certainly merited Tait the title of the "Rudyard Kipling of English Physics."

These two papers, which conclude the debate, may serve to remind us that though most of the papers in the debate discussed the question of *which* vectorial method should be used, the question whether *any* vectorial method should be used was hardly forgotten.

III. *Conclusions*

If the materials discussed previously in this chapter are viewed broadly, a number of important generalizations emerge. Altogether thirty-six publications (thirty-eight including the two Heaviside publications discussed in the last chapter) appeared in the years from 1890–1894. Eight leading scientific journals were involved, with *Nature* carrying twenty articles. Twelve scientists, writing from England, Scotland, Australia, and the United States, participated. Nearly all the articles referred to other articles in the debate, so that the debate should be viewed as a definite historical unit.

A high level of intensity and a certain fierceness characterized much of the debate and must have led many readers to follow it with interest. And the penchant of the participants for striking metaphors (at times by their absurdity) led them to compare quaternions with, among other things, a Highlander's musket, strong meat, archangels, a map, and an elephant's trunk! Such analogies did at least lead to readability, though only indirectly to meaningful discussion.

The ratio of heat to light was especially high in the writings of the quaternionists, and it could hardly have been otherwise. Gibbs and Heaviside must have appeared to the quaternionists as unwelcome intruders who had burst in upon the developing dialogue between the quaternionists and the scientists of the day to arrive at a moment when success seemed not far distant. Charging forth, these two vectorists, the one brash and sarcastic, the other spouting historical irrelevancies, had promised a bright new day for any who would accept their overtly pragmatic arguments for an algebraically crude and highly arbitrary system. And worst of all, the system they recommended was, not some new system (even if Gibbs called it "Grassmannian"), but only a perverted version of the quaternion system. Heretics are always more hated than infidels, and these two heretics had, with little understanding and less acknowledgment, wrenched major portions from the Hamiltonian system and then claimed that these parts surpassed the whole. So at least it must have appeared to the quaternionists, and if this description of their reaction is only half correct, still it should be sufficient to explain why there was so little communication between the contending parties. Darwin's remark in the final chapter of his *Origin of Species* may serve to remind us that the quaternionists were not unique in their reaction. Darwin wrote:

> Although I am fully convinced of the truth of the views given in this volume under the form of an abstract, I by no means expect to convince experienced naturalists whose minds are stocked with a multitude of facts all viewed, during a long course of years, from a point of view directly opposite to mine. . . . A few naturalists, endowed with much flexibility of mind, and who have already begun to doubt the immutability of species, may be influenced by this volume; but I look with confidence to the future,—to young and rising naturalists, who will be able to view both sides of the question with impartiality.[57]

Though all the quaternionists wrote with some bitterness, it should not be forgotten that Knott wrote with care and thoroughness, Tait with the prestige that came from his distinguished career in science,[58] and McAulay with some success in demonstrating the usefulness of quaternions in physical application. However, perhaps too frequently Knott, Tait, and Peddie wrote for the already favorable Edinburgh audience. Prophets may never be heard in their homeland, but patriots are seldom heard anywhere else.

Among the opponents of the quaternion system Gibbs stands out as having contributed timely, carefully reasoned, and persuasive articles. It is perhaps not too much to describe his papers as masterpieces of mathematical rhetoric and as capable of leading thoughtful readers to accept the Gibbs-Heaviside system. Their effect on the less careful reader certainly was less great. Heaviside's short polemical papers probably had a certain appeal to both types of readers. His greatest contribution was however his demonstration of the effectiveness of the vectorial approach by his use of vectorial methods in his important electrical publications. It is noteworthy too that it was in the 1890's that Gibbs and Heaviside were becoming recognized as extremely important physical scientists. There was no one in the quaternion camp who was to receive such acclaim except perhaps Tait. Gibbs and Heaviside wrote only eight of the papers, but these papers were especially effective since the authors had the advantage that they were fully experienced in the system advocated by their opponents, whereas only Knott was well acquainted with his opponent's system.

In 1890 the quaternion system, though not widely used, was at least widely known; such was not the case with the Gibbs-Heaviside system. There are two stages preliminary to the acceptance of any system: it must first become known and then be tried and discussed. The quaternion system had from 1844 to 1894 passed through the first stage and was well into the second. The Gibbs-

Heaviside system had made roughly the same progress in the period from 1881 to 1894. The publicity associated with the debate was very helpful to the Gibbs-Heaviside system but was unnecessary and perhaps even dangerous for the quaternion system. From this point of view the contribution of Macfarlane may be evaluated. Since he presented his system as resulting from defects in the quaternion system and since some of these defects were also noted by Gibbs and Heaviside, his five papers must have acted to support the vector analysis (Gibbs-Heaviside) cause. Not a great deal was written on behalf of the full Grassmannian system. What was written came from Gibbs, who used his discussion of it mainly in support of his own system, and from Ball and Genese, whose articles were extremely brief.

It is interesting that Gibbs, Heaviside, and to some extent Macfarlane took what may be described as a pragmatic approach to the question of which system was to be preferred. Many of their arguments were on grounds of expressiveness, congruity with physical relationships, and ease of understanding. The quaternionists, on the other hand, put somewhat greater stress on mathematical elegance and algebraic simplicity. Some quaternion advocates looked askance at their opponents' use of a number of products rather than the single quaternion product. That history had in this regard come full circle has previously been mentioned.

There was probably too much stress on the notation question (which was of little direct relevance) in the debate. This was natural however for two reasons. It was the common opinion of British mathematicians of the nineteenth century that continental mathematics had far surpassed English mathematics by 1800 in large part because of the superiority of the Leibnizian notation for calculus as compared to the Newtonian notation. Hence it would be natural for them to place an undue stress on notation. A second and more important reason is historically rooted in the fact that when Gibbs and Heaviside created their systems, they wished to dissociate their systems from the quaternion parent. One natural way to do this was through changes in symbolism. The notation question was further complicated by the fact that Grassmann's symbolism bore no relation to that of Hamilton, Gibbs, or Heaviside. Macfarlane too had suggested many changes in symbolization. Thus there were numerous, distinct sets on notation to be considered, and since the symbols themselves became symbolic of the differences between the systems, the whole question was hotly debated.

The notation problem reached major proportions in the first

decade of the twentieth century; committees were organized, which resulted in part in suggestions for new symbols but mainly in failure.[59] It is surprising that only two papers in the debate directly discussed the fundamental question of whether *any* vectorial system should be adopted, especially since vectorial systems were at that time still something of an innovation. Cayley was the only participant to argue that no vector system should be adopted, whereas a large number of the readers of the debate must, at least initially, have been of that persuasion. Cayley is also unique in that he was the only mathematician who contributed a major article to the debate. The vast majority of the participants were physicists who were interested primarily in the applications of vectorial methods to physics.

Some results of this debate may now be considered. In 1895 Shunkichi Kimura and Pieter Molenbroek published a notice in *Nature* addressed to "Friends and Fellow Workers in Quaternions."[60] Simultaneously they published a similar notice in *Science* entitled "To Those Interested in Quaternions and Allied Systems of Mathematics."[61] Kimura, a Japanese scientist then residing at Yale, and the Dutchman Molenbroek suggested in these notices that an association of those interested in various systems of vector analysis should be formed. Thus came about the International Association for Promoting the Study of Quaternions and Allied Systems of Mathematics. In March, 1900, the first issue of the *Bulletin* of the International Association was published, the last issue appearing in 1913. Tait was elected the first president, but declined, and Ball was elected with Macfarlane as general secretary. The membership as listed in the first issue included over sixty scientists from fifteen countries. One major result of the Association, primarily attributable to Macfarlane, who became its real leader, was the publication in 1904 (with supplements in the Bulletins until 1913) of a *Bibliography of Quaternions and Allied Systems of Mathematics* which contained (with the supplements) references to roughly twenty-five hundred articles in the vectorial tradition. This association, which must have advanced the vector cause considerably, came about in large part as a result of the "struggle for existence" of the early 1890's.[62]

Another major result of this series of articles of the early 1890's seems to have been a lessening of Hamilton's reputation. Tait and the other quaternionists had, as noted previously, frequently invoked Hamilton's name and reputation on behalf of the quaternion cause. Thus Gibbs and Heaviside, who were of course proponents of the system that eventually triumphed, tended to take their stance against Hamilton. The result was that the failure of the quaternion cause, which had become so closely linked to Hamilton's name, acted to diminish Hamilton's historical stature. This should not and probably need not have been the case. It is possible that if Gibbs and Heaviside had presented their system as a direct descendant of Hamilton's creation, if they had tried to capitalize on the successes of this tradition and had viewed themselves as moving within it rather than as breaking from it, the net result of the eventual triumph of their cause would have been an increase in Hamilton's reputation.

Macfarlane, it should be recalled, had taken this position in presenting his system. But intellectual debates tend to push the contending parties to extremes, and hence Gibbs and Heaviside have been viewed as the great revolutionaries; it should be clear however from what has been written that such a view does serious injustice to Hamilton and even to Tait. Gibbs and Heaviside were indeed heretics, but their cause was 90 percent in harmony with the Hamilton-Tait orthodoxy. Hamilton and Tait were their intellectual ancestors, but this genealogy was obscured in the heat of this debate.

Finally, it should be noted that in the course of the debate there was much speculation as to why the quaternion system had been as yet poorly received. In the light of this discussion and because of the relevance of this question to the general theme of this study some attempts to explain this phenomenon will now be made.

It is first of all important to note that the quaternionists painted the picture far too bleakly. By 1890 considerable recognition had been given to the quaternion system; as it was pointed out earlier, twenty-seven quaternion books and over four hundred articles had been published by 1890, and over half of these books and roughly a third of the articles were published in the 1880's. Moreover, to consider the question from a comparative point of view, there were few new mathematical or physical *systems* presented in the nineteenth century that were rapidly accepted. Appropriate comparisons might be non-Euclidean geometry, group theory, or Maxwell's theory. The quaternion system represented a real innovation; its fundamental laws broke with long-established traditions; its discovery was a key development in the birth of modern algebra; as an applied system its only traditional reference points were such things as the parallelogram of forces or velocities.

Other reasons for the slow acceptance are that Hamilton's style was unsuited to an introductory exposition and that mathematically anything that could be done by the application of quaternions in geometry and physics could also be done with the Cartesian methods, though usually by longer processes. The vectorial methods thus were not absolutely considered a *sine qua non* for progress. Also relevant is the fact that few important physical discoveries had been made by quaternion methods; indeed the physicist of the first two-thirds of the century had far less need for vectorial methods than his counterpart of the last third. It also took time to put the quaternion system into the most fruitful form for application to the needs of the physicist. This was the great achievement of Tait. Systems of vectorial analysis progressed not only in relation to physical developments but also in relation to mathematical developments; algebra changed greatly and matured rapidly from 1840 to 1890 to allow a perspective within which the quaternion system might be judged. There were of course innate mathematical difficulties within the quaternion system that hindered its acceptance, for example, the scalar product was negative. Finally, up to 1890 there had been no widespread published discussion of the merits of the quaternion system. There are certainly other reasons, but these are among the most important.

The question of the success of the quaternion system should be viewed as subservient to another question. This is the success of the vectorial approach in general. If this question is emphasized (as Gibbs stressed that it should be), then the situation appears far from bleak. The vectorial approach in its many forms was becoming evermore common. The challenge facing Gibbs and Heaviside was by no means the challenge that had faced Hamilton a half century earlier. The Gibbs-Heaviside system came into existence in an entirely different climate from that of the 1840's. Many of the forces which had earlier acted against the quaternion system were no longer present. For example, the maturation of algebra supplied a perspective that made the new system seem far less revolutionary. Moreover, there was a long tradition by then of work in vector analysis. Physics had changed; the physicist was daily forced to deal with vectorial entities. The Gibbs-Heaviside vector analysis system *was* closely associated with important new physical ideas in Heaviside's writings. If one wished to read Heaviside, one had to learn the language of vector analysis. Finally, the mathematical defects of the quaternion system as an applied system were in part eliminated in the new system, and the merits of systems of vector analysis had been widely discussed.

The times were thus in many ways disposed to the acceptance of some vector system; that system was to be the Gibbs-Heaviside system; that acceptance will be discussed in the next chapter.

Notes

[1] Peter Guthrie Tait, "On the Importance of Quaternions in Physics" in *Philosophical Magazine,* 5th Ser., 29 (January, 1890), 84–97.

[2] Peter Guthrie Tait, "Preface" in *An Elementary Treatise on Quaternions,* 3rd ed. (Cambridge, England, 1890), v-viii.

[3] Josiah Willard Gibbs, "On the Role of Quaternions in the Algebra of Vectors" in *Nature, 43* (April 2, 1891), 511–513.

[4] Peter Guthrie Tait, "The Role of Quaternions in the Algebra of Vectors" in *Nature, 43* (April 30, 1891), 608.

[5] Josiah Willard Gibbs, "Quaternions and the 'Ausdehnungslehre'" in *Nature, 44* (May 28, 1891), 79–82.

[6] Peter Guthrie Tait, "Quaternions and the Ausdehnungslehre" in *Nature, 44* (June 4, 1891), 105–106.

[7] Alexander McAulay, "Quaternions as a practical Instrument of Physical Research" in *Philosophical Magazine,* 5th Ser., *33* (June, 1892), 477–495.

[8] Alexander Macfarlane, "Principles of the Algebra of Physics" in *Proceedings of the American Association for the Advancement of Science, 40* (1891, published 1892), 65–117.

[9] [Alfred Lodge], "[Review of] 'Principles of the Algebra of Vectors' By Alexander Macfarlane" in *Nature, 47* (November 3, 1892), 3–5. Note that the author of the review used an incorrect title for Macfarlane's work.

[10] Oliver Heaviside, "On the Forces, Stresses, and Fluxes of Energy in the Electromagnetic Field" in *Philosophical Transactions of the Royal Society of London, 183* A (1892, read in 1891, published in 1893), 423–484. Published earlier in Oliver Heaviside, *Electrical Papers, 2* (London, 1892), 521–574. References will be given for the *Phil. Trans.* publication.

[11] Alexander McAulay, "Quaternions" in *Nature, 47* (December 15, 1892), 151.

[12] Alexander McAulay, "On the Mathematical Theory of Electromagnetism" in *Philosophical Transactions of the Royal Society of London, 183* A (1892, read in 1892, published in 1893), 685–779.

[13] Alexander McAulay, *Utility of Quaternions in Physics* (London, 1893).

[14] Alexander Macfarlane, "[Review of] *Utility of Quaternions in Physics.* By A. McAulay" in *Physical Review, 1* (1893), 387–390.

[15] Peter Guthrie Tait, "[Review of] *Utility of Quaternions in Physics.* By A. McAulay" in *Nature, 49* (December 28, 1893), 193–194.

[16] Peter Guthrie Tait, "Vector Analysis" in *Nature, 47* (January 5, 1893), 225–226.

[17] Josiah Willard Gibbs, "Quaternions and the Algebra of Vectors" in *Nature, 47* (March 16, 1893), 463–464.

[18] Oliver Heaviside, "Vectors *versus* Quaternions" in *Nature, 47* (April 6, 1893), 533–534.

[19] Cargill Gilston Knott, "Recent Innovations in Vector Theory" in *Proceedings of the Royal Society of Edinburgh, 19* (1892, read December 19, 1892, published 1893), 212–237.

[20] Cargill Gilston Knott, "Recent Innovations in Vector Theory" (An Abstract) in *Nature, 47* (April 20, 1893), 590–593.

[21] Alexander Macfarlane, "Vector *versus* Quaternions" in *Nature, 48* (May 25, 1893), 75–76.

[22] Cargill Gilston Knott, "Vectors and Quaternions" in *Nature, 48* (June 15, 1893), 148–149.

[23] Alfred Lodge, "Vectors and Quaternions" in *Nature, 48* (June 29, 1893), 198–199.

[24] Josiah Willard Gibbs, "Quaternions and Vector Analysis" in *Nature, 48* (August 17, 1893), 364–367.

[25] Robert S. Ball, "The Discussion on Quaternions" in *Nature, 48* (August 24, 1893), 391.

[26] Robert William Genese, "Grassmann's 'Ausdehnungslehre'" in *Nature, 48* (September 28, 1893), 517.

[27] Cargill Gilston Knott, "Quaternions and Vectors" in *Nature, 48* (September 28, 1893), 516–517.

[28] Alexander Macfarlane, "Vectors and Quaternions" in *Nature, 48* (October 5, 1893), 540–541.

[29] Oliver Heaviside, "Quaternionic Innovations" in *Nature, 49* (January 11, 1894), 246.

[30] Cargill Gilston Knott, "The Quaternion and its Depreciators" in *Proceedings of the Edinburgh Mathematical Society, 11* (1893, read January 13, 1893), 62–80.

[31] William Peddie, "On the Fundamental Principles of Quaternions and other Vector Analyses" in *Proceedings of the Edinburgh Mathematical Society, 11* (1893, read February 10, 1893), 84–92.

[32] William Peddie, "The Elements of Quaternions" (Abstract) in *Proceedings of the Edinburgh Mathematical Society, 11* (1893, read April 14, 1893), 130–136.

[33] Alexander Macfarlane, "[Review of] *Electromagnetic Theory.* Vol. I. By Oliver Heaviside" in *Physical Review, 2* (1894), 152–154.

[34] George M. Minchin, "[Review of] *Electromagnetic Theory.* By Oliver Heaviside" in *Philosophical Magazine,* 5th Ser., *38* (July, 1894), 146–156.

[35] Arthur Cayley, "Coordinates versus Quaternions" in *Proceedings of the Royal Society of Edinburgh, 20* (1893–1894 session, read July 2, 1894, published 1895), 271–275.

[36] Peter Guthrie Tait, "On the Intrinsic Nature of the Quaternion Method" in *Proceedings of the Royal Society of Edinburgh, 20* (1893–1894 session, read July 2, 1894, published 1895), 276–284.

[37] Lynde Phelps Wheeler, *Josiah Willard Gibbs* (New Haven, 1962), 115.

[38] E. T. Whittaker, "Oliver Heaviside" in *Calcutta Mathematical Society Bulletin, 20* (1930), 205.

[39] It is hardly necessary to mention that an important article of only mathematical content has a definite polemical force. For obvious reasons such articles have not been included.

[40] The reader may wish to consult the recent article by Alfred M. Bork, " 'Vectors Versus Quaternions'—The Letters in *Nature*" in *American Journal of Physics, 34* (1966), 202–211. In March, 1965, Professor Bork generously sent me a copy of his manuscript; I had written my study of the debate in the summer of 1964 and was pleased to find that we had reached many of the same conclusions. The chief methodological difference between the two studies is that Professor Bork limited his discussion to the articles in *Nature.*

[41] (2; vi). Tait either did not know Gibbs' notation or Grassmann's notation or knew neither of them; this is evident since his statement about notation is completely wrong.

[42] A. Macfarlane, *Vector Analysis and Quaternions* (New York, 1906), 50 pp. No second edition of the book appeared.

[43] It may be noted that Macfarlane delivered another paper, "On the Imaginary of Algebra" at the American Association for the Advancement of Science meeting of 1892, which was published (in December, 1892) in the *Proceedings* of that society, volume *41*, 33–55. This paper does not require further discussion, since it is essentially an extension of the earlier paper.

[44] The review was not signed. The evidence for attributing it to Alfred Lodge is that Macfarlane did so in his *Bibliography of Quaternions and Allied Systems of Mathematics* (Dublin, 1904), 51. Alfred Lodge should not be confused with the better-known scientist Oliver Lodge.

[45] (13; v). It seems quite clear that McAulay's essay did not win the prize.

[46] I do not believe it is necessary to discuss a lengthy, favorable, mainly summarizing review written by A. S. Hathaway, who taught at Rose Polytechnic Institute in Terre Haute, Indiana, and who was a quaternion advocate and author of an 1896 *Primer of Quaternions* (New York). Hathaway's review appeared in the *Bulletin of the American Mathematical Association*, 1st Ser., *3* (1893), 179–185.

[47] (15; 193). Heaviside never attained the doctoral degree (unless honorary and then certainly after 1893) since he never attended a university. It would seem that Tait would know this. It is doubtful then that Tait's bestowing of a "Dr." on Heaviside was a simple slip of the pen.

[48] From a letter dated August 6, 1894, in the "Scientific Correspondence" of Gibbs preserved at Yale University.

[49] The first abstract was part of a report on the Edinburgh Royal Society meeting of December 19, 1892. See *Nature*, *47* (January 19, 1893), 287.

[50] This information has been obtained mainly from Cargill Gilston Knott, *Life and Scientific Works of Peter Guthrie Tait* (Cambridge, England, 1911); see especially the title page.

[51] J. W. Gibbs, *The Scientific Papers of J. Willard Gibbs*, vol. II (New York, 1961), 44–50.

[52] (22; 149). Knott's quotation is from Milton's *Paradise Lost*, I, 591–593.

[53] William Peddie, "Elements of Quaternions" in *Proceedings of the Edinburgh Mathematical Society*, *11* (1893), 104.

[54] Cargill Gilston Knott published selections from this correspondence in his *Life of Tait*, 154–166.

[55] *Ibid.*, 159.

[56] *Ibid.*, 164–165.

[57] Charles Darwin, *Origin of Species*, 6th ed. (New York, 1963), 444.

[58] The vector versus quaternion dispute was not the only dispute in which Tait took part. Throughout his life he manifested a tendency to become embroiled in controversies. The most famous of these concerned the history of thermodynamics, and he has in this regard frequently been accused of chauvinism. It may be mentioned that Tait had argued in his 1880 article for the *Encyclopaedia Britannica*, entitled "Hamilton," that Hamilton was actually of Scotch ancestry.

[59] See on this point Felix Klein, *Elementary Mathematics from an Advanced Standpoint: Arithmetic, Algebra, Analysis* (New York, n. d.), 65, and Florian Cajori, *A History of Mathematical Notations*, vol. II (Chicago, 1952), 136–139.

[60] Shunkichi Kimura and Pieter Molenbroek, "Friends and Fellow Workers in Quaternions" in *Nature*, *52* (1895), 545–546.

[61] Shunkichi Kimura and Pieter Molenbroek, "To Those Interested in Quaternions and Allied Systems of Mathematics" in *Science*, 2nd Ser., *2* (1895), 524–525.

[62] Another event, almost certainly engendered by this debate, was the offering of a prize in 1894 by the "Dutch Society of Sciences" for a comparison of the methods of Grassmann, Hamilton, and Cauchy with emphasis on their applicability in physical science. This information is given in Victor Schlegel, "Die Grassmann'sche Ausdehnungslehre" in *Zeitschrift für Mathematik und Physik, 41* (1896), 53. I have not been able to determine to whom or even whether the prize was awarded.

CHAPTER SEVEN

The Emergence of the Modern System of Vector Analysis: 1894-1910

I. *Introduction*

In the last chapter the debate on vectorial methods that occurred in the period 1890 to 1894 was discussed, and it was suggested that as a result of this debate the Gibbs-Heaviside system became widely known. But by 1894 this system still had not been widely accepted by the scientific community. The present chapter will deal with the developments from 1894 to 1910. It will be argued that vectorial analysis came to be accepted widely during this period; that the form of vectorial analysis that came to be accepted was that in the tradition established by Hamilton, Tait, Gibbs, and Heaviside; and that the most influential force in producing acceptance stemmed from the association of vectorial analysis with electrical theory, an association to be credited to Maxwell and Heaviside.

The evidence employed to support the statement that vector analysis came to be accepted widely during this period consists in the establishment of the fact that a substantial number of major publications [14] presenting the now common system of vector analysis were published at this time. It has been assumed that the publication of such a work indicates a belief on the part of both author and publisher that such works would be widely read. When later editions of a book were called for, this is taken as evidence that the book was in fact widely read. To support the conclusion that the ancestry of the majority of these publications extends back to the Hamilton, Tait, Maxwell, Gibbs, and Heaviside tradition, each of the works has been analyzed in terms of its origin and content.

Much information concerning the origins of the works has been derived from biographical information concerning the authors and also from reviews of the books.

There were perhaps 1000 journal articles published during the period 1894 to 1910 in which some form of vectorial analysis was discussed or employed. These have not been analyzed since the evidence derived from the major publications has seemed sufficient unto the conclusions propounded.[15]

In a number of cases the analysis of twelve major publications has served to reveal important lines of development, and there has been no hesitation to discuss these broader aspects. In the conclusion to this chapter the fate of the quaternionic and the Grassmannian traditions is briefly treated.

II. *Twelve Major Publications in Vector Analysis from 1894 to 1910*

August Föppl's *Einfuhrung in die Maxwell'sche Theorie der Elektricität*[1] of 1894 is an important book not only in the history of vector analysis but also in the history of electricity. Föppl's book was one of the first expositions in German of electricity as presented in accordance with Maxwell's ideas. In his foreword Föppl stated: "The circle of ardent followers of Maxwell's electrical theories consisted until recently almost exclusively of English physicists. Earlier considerable attention had been accorded this theory in Germany, but scientists were still too biased by the ban on action at a distance to be able to become fully accustomed to it." (1; III) Föppl proceeded to cite the famous experiment performed by Heinrich Hertz in 1887 to verify Maxwell's statement that light is an electromagnetic wave, as the turning point in German hostility to Maxwell. After Hertz's experiment the interest in Maxwell's entire system became widespread, and Föppl stated that this interest led him to write his book. (1; III–IV)

Föppl went on: "In the mathematical formulation of the theories discussed I have throughout made use of the symbols and methods of vector calculus; these are discussed in the first chapter to the extent that they will be used. The manner of presentation is very simple and, as I dare to assume unconditionally, also very easy to understand." (1; V–VI) After discussing the advantages of vector analysis, Föppl revealed how he came to introduce vectorial methods in his book.

> In the presentation of vector techniques and on many other points, I followed the pattern set by O. Heaviside in his papers, which have re-

cently been collected into book form. The works of this author have in general influenced my presentation more than those of any other physicist with the obvious exception of Maxwell himself. I consider Heaviside to be the most eminent successor to Maxwell in regard to theoretical developments. . . . (1; VII)

Among Föppl's arguments for the adoption of vector analysis in Germany was the following: "The country which produced a Grassmann should no longer stand behind the country of Hamilton in regard to the introduction of these important improvements in the mathematical aids to theoretical physics." (1; VII)

Thus Föppl devoted his first three chapters (84 pages) to an explanation of vector analysis and employed the symbolisms of Maxwell and especially Heaviside in doing so. The Grassmannian and the Gibbsian systems of symbolism were not used.[16] Thus Föppl's book was the first written in German to present a detailed exposition of the modern system of vector analysis, and it was a very popular book: a second edition appeared in 1904, edited by Max Abraham,[17] and a third edition in 1907. In the 1904 Abraham edition of Föppl's book additional subjects in vector analysis were treated. These additions were based on Föppl's short publication of 1897, *Die Geometrie der Wirbelfelder,* in which Föppl had presented some of the more advanced parts of vector analysis, including the linear vector function and the vectorial treatment of potential theory. Here as before Föppl closely followed Heaviside.

The importance of Föppl's work was well summed up in a statement made in the 1910's by Felix Klein, who was a contemporary to Föppl and to the developments at that time. Klein wrote:

Heaviside is also linked with the first independent presentation which vector analysis has found in Germany. This is A. Föppl's "Geometrie der Wirbelfelder" (1897), which was a completion of the presentation in Föppl's "Einleitung in die Maxwellsche Theorie" (1894). From these two publications there later arose the two-volume "Theorie der Elektrizität" which was worked on by Abraham and which is now one of the most frequently used textbooks in electricity.[18]

Föppl also published a four-volume work entitled *Vorlesungen über technische Mechanik* (1897–1900) in which he used vector analysis extensively. This publication must have been very successful, since a second edition was immediately called for and appeared from 1900–1903.[19]

A book similar to Föppl's 1894 publication was published in 1899 by the Italian Galileo Ferraris under the title *Lezioni di Elettrotecnica,*[20] and Ferraris, like Föppl, followed Heaviside in both his presentation of electricity and of vector analysis.

Thus the first extensive published presentations of modern vector analysis to appear in England, Germany, and Italy were included in books on electricity presented from a Maxwellian point of view. One was written by Heaviside; the other two stemmed directly from him.

Edwin Bidwell Wilson's book of 1901, *Vector Analysis: A Text Book for the Use of Students of Mathematics and Physics Founded upon the Lectures of J. Willard Gibbs*,[2] was the first formally published book that was entirely devoted to presenting the modern system of vector analysis.[21]

Wilson's textbook on vector analysis was one of the longest and best books to be published on that subject. The story of how it came to be written is directly relevant to the present study, since it reinforces one of the more surprising and important conclusions reached concerning the manner in which modern vector analysis came to be developed and widely applied.

Wilson told the story in fullest detail in a publication of 1931.[13] As Wilson was completing his undergraduate work in mathematics at Harvard in 1899, the mathematician W. F. Osgood suggested to him that he do graduate work at Yale and study with Pierpont and Percy Smith. Of his teachers only B. O. Peirce mentioned Gibbs by referring to him as someone "whom some of us here think a rather able fellow." (13; 211) Thus Wilson went to Yale to study with Pierpont and Smith. At his first registration for courses Wilson planned to register for only three courses until Dean A. W. Phillips required that he take a fourth and suggested the course in vector analysis taught by Gibbs. To this Wilson objected that he had had a full year course in quaternions at Harvard from J. M. Peirce and that consequently the course would in large part serve only as a review. Somewhat unhappily Wilson followed the Dean's advice and joined the handful of others who had registered for the course. Wilson found the course easy, and soon after he had completed it, he was asked by Professor Morris, editor of the Yale Bicentennial Series, to prepare a book on vector analysis for that series. Wilson met with Gibbs, who was engaged in preparing a work on statistical mechanics and thus "would not have time to advise on the composition of 'Vector Analysis,' to read the manuscript or the proof. . . ." (13; 214) Wilson stated that his contact with Gibbs concerning the book was essentially limited to that meeting. Nevertheless Gibbs later mentioned to Wilson that he found the book satisfactory. (13; 214) It is interesting to note that Gibbs and Wilson corresponded in 1903 on the question of whether Wilson should write a shorter treatise or an abridgment of the vector analysis book. (13; 217–218) However no such book came into print.

That the primary source of Wilson's book was Gibbs' lectures is established by the following quotation from Wilson's general preface.

> By far the greater part of the material used in the following pages has been taken from the course of lectures on Vector Analysis delivered annually at the University by Professor Gibbs. Some use, however, has been made of the chapters on Vector Analysis in Mr. Heaviside's *Electromagnetic Theory* (Electrician Series, 1893) and in Professor Föppl's lectures on *Die Maxwell'sche Theorie der Electricität* (Teubner, 1894). My previous study of Quaternions has also been of great assistance. (2; ix)

The book was very clear and thorough in its presentation; it included, with abundant explanation, nearly all the material in Gibbs' earlier work. A small quantity of new material was added, for example, an introduction to barycentric calculus and some further developments concerning the linear vector function. The book received a number of very favorable reviews, one of nine pages by Alexander Ziwet in the *Bulletin of the American Mathematical Society*,[22] one by an anonymous reviewer in the *Bulletin des sciences mathématiques* which ran to ten pages,[23] and a shorter one by Victor Schlegel in *Fortschritte der Mathematik* in which Schlegel referred to the book as "essentially a working out of quaternion theory in the sense of the Ausdehnungslehre. . . ."[24] A very unfavorable review appeared in the *Philosophical Magazine.* The author was C. G. Knott; his criticism was that the book was not quaternionic.[25] The success of the book is perhaps best indicated by the fact that a second edition was published in 1909.[26]

In summary, the publication of Wilson's *Vector Analysis* constituted the appearance of the first full-length book presenting the modern system of vector analysis. In one way Gibbs had less to do with this book than one would expect, and some of Heaviside's ideas were included. The book was well received and popular. It is striking, but in one sense typical of developments that have been previously discussed, that the author of the book was a convert from the quaternion camp.

Alfred Heinrich Bucherer's book of 1903, *Elemente der Vektor-Analysis mit Beispielen aus der theoretischen Physik*,[3] was the first book on the modern system of vector analysis published in Germany. Bucherer, who was born in 1863, was in 1903 a Privatdozent at Bonn. He had in 1897 published a book relating to electricity[27] and was to publish another in 1904.[28] It is thus probable that his interest in vector analysis stemmed from his interest in Maxwellian electrical theory. Further support for this speculation is derived from the following statement from the foreword of his book: "In regard to the form of presentation I have on the whole followed Heaviside and used the same symbols as did A. Föppl in his excellent work 'Einführung in die Maxwellsche Theorie.' . . ." (3; IV) That Bucherer did not follow Föppl in all things is shown by the following statement: "I have nevertheless believed it to be expedient to make full use of Grassmann's association of vectors with surfaces. This serves to simplify the derivations of many theorems." (3; IV) At one point he also mentioned Wilson's recent book. (3; 12) Bucherer's book covered the major topics in vector analysis (addition, multiplication, and differentiation of vectors, potential theory, and the transformation theorems) with the exclusion of the linear vector function. It offered a number of applications chosen especially from mechanics, hydrodynamics, and electricity.

The reception given his book is indicated by the fact that a second edition appeared two years later.[29] In the foreword to this edition Bucherer stated: "When I wrote the first edition of this small work, the discussions and deliberations concerning a uniform symbolism for vector analysis were still in flux. Since that time through the adoption of a suitable method of designation by those working on the *Encyklopädie* an important system of symbolism has been put forward." [30] The encyclopedia referred to was the famous *Encyklopädie der mathematischen Wissenschaften* which was appearing in those years and which will be discussed later. Other than the change in notation, no major alterations were made for the second edition.

Thus Bucherer produced the first separately published treatment of modern vector analysis written in German; however in the year (1905) of the publication of his second edition two other such books were published, all three coming from the same publishing house, that of B. G. Teubner.

Dr. Richard Gans' *Einführung in die Vektoranalysis mit Anwendungen auf die mathematische Physik* [4] was published in 1905. Gans was born in 1880 and by 1905 had become Privatdozent at the University of Tübingen. Since his inaugural dissertation dealt with electricity [31] and since he published major writings on electrical theory in 1906 [32] and in 1908,[33] it is probable that Gans also came to vector analysis through electricity. Further support for this conjecture as well as other interesting information is given in the following quotation from his foreword:

> Since however through the development of the electrodynamics of moved bodies and of the theory of electron more demands are constantly being placed on readers in relation to their command of the methods of vector analysis, it seemed to me not out of place to write a book which would be aimed at fullfilling the needs of these branches of science, for it cannot be questioned that the important results attained in the above-mentioned fields will not safely be surveyed by many persons, since they are not sufficiently familiar with the mathematical techniques involved. It is also certain that those who have the intention of assimilating thoroughly and with the greatest possible ease the newer literature and that which is yet to appear, or of working independently in the field of theoretical electricity, must above all provide for the proper tools, i.e., for a knowledge of vector analysis. (4; v)

The notation used by Gans was based mainly on the system used by Lorentz and Abraham in the *Encyklopädie der mathematischen Wissenschaften.*[34] Gans treated such topics as vector addition and multiplication, the differential and integral calculus of vectors, the transformation theorems, and curvilinear co-ordinates; and he gave numerous applications to mechanics, hydrodynamics, and above all electrodynamics. Gans' book, like Bucherer's, was a short work composed primarily for physicists and engineers, with no discussion of such topics as Möbius', Grassmann's, or Hamilton's system. Like Bucherer's book, Gans' book met with success, for a second edition appeared in 1909.[35] The only major change introduced in the second edition was the inclusion of a seventeen-page chapter on tensors and the linear vector function.

The second vectorial book published in Germany in 1905 was Eugen Jahnke's *Vorlesungen über die Vektorenrechnung mit Anwendungen auf Geometrie, Mechanik und mathematische Physik.*[5] Jahnke (1863–1921) in 1905 held the academic position (as indicated on his title page) of "Etatsmässiger Professor an der Königl. Bergakademie zu Berlin." Unlike Föppl, Bucherer, and Gans, but like Wilson, Jahnke was primarily a mathematician; this is indicated by the nature of his other publications and from the approach taken in his vector analysis book.

In one sense Jahnke's book should not, despite its title, be included in this study of the first books presenting modern vector analysis, for its approach was primarily Grassmannian. The other books published in this period and written directly within the Grassmann tradition have been excluded on the grounds that they differ from modern books on vector analysis in regard to a number of fundamental ideas (for example, a slightly different definition of the vector product), and that their contents include many mathematical ideas (for example, point analysis) which are not found in most modern vector analysis books. Jahnke's book merits discussion since its origins lie in both the Gibbs-Heaviside tradition and the Grassmannian tradition. Jahnke, though primarily influenced by Grassmann's ideas, was also influenced by, and took advantage of, the ideas of the Hamilton-Tait-Maxwell-Heaviside tradition.

Jahnke began his foreword by quoting from a letter written by Gauss in 1843 in which Gauss, referring to Möbius' barycentric calculus, suggested that all new calculi are accepted only slowly, and accepted only when it is seen that they provided methods for dealing with problems too complicated for traditional methods. After referring to the slowness of the acceptance of vector analysis, Jahnke stated: "However the results which vector methods have recently achieved in the theory of electrons and in regard to the question of an electromagnetic foundation for mechanics have brought about a change." (5; III–IV) Jahnke then suggested that a major cause of the hesitation of the mathematical world was to be ascribed to the division in the historical development between the traditions stemming from Hamilton and from Grassmann. Jahnke proceeded to compare these traditions. After stating that some have argued that the Hamilton-Heaviside line of development is suited for physical applications but of limited value for geometrical applications, Jahnke revealed his own feelings by the statement: "On the other hand the direction established by Grassmann allows applications to geometry in the broadest sense of the word as well as to mathematical physics." (5; IV)

After describing the contents of his book, Jahnke briefly mentioned his system of symbolism. His symbolism was of course symbolic of his approach and was derived primarily from Grassmann but also partly from the tradition that led to Föppl's book. Jahnke mentioned sixteen men whose works had influenced his ideas; heading this list was Grassmann; mentioned in the list were such men as Schlegel, Kelland, and Tait, Hyde, Peano, Bucherer, Abraham, Föppl, and Gans; noteworthy by their absence from the list were the names of Hamilton, Gibbs, Heaviside, and Wilson.[36]

Jahnke's book was divided into two sections, the first was entitled "Vectors in the Plane," and the second "Vectors in Space." The latter section took up slightly less than two-thirds of the text. The first section began with a chapter on the addition and subtraction of point, i.e., the Möbius-Grassmann point analysis from the Grassmannian point of view. In the next two chapters Jahnke introduced free vectors and line-bound vectors (defined, as Grassmann had done, through the subtraction and multiplication of points respectively). In these first three chapters, as well as later, Jahnke presented numerous applications to physics and geometry; his intent was clearly to interest scientists and engineers as well as mathematicians in the contents of his book. The next three chapters dealt with vector multiplication and applications thereof to physics; the products defined were (1) the inner or scalar (dot) product common to modern vector analysis and to Grassmannian analysis, (2) the Grassmannian outer product which is (as discussed earlier) similar to the modern vector product, and (3) the Grassmannian regressive product which has no correlate in modern vector analysis.

Thus ended the first section; the second section dealt with the same topics as the first section but developed for three-dimensional space and with special attention to point relations. In addition two chapters on vector differentiation, the differential operator ∇, and a short section on tensors were included. It was especially in these two chapters that the influence of the Hamilton-Maxwell-Tait-Heaviside tradition was evident, for terms (and ideas) such as "Gradient," "Curl," "Divergenz," and "lineare Vektorfunktion" appeared, as well as applications to topics in physics such as Maxwell's equations and electron theory. To sum up, Jahnke's book was primarily, but not entirely, within the Möbius-Grassmann tradition. Numerous physical applications were included in the hope of interesting those who were in the process of adopting a system of vector analysis for physical application.

I know of four major reviews of Jahnke's book that shed light on the nature of the reception accorded it. The view of the book taken by E. B. Wilson writing in the *Bulletin of the American Mathematical Society* is summed up in the following statement: "These lectures are really lectures on multiple algebra and form an excellent introduction to the subject—[but] no more than an introduction. . . ." [37] In short, Wilson viewed it as a worthwhile book for the mathematician, but misleading in its title. Jules Tannery [38] writing from France, and O. Staude [39] writing in German, praised the book and described its contents in some detail. Emil Müller, who was a Grassmann enthusiast, criticized the work as giving a poor picture of Grassmann's ideas.[40] Much light is also shed on its reception by the fact that whereas Bucherer's book attained to a second edition, Gans' book to a seventh (1950) and to translations into English and Spanish, and Wilson's book had both a second edition and a major paperback reprinting, Jahnke's book was never reprinted, republished, or translated.[41] Thus Jahnke's book may be viewed as a carefully conceived but unsuccessful attempt to bring about the adoption of a somewhat revised Grassmannian system.

In 1906 Gibbs' original treatise on vector analysis was published as part of his collected works.[6] Published in the same volume were his six papers relating to vector analysis. If narrowly considered, this barely constitutes the publication of a vector analysis book; however if broadly considered, it constitutes an important publication for a number of reasons. First, since Gibbs' four polemical articles, his "Multiple Algebra" paper, and his presentation of a vectorial method for computing orbits were included in the volume, it offered the reader an array of important and valuable materials. Second, by 1906 Gibbs had become quite famous, certainly more famous than any other writer of a vector analysis book published before 1910. Thus it is probable that a substantial number of readers turned to his book and were influenced thereby. In one sense Gibbs' short treatise supplemented Wilson's longer treatise and provided the public with a compact treatment of vector analysis.

Pavel Osipovich Somoff (or Somov) published in 1907 one of the best of the early books on vector analysis; [7] with his book the modern system of vector analysis made its first appearance in Russia. Somoff's interest in vectorial analysis probably stemmed from his father, Joseph Somoff (Ossip Ivanovich Somoff, or Somov) [42] and from his own interests in mechanics and mathematics, which is witnessed by his publication of roughly twenty papers on these two subjects in Russian and German journals from the 1880's to 1907.

The more direct sources of Somoff's work on vector analysis are clear from the following quotation taken from the foreword of his book:

> In the study of this subject, it is possible to discern two directions, an older one connected with the names Grassmann ("Die lineale Ausdehnungslehre," 1844) and Hamilton ("Lectures on Quaternions," 1853). The other direction, which is more recent, appeared at the same time as the study of vector fields and developed through the work of Maxwell, Heaveside [sic], Gibbs, Föppl, and others. . . . We will hold to this second approach in our presentation of vector analysis. (7; iii–iv)

Somoff's book dealt with both the elementary and the advanced parts of vector analysis, including the linear vector function. It incorporated many examples selected primarily from mechanics, as well as brief explications of other systems of space analysis, such as those of Möbius and Hamilton. Somoff seems to have been acquainted with the majority of the existing works on vector analysis, and in the course of his book he mentioned such names as Resal, Tait, Heaviside, Henrici and Turner, Gibbs, Wilson, Fischer, Föppl, Bucherer, Gans, and Jahnke; correspondingly his presentation was eclectic, with the Gibbs-Wilson symbolism and techniques being the most frequently used. Thus with Somoff vector analysis entered Russia—and entered in a very respectable manner.

Siegfried Valentiner's *Vektoranalysis*[8] was published in 1907. Valentiner was born in 1876 and had by 1907 become Privatdozent for physics at the University of Berlin. Being a physicist, he wrote his book primarily for physicists, as Bucherer and Gans had done earlier. Thus nearly half of the book was devoted to the discussion of applications of vector analysis to electricity and mechanics. However Valentiner, unlike Bucherer and Gans, included in his book a treatment of the linear vector function presented from the Gibbsian point of view. Valentiner was in general more eclectic in his presentation than Bucherer or Gans, and his presentation, while decidedly influenced by the Heaviside-Föppl tradition, owed much to the Gibbs-Wilson tradition. Valentiner's book must have been well received, for a second edition appeared in 1912.[43]

Cesare Burali-Forti (1861–1931) and Roberto Marcolongo (1862–1943) published in 1909 their *Elementi di calcolo vettoriale con numerose applicazioni alla geometria, alla meccanica e alla Fisica-Matematica.* In the following year a French translation of this book was published.[9] At this time Burali-Forti, who was primarily a mathematician, was a professor at the Military Academy of Turin. Marcolongo was a physicist and professor of rational mechanics at the University of Naples.

Burali-Forti's first book relating to vectorial analysis was his 1897 *Introduction à la géometrie différentielle suivant la méthode de H. Grassmann.* It is highly probable that Burali-Forti's interest in Grassmann stemmed from another citizen of Turin who was an early advocate of Grassmann's system. This was the professor of infinitesimal calculus at the Royal University of Turin, Giuseppe Peano (1858–1932). Peano had received his doctorate in mathematics in 1880, and in 1887 he published a book that included much discussion of Grassmann's ideas; this was his *Applicazioni geometriche del calcolo infinitesimale.* This was followed by his *Calcolo geometrica secundo l'Ausdehnungslehre di H. Grassmann* (1888) and by a number of other works, including his famous work on the foundations of mathematics, *Formulaire de Mathématiques,* one part of which was devoted to vectors. Peano thus was an important proponent of the Grassmannian system, and in addition was a man who had the perspective provided by a knowledge of other systems, such as that of Hamilton.[44]

After his first Grassmannian publication of 1897 Burali-Forti continued to publish on vector analysis from the Grassmannian point of view. One of his articles (published with Marcolongo) deserves special mention in that the controversy provoked by it was both typical of the times and influential on the presentation and reception of their books of 1909–1910. There was during the first decade of this century a heated debate of considerable magnitude concerning the best symbolism for vector analysis.[45] A major cause of this debate and its intensity was the fact that the symbols used symbolized the essential mathematical differences between the still-contending systems—the Gibbs-Heaviside, the Grassmannian, the Hamiltonian, and others.[46]

At the Naturforscherversammlung at Kassel in 1903 a commission was set up to deal with the notation question. According to Felix Klein the only result of the activity of the commission was that "about three new notations came into existence!"[47] The Fourth International Congress of Mathematicians was scheduled for Rome in 1908, and in preparation for this meeting Burali-Forti and Marcolongo published an extensive study of the origins of the various notations. They also recommended the establishment of a new system of symbolism,[48] but the only major result of their recommendation seems to have been that mathematicians reviewed their books (which used the proposed symbolism) more in terms of their symbolism than their content.

In the preface to their book Burali-Forti and Marcolongo made the following statement which supports the conclusions given above concerning their interest in Grassmann's approach:

> Knowledge of vectorial methods is already being forced not only upon physicists and electricians but also on those who work in pure mathematics. May this book, prepared in the year when Germany celebrates the first centenary of the birth of Grassmann, contribute to spread and make known the methods of this great mathematician, an end on behalf of which we have labored with faith and patience for many years; may it crown the work begun among us with so much talent by Mr. PEANO! (9; v)

The authors divided their book into two parts: the first presenting vector methods and the second dealing with applications. The first two chapters were devoted to vector addition and the barycentric calculus. The title of their first chapter is symbolic of much else; it is "Prodotto vettoriale e intorno." Thus their scalar (dot) product was named in accordance with the Grassmannian tradition, whereas their vector (cross) product was named and defined in accordance with the Hamilton-Heaviside-Gibbs tradition, the product being another vector rather than a directed area. The final three chapters in the first part dealt with rotations in a plane (for which they introduced a new and controversial operator) and the differential calculus of vectors, including the operator ∇ (the symbol and usual presentation of which they rejected). In the second part, applications to geometry, mechanics, hydrodynamics, elasticity theory, and electrodynamics were given along with the development of the transformation theorems.[49] In the Italian edition a section of historical notes was appended at the end, and in the French edition the historical section was expanded, while some philosophical ideas as well as presentations of both quaternions and the original Grassmannian system were added.

Since many of the book reviews were unfavorable[50] and no second editions appeared,[41] it may be concluded that their books were not well received. The books of Burali-Forti and Marcolongo are, like Jahnke's book, especially interesting as representing partial departures from the Grassmannian system toward the system finally accepted. As compared to Jahnke's book, their books were somewhat more removed from the Grassmannian tradition than his; but because of their tendency toward the inclusion of original symbolisms and ideas, the authors cannot be viewed as nearer to the system that was eventually accepted than Jahnke.

Joseph George Coffin's book *Vector Analysis: An Introduction to Vector Methods and Their Various Applications to Physics and Mathematics*[10] was published in 1909. Coffin was born in 1877, and, after spending four years in Europe (Switzerland and France), he entered Massachusetts Institute of Technology in 1894 and graduated with a B.S. in 1898. Coffin completed his Ph.D. in physics at Clark University in 1903, and by 1909 he had taught physics at both Clark and the City College of New York.

His book requires little discussion; it may be described by recalling that Wilson corresponded with Gibbs about the advisability of publishing a shorter and more elementary work following the same format as Wilson's book. Wilson never published such a book; Coffin did. Coffin followed the Gibbs-Wilson tradition in both symbolism and in methods, though his explanations and proofs were less full and rigorous and his contents somewhat less inclusive. He did include all the major topics in vector analysis up to the linear vector function, which he treated rather briefly. The last two chapters, which compose well over a third of the book, were devoted to applications to electrical theory and to mechanics; and numerous applications were also given in the earlier chapters. Coffin, like nearly all the writers of works on vector analysis from the latter half of the first decade of this century, was acquainted with the majority of the earlier books. The reception accorded his book is indicated by the fact that a second edition was published in 1911 (which has been reprinted a large number of times and is still in print) and a French translation appeared in 1914.[10]

W. V. Ignatowsky published his book on vector analysis, *Die Vektoranalysis und ihre Anwendung in der theoretischen Physik,*[11] in two parts in 1909 and 1910. Dr. Ignatowsky had written an earlier book, entitled *Solution of Some Problems of Electrostatics and Electrodynamics with the Help of Vector Analysis,* published in 1902 and written in Russian. From the title of this book it seems very probable that his interest in vector analysis came from his interest in electricity, on which subject he had also published a number of papers in German journals.

Ignatowsky's vector analysis book was in the tradition established by Bucherer, Gans, and Valentiner, although his book was somewhat longer than theirs. His notation stemmed from Föppl, Lorentz, and Abraham; and his bibliography listed works by Abraham, Bucherer, Föppl, Gans, "Gibbs-Wilson," Jahnke, Jaumann,[51] and Valentiner. As he stated in his foreword, the book was intended for physicists (11,I; iii), and hence the entire second part dealt with applications. In his first part he presented the algebra and calculus of vectors, including special discussions of types of fields and curvilinear co-ordinates, and he concluded with a section on tensors. A number of his methods of presentation differed from the ordinary methods, for example, he made every effort to avoid decomposing vectors into their i, j, k components in proving theorems, and he defined the vector (cross) product through an integration. The applications in the second part were to nearly all the branches of physics where vector analysis was applicable. The book must have been well received, since by 1926 a third edition had appeared.

The final work to be discussed is, not a book on vector analysis, but the famous and influential *Encyklopädie der mathematischen Wissenschaften mit Einschluss ihrer Anwendungen.* The fact, form, and extent of the vectorial methods included in the *Encyklopädie* influenced the developments in the first decade of the twentieth century. The publication of the *Encyklopädie* began around 1898 and extended to 1935; a French translation and revision was begun about 1904.

Included in the part of the *Encyklopädie* devoted to geometry were extensive treatments of both the Grassmannian and the Hamiltonian systems, as well as shorter treatments of such systems as that of Möbius, but these sections were published only after 1915.[12] For the part of the *Encyklopädie* dealing with mechanics H. E. Timerding wrote a section entitled "Geometrical Basis of the Mechanics of a Rigid Body," which he finished in February, 1902, and in which he gave an elementary presentation of the Grassmannian system.[12] Max Abraham also wrote for the part of the mechanics section dealing with deformable bodies a section entitled "Fundamental Geometrical Concepts," which he finished in April, 1901.[12] Abraham included in this a presentation of vector analysis as applied to various types of fields. His presentation partly followed Grassmann and partly Gibbs and Heaviside. It is interesting to note that in 1899 Abraham wrote to Gibbs asking him to write a presentation of vector analysis for the mechanics section of the *Encyklopädie*.[52] Gibbs did not fulfill his request. Others made limited use of vector methods in the mechanics sections; more extensive use however is found in the section of the *Encyklopädie* entitled "Physik."

One of the major parts of this section on physics was entitled "Electricity and Optics." The first two parts of this section were mainly explanations of Maxwell's electrical ideas. These were written by the great Dutch physicist Hendrik Antoon Lorentz (1853–1928), and the Heaviside-Gibbs form of vector analysis was used throughout. Lorentz had finished these sections by late 1903.[12] He had used vector analysis at least as early as 1895, when his book *Versuch einer Theorie der elektrischen und optischen Erscheinungen in bewegten Korpern* had appeared. It is also known that he received a copy of Gibbs' original writing on vector analysis in 1884.[53] The section following on Lorentz' sections was entitled "Electrostatics and Magnetostatics"; its author was Richard Gans, who finished it in 1906.[12] Here and in the later sections on electricity by Lorentz, Abraham, and others, vector methods were extensively used. This fact must have led many scientists to an interest in vector analysis, for it seems probable that there were few parts of the *Encyklopädie* more carefully and more widely read than this section on electricity.

III. *Summary and Conclusion*

Since the detailed study of the reception of the modern system of vector analysis from 1894 to 1910 is now concluded, it will be profitable to set some of the facts within sharper perspective and to subsume them where possible into generalizations reflective of the times.

First of all, it should not be assumed that these were the only books presenting vectorial systems that were published at this time. Books on the quaternion system continued to appear; thus, for example, Hamilton's *Elements of Quaternion* was republished with annotations by Charles Jasper Joly in two volumes appearing in 1899 and 1901. Joly also published a book in 1905 entitled *Manual of Quaternions,* and in 1904 Knott edited and published the third edition of *Introduction to Quaternions* by Kelland and Tait. Similarly the Grassmannian tradition was represented to some degree in the book of Burali-Forti and Marcolongo and in that of Jahnke, as well as such books as Joseph V. Collins' *An Elementary Exposition of Grassmann's Ausdehnungslehre* (1901) and Edward Wyllys Hyde's *Grassmann's Space Analysis* (1906). Macfarlane's system was presented in his 1906 publication of *Vector Analysis and Quaternions.* None of the above however seems to have required a second edition.[54]

Thus in the first decade of this century other systems of vectorial analysis besides the modern system were in use, and books were published on them. Yet the rate of publication of books on the Grassmannian and the Hamiltonian systems was less during the interval 1901–1910 than during the interval 1891–1900, and the rate was to decrease further after 1910. That neither tradition has passed away even now is indicated by the publication of books within each tradition in the 1940's and 1950's.

Whereas interest in books presenting the Grassmannian forms of vectorial analysis decreased, interest in the books in the Gibbs-Heaviside tradition increased. This is evidenced by the following table in which some of the publication history of books in the Gibbs-Heaviside tradition is presented.

Author of Book and Year of First Publication	Some Translations and Republications
Föppl, 1894	2nd ed., 1904 3rd ed., 1907 4th ed., 1912 1st English ed., 1932 16th German ed., 1957
Wilson, 1901	2nd ed., 1900 8th reprinting, 1943 Paperback reprinting, 1960

Author of Book and Year of First Publication	Some Translations and Republications
Bucherer, 1903	2nd ed., 1905
Gans, 1905	2nd ed., 1909 3rd ed., 1913 1st English ed., 1931 2nd Spanish ed., 1940 7th German ed., 1950
Gibbs, 1906	Paperback reprinting, 1961
Valentiner, 1907	2nd ed., 1912 7th ed., 1950 Reprint of 7th ed., 1954
Coffin, 1909	2nd ed., 1911 1st French ed., 1914 2nd ed., 6th impression, 1923 9th reprinting, 1959
Ignatowsky, 1909–1910	3rd ed., 1926

From this table three generalizations may be made: first, the books listed were indeed in the tradition that survives and flourishes at present; second, many of these books were to play a part in the history of vector analysis after 1910, and indeed most are still in print; third, the books were widely read from 1895 to 1915, which is indicated by the fact that new editions were so frequently necessary.

The books in the Grassmannian tradition differed from these books in two ways: first, the Grassmannian books contained some ideas that varied from corresponding ideas in vector analysis books —for example, the Grassmannian outer product—and, second, the Grassmannian books contained many ideas and methods for which there was no correlate in the vector analysis books, for example, the regressive product and the system of point analysis. These differences seem to have been sufficient to deter many readers, for none of the books using the Grassmannian approach or the modified Grassmannian approach (Jahnke, Burali-Forti and Marcolongo) required second editions.

Thus it may be concluded that the Grassmannian tradition played no major role in the *acceptance* of the modern system of vector analysis. It was shown previously that the Grassmannian tradition played no major role in the *development* of the Gibbs-Heaviside system, that is, the modern system of vector analysis. These two conclusions are particularly striking when they are set beside two other conclusions: the Grassmannian system *could have* played a major role in both the *development* and the *acceptance* of vector analysis.

The traditions that *did* play a major role in the acceptance of the modern form of vector analysis are the following. The tradition that began with Gibbs and led to the books of Wilson and Coffin played a large part. It should be noted however that Wilson's interest had roots directly in the quaternion system and that Coffin as a physicist probably attained some of his interest and information from the Heaviside tradition. The same remark applies in stronger form to P. O. Somoff. The most influential tradition was that which had its beginning in Heaviside and in Heaviside's development of Maxwell's ideas. The source of the interest and information attained by Lorentz, Föppl, Abraham, Bucherer, Gans, Valentiner, and Ignatowsky has been traced to Heaviside and located specifically in his association of modern vector analysis with Maxwell's electrical ideas.

This fact is surprising in at least one way, for Gibbs' presentation was fuller than Heaviside's and included ideas that were original with Gibbs and more or less unique in his system. Furthermore, on the criterion that what has survived is what was fittest, Gibbs must be given credit for having developed a better notation than Heaviside. The conclusion that the Heaviside tradition was more influential than the Gibbsian tradition must also be qualified by the statement that many who became interested in vector analysis within the Heaviside tradition were hence led to the Gibbsian tradition and profited from the greater riches therein.

Finally it may be noted that the vast majority of the authors of the books presenting the modern form of vector analysis were physicists. This is appropriate in that the great future for vector analysis lay in physical science; at present nearly all books on electricity and mechanics use vector analysis, and it appears not infrequently in books on optics and heat conduction. It is also used in many parts of modern physics, and its applications for the engineer are legion. Vector analysis has been of great value to the geometer, but geometers are few in number among modern mathematicians. Such mathematical creations as matrices, vector spaces, groups, and fields are associated only indirectly with vector analysis in the traditional sense. In many cases however their roots extend back historically to the broad stream of development that culminated in the first decade of this century.

Notes

[1] August Föppl, *Einführung in die Maxwell'sche Theorie der Elektricität* (Leipzig, 1894).

[2] Edwin Bidwell Wilson, *Vector Analysis: A Text Book for the Use of Students of Mathematics and Physics Founded upon the Lectures of J. Willard Gibbs* (New York, 1901). References are to the reprint of the slightly revised second edition (New York, 1960).

[3] Alfred Heinrich Bucherer, *Elemente der Vektor-Analysis mit Beispielen aus der theoretischen Physik* (Leipzig, 1903).

[4] Richard Gans, *Einführung in die Vektoranalysis mit Anwendungen auf die mathematische Physik* (Leipzig, 1905).

[5] Eugen Jahnke, *Vorlesungen über die Vektorenrechnung mit Anwendungen auf Geometrie, Mechanik und mathematische Physik* (Leipzig, 1905).

[6] Josiah Willard Gibbs, *Elements of Vector Analysis Arranged for the Use of Students in Physics* in Gibbs, *Scientific Papers of J. Willard Gibbs,* vol. II (London, 1906), 17–90.

[7] Pavel Osipovich Somoff, *Vector Analysis and Its Applications* (in Russian), (Saint Petersberg, 1907).

[8] Siegfried Valentiner, *Vektoranalysis* (Leipzig, 1907).

[9] Cesare Burali-Forti and Roberto Marcolongo, *Elementi di calcolo vettoriale con numerose applicazioni alla geometria, alla meccanica e alla Fisica-Matematica* (Bologna, 1909). French translation: *Eléments de calcul vectoriel avec de nombreuses applications à la géométrie, à la mécanique et à la physique mathématique,* trans. S. Lattes (Paris, 1910).

[10] Joseph George Coffin, *Vector Analysis: An Introduction to Vector Methods and Their Various Applications to Physics and Mathematics* (New York, 1909). French translation: *Calcul Vectoriel,* trans. Alex Veronnet (Paris, 1914).

[11] W. v. Ignatowsky, *Die Vektoranalysis und ihre Anwendung in der theoretischen Physik,* Teil I (Leipzig and Berlin, 1909); Teil II (Leipzig and Berlin, 1910).

[12] *Encyklopädie der mathematischen Wissenschaften mit Einschluss ihrer Anwendungen* (Leipzig, 1898–1935). The following articles in the *Encyklopädie* are especially important for the history of vector analysis: (1) Hermann Rothe, "Systeme geometrischer Analyse, Erster Teil," vol. III, pt. I (Leipzig, 1914–1931), 1277–1423; (2) Alfred Lotze and Chr. Betsch, "Systeme geometrischer Analyse, Zweiter Teil," vol. III, pt. I (Leipzig, 1914–1931), 1425–1595; (3) H. E. Timerding, "Geometrische Grundlegung der Mechanik eines starren Körpers," vol. IV, pt. I (Leipzig, 1901–1908), 125–189; (4) Max Abraham, "Geometrische Grundbegriffe," vol. IV, pt. 3 (Leipzig, 1901-1908), 3–47; (5) Hendrik Antoon Lorentz, "Maxwells elektromagnetische Theorie" and "Weiterbildung der Maxwellschen Theorie. Elektronentheorie," vol. V, pt. II (Leipzig, 1904–1922), 63–280; (6) Richard Gans, "Elektrostatik und Magnetostatik," vol. V, pt. II (Leipzig, 1904–1922), 289–349.

[13] Edwin Bidwell Wilson, "Reminiscences of Gibbs by a Student and Colleague" in *Scientific Monthly, 32* (1931), 211–227.

[14] By a "major publication in vector analysis" I mean a relatively long work presenting a system of vectorial analysis similar or identical to the now-common system.

The presentation had to include exposition of some of the advanced parts of vector analysis and be published as part or all of a book. Thus all expositions presented in journals have been excluded. The majority of the works to be discussed are books on vector analysis. The two exceptions are the works listed in notes (1) and (12) above. These two works *included* extensive presentations of vector analysis and were especially influential. Five books that a reader knowledgeable in the vectorial literature of the period might expect to be discussed have not been treated. These are (1) Alfred North Whitehead, *A Treatise on Universal Algebra*, vol. I (only volume published), (Cambridge, England, 1898); (2) O. Henrici and G. C. Turner, *Vectors and Rotors* (London, 1903?); (3) Victor Fischer, *Vektordifferentiation und Vektorintegration* (Leipzig, 1904); (4) William Cramp and Charles F. Smith, *Vectors and Vector Diagrams* (London, 1909); (5) C. Fortin, *Théorie et applications elêmentaires des vecteurs* (Paris, 1910).
Whitehead's book was aimed at presenting "a thorough investigation of the various systems of Symbolic Reasoning allied to ordinary Algebra." Hence the majority of the work dealt with material extraneous to vector analysis, and that part dealing with vector analysis was primarily based on Grassmann. Henrici's book was a very elementary work written mainly under the influence of Heaviside. Fortin's book was also very elementary and was in the tradition begun by Bellavitis. Fischer's book was a specialized monograph and, according to the reviews, one of very poor quality. The Cramp-Smith book was an exposition of Steinmetz' methods for treating alternating currents by means of complex numbers and elementary vector analysis.

[15] I have of course read many of these journal publications, and in all cases they support the conclusions arrived at through the more systematic mode of analysis.

[16] For a table which gives the systems of symbolism used by nearly all the authors discussed in this chapter see James Byrnie Shaw, "Comparative Notation for Vector Expressions" in *Bulletin of the International Association for Promoting the Study of Quaternions and Allied Systems of Mathematics* (1912), 18–29.

[17] August Föppl, *Theorie der Elektricität*, ed. Max Abraham (Leipzig, 1904). The third edition appeared in 1907 and a sixteenth edition appeared in 1957.

[18] Felix Klein, *Vorlesungen über die Entwicklung der Mathematik im 19. Jahrhundert*, pt. 2., ed. R. Courant and St. Cohn-Vossen (New York, 1956), 47.

[19] A fourth edition appeared in 1911, and a fifteenth in 1951.

[20] A German translation of Ferraris' book appeared in 1901; Galileo Ferraris, *Wissenschaftliche Grundlagen der Elektrotechnik*, trans. Leo Finzi (Leipzig, 1901).

[21] Gibbs repeatedly insisted that his short printed work of 1881–1884 was not "formally published." It had in any case a very small printing.

[22] Alexander Ziwet, "[Review of] *Vector Analysis.* By Edwin Bidwell Wilson" in *Bulletin of the American Mathematical Association*, 8 (1901–1902), 207–215.

[23] Anonymous, "[Review of] Gibbs (I.-W.) et Wilson (E.-B.).—Vector analysis" in *Bulletin des sciences mathématiques*, 2nd Ser., 26 (1902), 21–30.

[24] Victor Schlegel, "[Review of] E. B. Wilson, *Vector Analysis*" in *Jahrbuch über die Fortschritte der Mathematik*, 33 (1902), 96–97.

[25] Cargill Gilston Knott, "[Review of] *Vector Analysis.* By Dr. Edwin Bidwell Wilson" in *Philosophical Magazine*, 6th Ser., 4 (1902), 614–622.

[26] The second edition was in its eighth reprinting in 1943, and a paperback reprint appeared in 1960.

[27] Alfred Heinrich Bucherer, *Grundzüge einer thermodynamischen Theorie electrochemischer Kräfte* (Freiberg, 1897).

[28] Alfred Heinrich Bucherer, *Mathematische Einführung in die Elektronentheorie* (Leipzig, 1904).

[29] Alfred Heinrich Bucherer, *Elemente der Vektor-Analysis mit Beispielen aus der theoretischen Physik,* 2nd ed. (Leipzig, 1905).

[30] *Ibid.*, V.

[31] Richard Gans, *Über Induction in rotierenden Leitern* (Leipzig, 1902), 33 pp.

[32] Richard Gans, "Elektrostatik und Magnetostatik" in *Encyklopädie der mathematischen Wissenschaften,* vol. IV, pt. 2 (Leipzig, finished in 1906), 289–249.

[33] Richard Gans, *Einführung in die Theorie des Magnetismus* (Leipzig and Berlin, 1908).

[34] For a comparison see Shaw, "Comparative Notation," 18–29.

[35] Richard Gans, *Einführung in die Vektoranalysis mit Anwendungen auf die mathematische Physik,* 2nd ed. (Leipzig and Berlin, 1909). It may be noted that a third edition came in 1913, a seventh in 1950, and translations into English and into Spanish in the interim.

[36] Hamilton, Gibbs, and Heaviside are mentioned in the book itself, but in such a way that there is no indication that Jahnke had read their works. The ideas of Hamilton and Heaviside of course influenced him (indirectly through their followers).

[37] Edwin Bidwell Wilson, "[Review of] *Vorlesungen über die Vektorenrechnung.* Von E. Jahnke" in *Bulletin of the American Mathematical Society,* 2nd Ser., *12* (1905–1906), 354.

[38] Jules Tannery, "[Review of] Gans (R.)—*Einführung in die Vektoranalysis . . .* Jahnke (E.)—*Vorlesungen über die Vektorenrechnung*" in *Bulletin des sciences mathématiques,* 2nd Ser., 29 (1905), 318–322.

[39] O. Staude, "[Review of] E. Jahnke, *Vorlesungen über die Vektorenrechnung*" in *Archive der Mathematik und Physik, 11* (1907), 268–275.

[40] Emil Müller, "*Vorlesungen über die Vektorenrechnung.* Von Dr. E. Jahnke" in *Monatshefte für Mathematik und Physik, 17* (1906), 56–57.

[41] I have based this statement on searches of such works as the published catalogues of the Library of Congress, British Museum, and Bibliothèque Nationale.

[42] J. Somoff was discussed previously in connection with Henri Resal. I have asserted that P. O. Somoff is the son of J. Somoff on the basis of the following facts: (1) Osipovich means "son of Joseph" and (2) J. Somoff had a son Pavel who worked in mechanics; concerning this see Alexander Ziwet's foreword to J. Somoff, *Theoretische Mechanik,* vol. I (Leipzig, 1878), v–vi.

[43] In 1954 the seventh edition of Valentiner's book was printed for a second time.

[44] In the preface to his *Calcolo infinitesmale* (Turin, 1887), v–vi, Peano mentioned, besides Hamilton and Grassmann, Möbius, Bellavitis, Resal, and J. Somoff.

[45] I would estimate that over twenty papers were devoted entirely to the notation question. If however papers that were only partly or indirectly concerned with the question were to be included, the number would probably triple. Many of the book reviews of the vector analysis books published at that time were little more than attacks on notation.

[46] For example, Alexander Macfarlane continued to defend his system and symbols. Also within each tradition listed above there were different symbolisms; thus Heaviside's symbolism differed from Gibbs', and Grassmann's earlier symbols differed in some cases from his later ones.

[47] Felix Klein, *Elementary Mathematics from an Advanced Standpoint: Arithmetic. Algebra. Analysis,* trans. E. R. Hedrick and C. A. Noble (New York, n.d.), 65.

[48] Cesare Burali-Forti and Roberto Marcolongo, "Per l'unificazione delle notazioni vettoriale" in *Rendiconti del Circolo Matematico di Palermo, 23* (1907), 324–328; *24* (1907), 65–80 and 318–332; *25* (1908), 352–375; *26* (1908), 369–377. See also Roberto Marcolongo, "Per l'unificazione delle notazioni vettoriali" in *Atti del IV*

Congresso Internazionale dei matematici. Roma (1908), vol. III (Rome, 1909), 191–197. The controversy is discussed in greater detail in Florian Cajori, *A History of Mathematical Notations*, vol. II (Chicago, 1952), 136–139.

[49] In 1909 Burali-Forti and Marcolongo published a short book in which they dealt with some of the more advanced topics relevant to vector analysis; this was their *Omografie vettoriali con applicazioni alle derivate rispetto ad un punto e alla matematica* (Turin, 1909).

[50] See for example Edwin Bidwell Wilson, "The Unification of Vectorial Notations," [Review of C. Burali-Forti and R. Marcolongo], *Elementi di calcolo vettoriale con numerose applicazioni* and *Omografie vettoriale con applicazioni* in *Bulletin of the American Mathematical Society, 16* (1909–1910), 415–436, and G. B. M., "[Review of] *Eléments de Calcul vectoriel.* . . . By Prof. C. Burali-Forti and Prof. R. Marcolongo" in *Nature, 86* (1911), 75.

[51] G. Jaumann, *Die Grundlagen der Bewegungslehre von einen modernen Standpunkt aus dargestellt* (Leipzig, 1905). In this book Jaumann made extensive use of vector methods and followed especially the Gibbs-Wilson tradition at least in notation.

[52] Lynde Phelps Wheeler, *Josiah Willard Gibbs* (New Haven, 1962), 231.

[53] *Ibid.*, 222.

[54] This statement requires a partial qualification. The books of Hyde and Macfarlane had originally appeared as chapters in Mansfield Merriam and Robert Simpson Woodward's *Higher Mathematics*, first published in 1896. This book consisted of eleven chapters surveying parts of higher mathematics and must have been quite popular, since it was in a third edition by 1902. About 1905 it was decided to print the chapters as separate books, and thus the title pages of Hyde's and Macfarlane's books contain the words "Fourth Edition." It seems that these two chapters were not among the most popular, for no "fifth edition" of them appeared.

CHAPTER EIGHT

Summary and Conclusions

This chapter focuses on a summary view of the history of the idea of a vectorial system. As in geography, so in history, a view from afar is at times a clearer view.

Concerning the early history of vectorial concepts it was suggested that the idea of a parallelogram of physical entities was indirectly influential. Although this concept does not necessarily involve the idea of a vector and did not lead directly to any vectorial systems, it did provide the early vectorists with an area in which the usefulness of elementary vectorial methods could convincingly be illustrated. To Leibniz belongs credit for having seen the desirability, and to a limited extent the nature, of a system for the analysis of three-dimensional space. Leibniz had no direct and major effect on the later history of vectorial analysis, but his indirect effect was noteworthy.

A tradition that led directly to Hamilton's discovery of quaternions (and to the less successful searches of others for a system of space analysis) was the complex number tradition. Within this tradition two branches are discernible: the first is that of the representation of complex numbers as ordered pairs of real numbers and the second is that of the representation of complex numbers through geometrical lines. The significance of the first branch was less than that of the second; indeed the first branch was only significant in that it provided Hamilton with an epistemological justification for his "four-dimensional" quaternions. Within the second tradition, Wessel, Argand, Français, Servois, Mourey, J. T. Graves, De Morgan, Bellavitis, Hamilton, and perhaps others sought a three-dimensional vectorial system. Wessel was the first to add vectors in three-dimensional space, and Hamilton was the first to publish an important type of multiplication of entities in three-dimensional space.

Within another tradition, which may be called the geometrical tradition, authors focused on geometrical entities and devised methods for operating directly with them. Möbius and in a restricted sense Bellavitis, Justus Günther Grassmann, Hermann Günther Grassmann, and Saint-Venant may be located within this tradition. An algebraic tradition is discernible within which authors focused on relations between mathematical entities or algebraic forms and hence devised more general meanings for such operations as multiplication. The spirit of this tradition stimulated many authors, above all Grassmann. It was noted that Grassmann's major work was done while he had no knowledge of the geometrical representation of complex numbers. Also intermingled with these traditions was the traditional quest for more natural, more compact, and more powerful mathematical methods for physical science.

These many traditions led to the creation of many systems of vectorial character. It was argued that the fact that a large number of men sought for (and some found) such systems indicates that this quest should be viewed as a definite movement within early nineteenth-century mathematical thought and that the existence of this movement augered well for the eventual widespread acceptance of such a system. On the basis of comparisons with other similar systems and their acceptance it was suggested that the acceptance of vectorial methods did not take an inordinately long time.

Of the many systems created, two were predominant; these were the systems of Hamilton and of Grassmann. If these two systems are compared, and modern vector analysis is taken as the standard of comparison, it is readily apparent that the Hamiltonian and Grassmannian systems have many similarities. Both were vectorial in character; both contained vector addition and subtraction and operations similar to the modern scalar (dot) and vector (cross) product. Hamilton's system however included these two products as parts of the quaternion product; the product of two vectors $\alpha\beta$ was for him composed of a scalar part $S\alpha\beta$ and a vector part $V\alpha\beta$. The former of these is the negative of the modern scalar product; the latter is the modern vector product. Among Grassmann's products the inner product is equivalent to the modern scalar product, while his outer product yields a directed and oriented area rather than another vector.

Both Hamilton and Grassmann dealt with vector differentiation, and both created full-blown systems rich in content and application. Hamilton's system was a vectorial system of such a character that by certain deletions, simplifications, and redefinitions it could be transformed into the modern vector analysis system. Grassmann's system was also in many ways similar to modern vector analysis; there were however two important types of differences. Grassmann's system contained some elements which were definitely not equivalent to their correlates in the modern system, for example, his outer product. Moreover there were many elements in Grassmann's system for which no correlates can be found in the modern system, for example, his point system and his other products. Broadly considered, the structure of his system was very different from the structure of the modern system, for example, vectors were derived from points.

Thus it was concluded that Grassmann's system, like Hamilton's, *could* have led to modern vector analysis through a process of deletion and alteration. Concerning priority in relation to the elements common to these systems Hamilton deserves credit for priority of publication, whereas Grassmann deserves credit for being the first to create his system and the first to publish an extensive exposition of it.

It is worthwhile now to compare these two men, especially in regard to factors influential on the history of their systems. Hamilton was born in 1805 and died in 1865; Grassmann was born in 1809 and died in 1877; they were thus close contemporaries. Hamilton was a prodigy who before he was thirty had made a number of very important discoveries which had brought him a large measure of fame; by comparison Grassmann's youth was undistinguished, and he published his first scientific paper only when he was thirty. Hamilton attained a university professorship while he was still an undergraduate; Grassmann never attained this distinction despite numerous efforts. Thus when in the 1840's their systems were being offered to the public, Hamilton had as he said a "capital" of personal fame that aided him in gaining recognition for his system, whereas Grassmann was then and long remained almost unknown.

Both men had strong interests outside mathematics, particularly in regard to languages, philosophy, and religion. Within mathematics both had a similar style that led them to write long, scarcely readable works embodying philosophical ideas. Both were isolated geographically from their important contemporaries: Hamilton lived in Dublin at a distance from Trinity College, and Grassmann lived at Stettin, from which he had to travel to Berlin to read the *Comptes rendus* of the Paris Academy. When each made his discovery, he expressed a readiness and a desire to spend a major part of his life developing his system.

Both also had one especially important follower, neither of whom had been a student of his master. Grassmann waited nearly until his death for a measure of fame; Hamilton's fame came early but probably did not increase during his later years or in the decades immediately after his death. Both were geniuses and alike in that the fate of their systems was to depend on the development and acceptance of another system, that of modern vector analysis. Correspondingly, judgments on the significance of these two men as creators of vectorial systems are complicated by the fact that their systems have not by and large survived in tact. Their systems were ancestors of the modern system, but this ancestry they themselves might well have disclaimed.

Concerning the reception accorded their systems a statistical study was made of the publications relating to each system from 1841 to 1900. This study revealed that roughly 594 publications relating to quaternions appeared from 1841 to 1900, whereas there were roughly 217 publications relating to Grassmann's system during this time span. Thus there were something on the order of 2.73 quaternion publications for each Grassmannian publication. There were 38 quaternion books as compared to 16 within the Grassmannian tradition, or 2.37 quaternion books for each book in the Grassmannian tradition. Thus interest in the quaternion system was on the order of two and one-half times as great as that in the Grassmannian system. This generalization was qualified by the observation that during the last decade of the century the number of publications within the Grassmannian tradition was nearly equal to the number published within the Hamiltonian tradition.

In relation to geographic distribution of interest it was observed that among the British, American, French, and German peoples interest in each system was strongest in the land of its origin, next strongest (for both systems) in America, and substantial in the other two countries. There was sufficiently widespread interest in quaternions outside these four countries that ten quaternion books were published in seven of the less intellectually productive countries of the late nineteenth-century world. Another major conclusions from the statistical study was that interest in vectorial systems was increasing rapidly during the latter third of the nineteenth century.

After this quantitative study established the degree of success of the quaternion system, the relation of Tait, Peirce, Maxwell, and Clifford to the quaternion tradition was analyzed. Tait was Hamilton's leading disciple and did much to make Hamilton's ideas better known. In addition to this he developed many parts of quaternion analysis, and some of these developments were later transferred into modern vector analysis. Most important was Tait's development of the operator ∇; as Maxwell put it, Tait was the "Chief Musician upon Nabla." Thus the very important transformation theorems associated with ∇ first appeared in vectorial form in

Tait's writings. Moreover Tait repeatedly stressed (as Hamilton had not) the applications of quaternion methods in physical science.

During the nineteenth century, physical science (above all electricity) developed in such a way that the need for a vectorial system increased. Thus for example emphasis was placed on the field concept, and potential theory was developed. Maxwell, partly through Tait, became interested in quaternion analysis some time around 1870. His enthusiasm for the "ideas," but not the "methods," of quaternion analysis was great, and his critical analysis of the quaternion system led him to definite views as to the defects in this system. Most of these defects were eliminated in the Gibbs-Heaviside system. The chief significance of Maxwell is that he associated certain parts of quaternion analysis with important ideas presented in his *Treatise on Electricity and Magnetism.* Though the use of quaternion methods in this influential book was very limited, it was sufficient to lead other scientists, most notably Gibbs and Heaviside, to the study of quaternions.

Certainly not all, probably not even a majority, of the physicists of the 1870's and 1880's were favorably disposed toward vectorial methods. Lord Kelvin was not only the most influential British physicist at this time; he was also an outspoken critic of quaternion methods as well as vectorial methods in general. Tait was not the only vigorous advocate of quaternions. Benjamin Peirce was an early and influential proponent of Hamilton's ideas, and much of the interest in quaternions in the United States can be traced to him. Peirce was also discussed as illustrative of the not small number of men who were led from quaternions to mathematical ideas of great importance, but ideas outside of the quaternion system. William Kingdon Clifford was seen as a very capable mathematician who was knowledgeable in both the Hamiltonian and Grassmannian traditions. Clifford's presentation of a vectorial system in his *Elements of Dynamic* was viewed as prophetic in the sense that it foreshadowed the presentations of Gibbs and Heaviside. Thus the period from 1865 to 1880 was characterized as a time of "realizations" as opposed to "discoveries." During this period many scientists came to realize that vectorial systems were important and useful, and a few scientists began to discern which elements in the existing systems were most significant. Thus the stage was set for Gibbs and Heaviside.

The modern system of vector analysis originated with Josiah Willard Gibbs and Oliver Heaviside. These two great physicists, working independently of each other, constructed their systems (which were essentially identical) during the late 1870's and early 1880's.

The position of both in relation to previous traditions was identical. It was demonstrated that each began to study the quaternion system under the stimulus supplied by Maxwell's electromagnetic writings. Both were exposed to Maxwell's critical remarks in his *Treatise on Electricity and Magnetism* and to the practice followed by Maxwell in that work. Thus Gibbs and Heaviside created a system that successfully avoided the defects ascribed to the quaternion system by Maxwell and simultaneously salvaged those parts of the quaternion approach that Maxwell praised. Though neither was influenced by Grassmann, Gibbs did become an advocate of some of Grassmann's ideas in multiple algebra.

During the 1880's the Gibbs-Heaviside system started to become known through Gibbs' selectively distributed booklet and through his lectures at Yale, and through Heaviside's publications in the *Electrician* and elsewhere. Gibbs' creative contribution to vector analysis was greater than Heaviside's, though Heaviside played a greater role in gaining acceptance for their system, since he set forth vector analysis in his very important electrical publications. It was pointed out that both Gibbs and Heaviside made efforts from a number of directions to make their systems better known and that one factor that must have influenced many readers was the rapidly increasing fame of each as a physical scientist. Thus by 1890 the modern system of vector analysis had been created and offered, as it were, to the public. It was not as yet widely known, but this situation was soon to change, for a "struggle for existence" was about to begin.

This "struggle for existence" took place from 1890 to 1894; thirty-eight publications written during that time by more than twelve scientists were discussed. It was stressed that the debate was widespread, conducted with vehemence, participated in mainly by physicists, and presented in a very readable manner. The main question debated was, Which vectorial system is best? All the participants except Cayley were convinced that a vectorial approach was desirable. The main contesting systems were the Hamilton-Tait and the Gibbs-Heaviside systems; the merits of the systems of Grassmann and Macfarlane were debated, but to a lesser extent. A major result of the debate was that the quaternion system, which was well known, was forcefully attacked; and the Gibbs-Heaviside system, which until then was not well known, received much publicity, though certainly not always of the most favorable kind.

It was suggested that the Gibbs-Heaviside system emerged from the debate with improved chances for widespread acceptance; this was partly due to the fact that those who wrote on its behalf presented a unified front, were more tactful in their presentation, were more knowledgeable of the opponents' system, and (judging by later history) were right in their arguments. Finally it was suggested that an unnecessary, but nonetheless real, result of the debate was that Hamilton's fame suffered. This was unnecessary in that, rhetorical considerations aside, Gibbs and Heaviside could have with justice presented themselves as descendents from Hamilton and Tait rather than as opponents of them. That they did not do this is certainly no cause for censure; that they chose not to do this was a result of the natural tendency of the quaternionists to draw on the capital of fame that had accrued to Hamilton. Thus by 1894 the Gibbs-Heaviside system had become known to a substantial number of scientists; it was not as yet widely used.

The methodology selected for the discussion of the acceptance of the Gibbs-Heaviside system was to analyze the books published from 1894 to 1910 that presented the now-common system of vector analysis in order to determine their origin, their nature, and their success in attracting readers. It was concluded that it was primarily through Heaviside and his association of vector analysis with Maxwell's electrical ideas that the modern system entered into the German-speaking lands, especially through the writings of such men as Föppl, Lorentz, Abraham, Bucherer, Gans, Valentiner, and Ignatowsky. Through Ferraris it became available in Italy, and through Heaviside himself it had already become available in Britain. Through Gibbs and his pupil Wilson an excellent presentation of the modern system was published in America. Coffin's shorter book further helped to make this system available. Through P. O. Somoff, who was acquainted with both the Gibbs and the Heaviside traditions, the modern system was presented in Russia.

The Grassmannian tradition was kept alive in the writings of Burali-Forti, Marcolongo, Timerding, and Jahnke. In the writings of these men Grassmann's ideas were dominant, though the methods and style of the Gibbs-Heaviside tradition were not absent. To a far lesser degree the Grassmannian tradition entered into such works as those of Wilson and Valentiner. However, it was concluded on the bases of republication and translation information that the books directly within the Gibbs-Heaviside tradition were the ones that were widely used by the scientific world. Thus, in short, the Heaviside tradition was the most important, the Gibbsian tradition second in importance, and the Grassmannian tradition a distant third. Books within other traditions had not ceased to appear, but their number had decreased greatly. The pure Hamiltonian and the pure Grassmannian systems are not yet mathematical antiques; thus, for example, Henry George Forder published in 1941 a book entitled *The Calculus of Extension* [1] which is in the Grassmannian tradition, and Otto F. Fischer published in the 1950's two long books within the Hamiltonian tradition.[2]

Heroes the history of science must have, though within the scope of the present study so many scientists have contributed significantly that the discernment of an order of heroic accomplishment is far from easy. Nonetheless, if two of the men discussed are deserving of last mention, they are certainly Grassmann and Hamilton. Each made a prophetic statement, and these statements will be quoted, but prefaced by the remark that the present almost universal adoption of vector analysis has fulfilled their prophecies, not to the letter of the statements, but to the spirit thereof.

Writing in 1861, Grassmann concluded the foreword of his second *Ausdehnungslehre* with the following statement:

> For I remain completely confident that the labor which I have expanded on the science presented here and which has demanded a significant part of my life as well as the most strenuous application of my powers will not be lost. It is true that I am aware that the form which I have given the science is imperfect and must be imperfect. But I know and feel obliged to state (though I run the risk of seeming arrogant) that even if this work should again remain unused for another seventeen years or even longer, without entering into the actual development of science, still that time will come when it will be brought forth from the dust of oblivion, and when ideas now dormant will bring forth fruit. I know that if I also fail to gather around me in a position (which I have up to now desired in vain) a circle of scholars, whom I could fructify with these ideas, and whom I could stimulate to develop and enrich further these ideas, nevertheless there will come a time when these ideas, perhaps in a new form, will arise anew and will enter into living communication with contemporary developments. For truth is eternal and divine, and no phase in the development of truth, however small may be the region encompassed, can pass on without leaving a trace; truth remains, even though the garment in which poor mortals clothe it may fall to dust.[3]

Tait in the preface to the third edition of his *Treatise on Quaternions* quoted from a letter written to him in 1859 by Hamilton. Hamilton had asked Tait:

> *Could* anything be simpler or more satisfactory? Don't you *feel*, as well as think, that we are on a *right track*, and shall be *thanked* hereafter. Never mind when.[4]

Notes

[1] Published at Cambridge, England, xvi + 490 pp.

[2] See Otto F. Fischer, *Universal Mechanics and Hamilton's Quaternions, A Cavalcade* (Stockholm, 1951), and Fischer, *Five Mathematical Structural Models in Natural Philosophy with Technical Physical Quaternions* (Stockholm, 1957).

[3] Hermann Günther Grassmann, *Gesammelte mathematische und physikalische Werke*, vol. I, pt. II (Leipzig, 1896), 10.

[4] As quoted in Peter Guthrie Tait, *An Elementary Treatise on Quaternions*, 3rd ed. (Cambridge, England, 1890), viii. The date of the letter (April 12, 1859) was given by Cargill Gilston Knott, *Life and Scientific Work of Peter Guthrie Tait* (Cambridge, England, 1911), 134.

Chronology

1673	Wallis' treatment of complex numbers
1679	Leibniz' letter to Huygens on a geometry of situation
1799	Wessel publishes the first 2-dimensional and first 3-dimensional vectorial system
by 1799	Gauss discovers geometrical representation of complex numbers
by 1800	Parallelogram of forces, velocities . . . becomes common in physical science books
1805	Birth of Hamilton
1806	Argand publishes the geometrical representation of complex numbers
1806	Buée publishes paper on complex numbers
1809	Birth of Grassmann
1814–1815	Argand, Français, and Servois publish ideas on 3-dimensional vectorial systems
1823	Möbius publishes his first treatment of his barycentric calculus
1824	Grassmann's father (J. G. Grassmann) publishes his idea of a geometrical product
1826	Hamilton becomes Andrews' Professor of Astronomy at Trinity College, Dublin, and Royal Astronomer of Ireland
1827	Möbius' *Der barycentrische Calcul*
1828	Warren publishes the geometrical representation of complex numbers
1828	Mourey publishes the geometrical representation of complex numbers
by 1830	Four men had published geometrical representation of complex numbers, but all are neglected
1830	Hamilton begins searching for a 3-dimensional vectorial system
1831	Gauss publishes the geometrical representation of complex numbers

1831	Birth of Tait and Maxwell
1832	Grassmann gets first ideas for his calculus of extension
1835	Bellavitis publishes his first major paper on his calculus of equipollences
1837	Hamilton publishes representation of complex numbers as "couples" of numbers
1839	Birth of Gibbs
by 1840	At least ten men have searched for a "triple" algebra
1840	Grassmann submits his dissertation on tidal theory containing the first presentation of his calculus of extension
1843	Hamilton discovers quaternions
1844	Grassmann publishes his system in his *Ausdehnungslehre*
1845	Saint-Venant publishes his vectorial system
1846	Hamilton introduces $S\alpha\beta$ and $V\alpha\beta$
1850	Birth of Heaviside
1852	O'Brien publishes his most complete treatment of his vectorial system
1853	Hamilton's *Lectures on Quaternions*
1853	Cauchy's "Sur les clefs algébriques"
1855	Peirce strongly recommends quaternions in his *A System of Analytical Mechanics*
1858	Tait-Hamilton correspondence begins
1862	Grassmann's second *Ausdehnungslehre*
1865	Death of Hamilton
by 1865	Grassmann's publications still totally neglected
1866	Hamilton's *Elements of Quaternions*
1867	Tait's *Elementary Treatise on Quaternions*
1867	Thomson's and Tait's *Treatise on Natural Philosophy*
1867	Hankel praises Grassmann's ideas in his *Theorie der complexen Zahlensysteme*
1871	Maxwell's "Mathematical Classification of Physical Quantities"
1872–1875	Schlegel's *System der Raumlehre*
1873	Tait's *Elementary Treatise on Quaternions* (2nd ed.)
1873	Kelland's and Tait's *Introduction to Quaternions*
1873	Maxwell's *Treatise on Electricity and Magnetism*

1877	Death of Grassmann
1878	Second edition of Grassmann's 1844 *Ausdehnungslehre*
1878	Schlegel's *Hermann Grassmann*
1878–1887	Clifford's *Elements of Dynamic*
1879	Death of Maxwell and Clifford
1880	Death of Peirce
by 1880	Gibbs begins teaching vector analysis courses
1880	German translation of Tait's *Treatise on Quaternions*
1881	Peirce's "Linear Associative Algebra" is published
1881–1884	Gibbs' *Elements of Vector Analysis*
1881–1884	German translation of Hamilton's *Elements of Quaternions*
1882–1884	French translation of Tait's *Treatise on Quaternions*
1882	Kelland's and Tait's *Introduction to Quaternions* (2nd ed.)
1882–1889	Grave's *Life of Sir William Rowan Hamilton*
1883	Heaviside begins using his system in his papers in the *Electrician*
1886	Gibbs' paper "On Multiple Algebra"
1890	Tait's *Elementary Treatise on Quaternions* (3rd ed.)
1890–1894	Thirty-eight publications discussing the merits of the various vectorial systems
1891	Maxwell's *Treatise on Electricity and Magnetism* (3rd ed.)
1891	Macfarlane's first presentation of his system
1892	Heaviside's *Electrical Paper*
1893	Heaviside publishes a lengthy exposition of his vectorial system in his *Electromagnetic Theory*
1894–1911	Grassmann's collected scientific papers published
1894	Föppl publishes first presentation of modern vector analysis in German
1895	Kimura and Molenbroek propose that advocates of vectorial methods form a society
1898–1900	Tait's *Scientific Papers*
1899	Ferraris publishes first long exposition of modern vector analysis in Italian
1899–1901	Hamilton's *Elements of Quaternions* (2nd ed.)
by 1900	Over 1000 vectorial publications: 594 quaternionic; 217 Grassmannian

1900	First issue of the *Bulletin of the International Association for Promoting the Study of Quaternions and Allied Systems of Mathematics*
1901	Death of Tait
1901–1909	Modern vector system used in important articles in the *Encyklopädie der mathematischen Wissenschaften*
1901	Wilson publishes first book-length presentation of modern vector analysis in English
1903	Bucherer publishes first German book on modern vector analysis
1903	Death of Gibbs
1904	Second edition of Föppl's book
1905	Gans' *Einführung in die Vektoranalysis*
1905	Jahnke's *Vorlesungen über die Vektorenrechnung*
1905	Bucherer's *Elemente der Vektor-Analysis* (2nd ed.)
1906	Gibbs' collected *Scientific Papers* containing his *Elements of Vector Analysis*
1907	Valentiner's *Vektoranalysis*
1907	Somoff publishes first book on modern vector analysis in Russian
1907	Third edition of Föppl's book
1909	Coffin's *Vector Analysis*
1909–1910	Ignatowsky's *Die Vektoranalysis*
1909	Wilson's *Vector Analysis* (2nd ed.)
1909	Gans' *Einführung in die Vektoranalysis* (2nd ed.)
1909	Burali-Forti and Marcolongo's *Elementi di calcolo vettoriale*
1910	French translation of Burali-Forti and Marcolongo's book of 1909
by 1910	Interest in quaternions and Grassmann's system declining; interest in Gibbs-Heaviside system rapidly increasing
1913	Last issue of the *Bulletin of the International Association for Promoting the Study of Quaternions and Allied Systems of Mathematics*
1925	Death of Heaviside

Index

A

B

W

Y

Z

A CATALOG OF SELECTED

DOVER BOOKS

IN SCIENCE AND MATHEMATICS

Astronomy

CHARIOTS FOR APOLLO: The NASA History of Manned Lunar Spacecraft to 1969, Courtney G. Brooks, James M. Grimwood, and Loyd S. Swenson, Jr. This illustrated history by a trio of experts is the definitive reference on the Apollo spacecraft and lunar modules. It traces the vehicles' design, development, and operation in space. More than 100 photographs and illustrations. 576pp. 6 3/4 x 9 1/4. 0-486-46756-2

EXPLORING THE MOON THROUGH BINOCULARS AND SMALL TELESCOPES, Ernest H. Cherrington, Jr. Informative, profusely illustrated guide to locating and identifying craters, rills, seas, mountains, other lunar features. Newly revised and updated with special section of new photos. Over 100 photos and diagrams. 240pp. 8 1/4 x 11. 0-486-24491-1

WHERE NO MAN HAS GONE BEFORE: A History of NASA's Apollo Lunar Expeditions, William David Compton. Introduction by Paul Dickson. This official NASA history traces behind-the-scenes conflicts and cooperation between scientists and engineers. The first half concerns preparations for the Moon landings, and the second half documents the flights that followed Apollo 11. 1989 edition. 432pp. 7 x 10. 0-486-47888-2

APOLLO EXPEDITIONS TO THE MOON: The NASA History, Edited by Edgar M. Cortright. Official NASA publication marks the 40th anniversary of the first lunar landing and features essays by project participants recalling engineering and administrative challenges. Accessible, jargon-free accounts, highlighted by numerous illustrations. 336pp. 8 3/8 x 10 7/8. 0-486-47175-6

ON MARS: Exploration of the Red Planet, 1958-1978--The NASA History, Edward Clinton Ezell and Linda Neuman Ezell. NASA's official history chronicles the start of our explorations of our planetary neighbor. It recounts cooperation among government, industry, and academia, and it features dozens of photos from Viking cameras. 560pp. 6 3/4 x 9 1/4. 0-486-46757-0

ARISTARCHUS OF SAMOS: The Ancient Copernicus, Sir Thomas Heath. Heath's history of astronomy ranges from Homer and Hesiod to Aristarchus and includes quotes from numerous thinkers, compilers, and scholasticists from Thales and Anaximander through Pythagoras, Plato, Aristotle, and Heraclides. 34 figures. 448pp. 5 3/8 x 8 1/2. 0-486-43886-4

AN INTRODUCTION TO CELESTIAL MECHANICS, Forest Ray Moulton. Classic text still unsurpassed in presentation of fundamental principles. Covers rectilinear motion, central forces, problems of two and three bodies, much more. Includes over 200 problems, some with answers. 437pp. 5 3/8 x 8 1/2. 0-486-64687-4

BEYOND THE ATMOSPHERE: Early Years of Space Science, Homer E. Newell. This exciting survey is the work of a top NASA administrator who chronicles technological advances, the relationship of space science to general science, and the space program's social, political, and economic contexts. 528pp. 6 3/4 x 9 1/4. 0-486-47464-X

STAR LORE: Myths, Legends, and Facts, William Tyler Olcott. Captivating retellings of the origins and histories of ancient star groups include Pegasus, Ursa Major, Pleiades, signs of the zodiac, and other constellations. "Classic." – *Sky & Telescope.* 58 illustrations. 544pp. 5 3/8 x 8 1/2. 0-486-43581-4

A COMPLETE MANUAL OF AMATEUR ASTRONOMY: Tools and Techniques for Astronomical Observations, P. Clay Sherrod with Thomas L. Koed. Concise, highly readable book discusses the selection, set-up, and maintenance of a telescope; amateur studies of the sun; lunar topography and occultations; and more. 124 figures. 26 halftones. 37 tables. 335pp. 6 1/2 x 9 1/4. 0-486-42820-6

Chemistry

MOLECULAR COLLISION THEORY, M. S. Child. This high-level monograph offers an analytical treatment of classical scattering by a central force, quantum scattering by a central force, elastic scattering phase shifts, and semi-classical elastic scattering. 1974 edition. 310pp. 5 3/8 x 8 1/2. 0-486-69437-2

HANDBOOK OF COMPUTATIONAL QUANTUM CHEMISTRY, David B. Cook. This comprehensive text provides upper-level undergraduates and graduate students with an accessible introduction to the implementation of quantum ideas in molecular modeling, exploring practical applications alongside theoretical explanations. 1998 edition. 832pp. 5 3/8 x 8 1/2. 0-486-44307-8

RADIOACTIVE SUBSTANCES, Marie Curie. The celebrated scientist's thesis, which directly preceded her 1903 Nobel Prize, discusses establishing atomic character of radioactivity; extraction from pitchblende of polonium and radium; isolation of pure radium chloride; more. 96pp. 5 3/8 x 8 1/2. 0-486-42550-9

CHEMICAL MAGIC, Leonard A. Ford. Classic guide provides intriguing entertainment while elucidating sound scientific principles, with more than 100 unusual stunts: cold fire, dust explosions, a nylon rope trick, a disappearing beaker, much more. 128pp. 5 3/8 x 8 1/2. 0-486-67628-5

ALCHEMY, E. J. Holmyard. Classic study by noted authority covers 2,000 years of alchemical history: religious, mystical overtones; apparatus; signs, symbols, and secret terms; advent of scientific method, much more. Illustrated. 320pp. 5 3/8 x 8 1/2. 0-486-26298-7

CHEMICAL KINETICS AND REACTION DYNAMICS, Paul L. Houston. This text teaches the principles underlying modern chemical kinetics in a clear, direct fashion, using several examples to enhance basic understanding. Solutions to selected problems. 2001 edition. 352pp. 8 3/8 x 11. 0-486-45334-0

PROBLEMS AND SOLUTIONS IN QUANTUM CHEMISTRY AND PHYSICS, Charles S. Johnson and Lee G. Pedersen. Unusually varied problems, with detailed solutions, cover of quantum mechanics, wave mechanics, angular momentum, molecular spectroscopy, scattering theory, more. 280 problems, plus 139 supplementary exercises. 430pp. 6 1/2 x 9 1/4. 0-486-65236-X

ELEMENTS OF CHEMISTRY, Antoine Lavoisier. Monumental classic by the founder of modern chemistry features first explicit statement of law of conservation of matter in chemical change, and more. Facsimile reprint of original (1790) Kerr translation. 539pp. 5 3/8 x 8 1/2. 0-486-64624-6

MAGNETISM AND TRANSITION METAL COMPLEXES, F. E. Mabbs and D. J. Machin. A detailed view of the calculation methods involved in the magnetic properties of transition metal complexes, this volume offers sufficient background for original work in the field. 1973 edition. 240pp. 5 3/8 x 8 1/2. 0-486-46284-6

GENERAL CHEMISTRY, Linus Pauling. Revised third edition of classic first-year text by Nobel laureate. Atomic and molecular structure, quantum mechanics, statistical mechanics, thermodynamics correlated with descriptive chemistry. Problems. 992pp. 5 3/8 x 8 1/2. 0-486-65622-5

ELECTROLYTE SOLUTIONS: Second Revised Edition, R. A. Robinson and R. H. Stokes. Classic text deals primarily with measurement, interpretation of conductance, chemical potential, and diffusion in electrolyte solutions. Detailed theoretical interpretations, plus extensive tables of thermodynamic and transport properties. 1970 edition. 590pp. 5 3/8 x 8 1/2. 0-486-42225-9

Engineering

FUNDAMENTALS OF ASTRODYNAMICS, Roger R. Bate, Donald D. Mueller, and Jerry E. White. Teaching text developed by U.S. Air Force Academy develops the basic two-body and n-body equations of motion; orbit determination; classical orbital elements, coordinate transformations; differential correction; more. 1971 edition. 455pp. 5 3/8 x 8 1/2. 0-486-60061-0

INTRODUCTION TO CONTINUUM MECHANICS FOR ENGINEERS: Revised Edition, Ray M. Bowen. This self-contained text introduces classical continuum models within a modern framework. Its numerous exercises illustrate the governing principles, linearizations, and other approximations that constitute classical continuum models. 2007 edition. 320pp. 6 1/8 x 9 1/4. 0-486-47460-7

ENGINEERING MECHANICS FOR STRUCTURES, Louis L. Bucciarelli. This text explores the mechanics of solids and statics as well as the strength of materials and elasticity theory. Its many design exercises encourage creative initiative and systems thinking. 2009 edition. 320pp. 6 1/8 x 9 1/4. 0-486-46855-0

FEEDBACK CONTROL THEORY, John C. Doyle, Bruce A. Francis and Allen R. Tannenbaum. This excellent introduction to feedback control system design offers a theoretical approach that captures the essential issues and can be applied to a wide range of practical problems. 1992 edition. 224pp. 6 1/2 x 9 1/4. 0-486-46933-6

THE FORCES OF MATTER, Michael Faraday. These lectures by a famous inventor offer an easy-to-understand introduction to the interactions of the universe's physical forces. Six essays explore gravitation, cohesion, chemical affinity, heat, magnetism, and electricity. 1993 edition. 96pp. 5 3/8 x 8 1/2. 0-486-47482-8

DYNAMICS, Lawrence E. Goodman and William H. Warner. Beginning engineering text introduces calculus of vectors, particle motion, dynamics of particle systems and plane rigid bodies, technical applications in plane motions, and more. Exercises and answers in every chapter. 619pp. 5 3/8 x 8 1/2. 0-486-42006-X

ADAPTIVE FILTERING PREDICTION AND CONTROL, Graham C. Goodwin and Kwai Sang Sin. This unified survey focuses on linear discrete-time systems and explores natural extensions to nonlinear systems. It emphasizes discrete-time systems, summarizing theoretical and practical aspects of a large class of adaptive algorithms. 1984 edition. 560pp. 6 1/2 x 9 1/4. 0-486-46932-8

INDUCTANCE CALCULATIONS, Frederick W. Grover. This authoritative reference enables the design of virtually every type of inductor. It features a single simple formula for each type of inductor, together with tables containing essential numerical factors. 1946 edition. 304pp. 5 3/8 x 8 1/2. 0-486-47440-2

THERMODYNAMICS: Foundations and Applications, Elias P. Gyftopoulos and Gian Paolo Beretta. Designed by two MIT professors, this authoritative text discusses basic concepts and applications in detail, emphasizing generality, definitions, and logical consistency. More than 300 solved problems cover realistic energy systems and processes. 800pp. 6 1/8 x 9 1/4. 0-486-43932-1

THE FINITE ELEMENT METHOD: Linear Static and Dynamic Finite Element Analysis, Thomas J. R. Hughes. Text for students without in-depth mathematical training, this text includes a comprehensive presentation and analysis of algorithms of time-dependent phenomena plus beam, plate, and shell theories. Solution guide available upon request. 672pp. 6 1/2 x 9 1/4. 0-486-41181-8

HELICOPTER THEORY, Wayne Johnson. Monumental engineering text covers vertical flight, forward flight, performance, mathematics of rotating systems, rotary wing dynamics and aerodynamics, aeroelasticity, stability and control, stall, noise, and more. 189 illustrations. 1980 edition. 1089pp. 5 5/8 x 8 1/4. 0-486-68230-7

MATHEMATICAL HANDBOOK FOR SCIENTISTS AND ENGINEERS: Definitions, Theorems, and Formulas for Reference and Review, Granino A. Korn and Theresa M. Korn. Convenient access to information from every area of mathematics: Fourier transforms, Z transforms, linear and nonlinear programming, calculus of variations, random-process theory, special functions, combinatorial analysis, game theory, much more. 1152pp. 5 3/8 x 8 1/2. 0-486-41147-8

A HEAT TRANSFER TEXTBOOK: Fourth Edition, John H. Lienhard V and John H. Lienhard IV. This introduction to heat and mass transfer for engineering students features worked examples and end-of-chapter exercises. Worked examples and end-of-chapter exercises appear throughout the book, along with well-drawn, illuminating figures. 768pp. 7 x 9 1/4. 0-486-47931-5

BASIC ELECTRICITY, U.S. Bureau of Naval Personnel. Originally a training course; best nontechnical coverage. Topics include batteries, circuits, conductors, AC and DC, inductance and capacitance, generators, motors, transformers, amplifiers, etc. Many questions with answers. 349 illustrations. 1969 edition. 448pp. 6 1/2 x 9 1/4. 0-486-20973-3

BASIC ELECTRONICS, U.S. Bureau of Naval Personnel. Clear, well-illustrated introduction to electronic equipment covers numerous essential topics: electron tubes, semiconductors, electronic power supplies, tuned circuits, amplifiers, receivers, ranging and navigation systems, computers, antennas, more. 560 illustrations. 567pp. 6 1/2 x 9 1/4. 0-486-21076-6

BASIC WING AND AIRFOIL THEORY, Alan Pope. This self-contained treatment by a pioneer in the study of wind effects covers flow functions, airfoil construction and pressure distribution, finite and monoplane wings, and many other subjects. 1951 edition. 320pp. 5 3/8 x 8 1/2. 0-486-47188-8

SYNTHETIC FUELS, Ronald F. Probstein and R. Edwin Hicks. This unified presentation examines the methods and processes for converting coal, oil, shale, tar sands, and various forms of biomass into liquid, gaseous, and clean solid fuels. 1982 edition. 512pp. 6 1/8 x 9 1/4. 0-486-44977-7

THEORY OF ELASTIC STABILITY, Stephen P. Timoshenko and James M. Gere. Written by world-renowned authorities on mechanics, this classic ranges from theoretical explanations of 2- and 3-D stress and strain to practical applications such as torsion, bending, and thermal stress. 1961 edition. 560pp. 5 3/8 x 8 1/2. 0-486-47207-8

PRINCIPLES OF DIGITAL COMMUNICATION AND CODING, Andrew J. Viterbi and Jim K. Omura. This classic by two digital communications experts is geared toward students of communications theory and to designers of channels, links, terminals, modems, or networks used to transmit and receive digital messages. 1979 edition. 576pp. 6 1/8 x 9 1/4. 0-486-46901-8

LINEAR SYSTEM THEORY: The State Space Approach, Lotfi A. Zadeh and Charles A. Desoer. Written by two pioneers in the field, this exploration of the state space approach focuses on problems of stability and control, plus connections between this approach and classical techniques. 1963 edition. 656pp. 6 1/8 x 9 1/4. 0-486-46663-9

Mathematics–Bestsellers

HANDBOOK OF MATHEMATICAL FUNCTIONS: with Formulas, Graphs, and Mathematical Tables, Edited by Milton Abramowitz and Irene A. Stegun. A classic resource for working with special functions, standard trig, and exponential logarithmic definitions and extensions, it features 29 sets of tables, some to as high as 20 places. 1046pp. 8 x 10 1/2. 0-486-61272-4

ABSTRACT AND CONCRETE CATEGORIES: The Joy of Cats, Jiri Adamek, Horst Herrlich, and George E. Strecker. This up-to-date introductory treatment employs category theory to explore the theory of structures. Its unique approach stresses concrete categories and presents a systematic view of factorization structures. Numerous examples. 1990 edition, updated 2004. 528pp. 6 1/8 x 9 1/4. 0-486-46934-4

MATHEMATICS: Its Content, Methods and Meaning, A. D. Aleksandrov, A. N. Kolmogorov, and M. A. Lavrent'ev. Major survey offers comprehensive, coherent discussions of analytic geometry, algebra, differential equations, calculus of variations, functions of a complex variable, prime numbers, linear and non-Euclidean geometry, topology, functional analysis, more. 1963 edition. 1120pp. 5 3/8 x 8 1/2. 0-486-40916-3

INTRODUCTION TO VECTORS AND TENSORS: Second Edition--Two Volumes Bound as One, Ray M. Bowen and C.-C. Wang. Convenient single-volume compilation of two texts offers both introduction and in-depth survey. Geared toward engineering and science students rather than mathematicians, it focuses on physics and engineering applications. 1976 edition. 560pp. 6 1/2 x 9 1/4. 0-486-46914-X

AN INTRODUCTION TO ORTHOGONAL POLYNOMIALS, Theodore S. Chihara. Concise introduction covers general elementary theory, including the representation theorem and distribution functions, continued fractions and chain sequences, the recurrence formula, special functions, and some specific systems. 1978 edition. 272pp. 5 3/8 x 8 1/2. 0-486-47929-3

ADVANCED MATHEMATICS FOR ENGINEERS AND SCIENTISTS, Paul DuChateau. This primary text and supplemental reference focuses on linear algebra, calculus, and ordinary differential equations. Additional topics include partial differential equations and approximation methods. Includes solved problems. 1992 edition. 400pp. 7 1/2 x 9 1/4. 0-486-47930-7

PARTIAL DIFFERENTIAL EQUATIONS FOR SCIENTISTS AND ENGINEERS, Stanley J. Farlow. Practical text shows how to formulate and solve partial differential equations. Coverage of diffusion-type problems, hyperbolic-type problems, elliptic-type problems, numerical and approximate methods. Solution guide available upon request. 1982 edition. 414pp. 6 1/8 x 9 1/4. 0-486-67620-X

VARIATIONAL PRINCIPLES AND FREE-BOUNDARY PROBLEMS, Avner Friedman. Advanced graduate-level text examines variational methods in partial differential equations and illustrates their applications to free-boundary problems. Features detailed statements of standard theory of elliptic and parabolic operators. 1982 edition. 720pp. 6 1/8 x 9 1/4. 0-486-47853-X

LINEAR ANALYSIS AND REPRESENTATION THEORY, Steven A. Gaal. Unified treatment covers topics from the theory of operators and operator algebras on Hilbert spaces; integration and representation theory for topological groups; and the theory of Lie algebras, Lie groups, and transform groups. 1973 edition. 704pp. 6 1/8 x 9 1/4. 0-486-47851-3

Mathematics–Logic and Problem Solving

PERPLEXING PUZZLES AND TANTALIZING TEASERS, Martin Gardner. Ninety-three riddles, mazes, illusions, tricky questions, word and picture puzzles, and other challenges offer hours of entertainment for youngsters. Filled with rib-tickling drawings. Solutions. 224pp. 5 3/8 x 8 1/2. 0-486-25637-5

MY BEST MATHEMATICAL AND LOGIC PUZZLES, Martin Gardner. The noted expert selects 70 of his favorite "short" puzzles. Includes The Returning Explorer, The Mutilated Chessboard, Scrambled Box Tops, and dozens more. Complete solutions included. 96pp. 5 3/8 x 8 1/2. 0-486-28152-3

THE LADY OR THE TIGER?: and Other Logic Puzzles, Raymond M. Smullyan. Created by a renowned puzzle master, these whimsically themed challenges involve paradoxes about probability, time, and change; metapuzzles; and self-referentiality. Nineteen chapters advance in difficulty from relatively simple to highly complex. 1982 edition. 240pp. 5 3/8 x 8 1/2. 0-486-47027-X

SATAN, CANTOR AND INFINITY: Mind-Boggling Puzzles, Raymond M. Smullyan. A renowned mathematician tells stories of knights and knaves in an entertaining look at the logical precepts behind infinity, probability, time, and change. Requires a strong background in mathematics. Complete solutions. 288pp. 5 3/8 x 8 1/2. 0-486-47036-9

THE RED BOOK OF MATHEMATICAL PROBLEMS, Kenneth S. Williams and Kenneth Hardy. Handy compilation of 100 practice problems, hints and solutions indispensable for students preparing for the William Lowell Putnam and other mathematical competitions. Preface to the First Edition. Sources. 1988 edition. 192pp. 5 3/8 x 8 1/2. 0-486-69415-1

KING ARTHUR IN SEARCH OF HIS DOG AND OTHER CURIOUS PUZZLES, Raymond M. Smullyan. This fanciful, original collection for readers of all ages features arithmetic puzzles, logic problems related to crime detection, and logic and arithmetic puzzles involving King Arthur and his Dogs of the Round Table. 160pp. 5 3/8 x 8 1/2. 0-486-47435-6

UNDECIDABLE THEORIES: Studies in Logic and the Foundation of Mathematics, Alfred Tarski in collaboration with Andrzej Mostowski and Raphael M. Robinson. This well-known book by the famed logician consists of three treatises: "A General Method in Proofs of Undecidability," "Undecidability and Essential Undecidability in Mathematics," and "Undecidability of the Elementary Theory of Groups." 1953 edition. 112pp. 5 3/8 x 8 1/2. 0-486-47703-7

LOGIC FOR MATHEMATICIANS, J. Barkley Rosser. Examination of essential topics and theorems assumes no background in logic. "Undoubtedly a major addition to the literature of mathematical logic." – *Bulletin of the American Mathematical Society.* 1978 edition. 592pp. 6 1/8 x 9 1/4. 0-486-46898-4

INTRODUCTION TO PROOF IN ABSTRACT MATHEMATICS, Andrew Wohlgemuth. This undergraduate text teaches students what constitutes an acceptable proof, and it develops their ability to do proofs of routine problems as well as those requiring creative insights. 1990 edition. 384pp. 6 1/2 x 9 1/4. 0-486-47854-8

FIRST COURSE IN MATHEMATICAL LOGIC, Patrick Suppes and Shirley Hill. Rigorous introduction is simple enough in presentation and context for wide range of students. Symbolizing sentences; logical inference; truth and validity; truth tables; terms, predicates, universal quantifiers; universal specification and laws of identity; more. 288pp. 5 3/8 x 8 1/2. 0-486-42259-3

Mathematics-Algebra and Calculus

VECTOR CALCULUS, Peter Baxandall and Hans Liebeck. This introductory text offers a rigorous, comprehensive treatment. Classical theorems of vector calculus are amply illustrated with figures, worked examples, physical applications, and exercises with hints and answers. 1986 edition. 560pp. 5 3/8 x 8 1/2. 0-486-46620-5

ADVANCED CALCULUS: An Introduction to Classical Analysis, Louis Brand. A course in analysis that focuses on the functions of a real variable, this text introduces the basic concepts in their simplest setting and illustrates its teachings with numerous examples, theorems, and proofs. 1955 edition. 592pp. 5 3/8 x 8 1/2. 0-486-44548-8

ADVANCED CALCULUS, Avner Friedman. Intended for students who have already completed a one-year course in elementary calculus, this two-part treatment advances from functions of one variable to those of several variables. Solutions. 1971 edition. 432pp. 5 3/8 x 8 1/2. 0-486-45795-8

METHODS OF MATHEMATICS APPLIED TO CALCULUS, PROBABILITY, AND STATISTICS, Richard W. Hamming. This 4-part treatment begins with algebra and analytic geometry and proceeds to an exploration of the calculus of algebraic functions and transcendental functions and applications. 1985 edition. Includes 310 figures and 18 tables. 880pp. 6 1/2 x 9 1/4. 0-486-43945-3

BASIC ALGEBRA I: Second Edition, Nathan Jacobson. A classic text and standard reference for a generation, this volume covers all undergraduate algebra topics, including groups, rings, modules, Galois theory, polynomials, linear algebra, and associative algebra. 1985 edition. 528pp. 6 1/8 x 9 1/4. 0-486-47189-6

BASIC ALGEBRA II: Second Edition, Nathan Jacobson. This classic text and standard reference comprises all subjects of a first-year graduate-level course, including in-depth coverage of groups and polynomials and extensive use of categories and functors. 1989 edition. 704pp. 6 1/8 x 9 1/4. 0-486-47187-X

CALCULUS: An Intuitive and Physical Approach (Second Edition), Morris Kline. Application-oriented introduction relates the subject as closely as possible to science with explorations of the derivative; differentiation and integration of the powers of x; theorems on differentiation, antidifferentiation; the chain rule; trigonometric functions; more. Examples. 1967 edition. 960pp. 6 1/2 x 9 1/4. 0-486-40453-6

ABSTRACT ALGEBRA AND SOLUTION BY RADICALS, John E. Maxfield and Margaret W. Maxfield. Accessible advanced undergraduate-level text starts with groups, rings, fields, and polynomials and advances to Galois theory, radicals and roots of unity, and solution by radicals. Numerous examples, illustrations, exercises, appendixes. 1971 edition. 224pp. 6 1/8 x 9 1/4. 0-486-47723-1

AN INTRODUCTION TO THE THEORY OF LINEAR SPACES, Georgi E. Shilov. Translated by Richard A. Silverman. Introductory treatment offers a clear exposition of algebra, geometry, and analysis as parts of an integrated whole rather than separate subjects. Numerous examples illustrate many different fields, and problems include hints or answers. 1961 edition. 320pp. 5 3/8 x 8 1/2. 0-486-63070-6

LINEAR ALGEBRA, Georgi E. Shilov. Covers determinants, linear spaces, systems of linear equations, linear functions of a vector argument, coordinate transformations, the canonical form of the matrix of a linear operator, bilinear and quadratic forms, and more. 387pp. 5 3/8 x 8 1/2. 0-486-63518-X

Mathematics–Probability and Statistics

BASIC PROBABILITY THEORY, Robert B. Ash. This text emphasizes the probabilistic way of thinking, rather than measure-theoretic concepts. Geared toward advanced undergraduates and graduate students, it features solutions to some of the problems. 1970 edition. 352pp. 5 3/8 x 8 1/2. 0-486-46628-0

PRINCIPLES OF STATISTICS, M. G. Bulmer. Concise description of classical statistics, from basic dice probabilities to modern regression analysis. Equal stress on theory and applications. Moderate difficulty; only basic calculus required. Includes problems with answers. 252pp. 5 5/8 x 8 1/4. 0-486-63760-3

OUTLINE OF BASIC STATISTICS: Dictionary and Formulas, John E. Freund and Frank J. Williams. Handy guide includes a 70-page outline of essential statistical formulas covering grouped and ungrouped data, finite populations, probability, and more, plus over 1,000 clear, concise definitions of statistical terms. 1966 edition. 208pp. 5 3/8 x 8 1/2. 0-486-47769-X

GOOD THINKING: The Foundations of Probability and Its Applications, Irving J. Good. This in-depth treatment of probability theory by a famous British statistician explores Keynesian principles and surveys such topics as Bayesian rationality, corroboration, hypothesis testing, and mathematical tools for induction and simplicity. 1983 edition. 352pp. 5 3/8 x 8 1/2. 0-486-47438-0

INTRODUCTION TO PROBABILITY THEORY WITH CONTEMPORARY APPLICATIONS, Lester L. Helms. Extensive discussions and clear examples, written in plain language, expose students to the rules and methods of probability. Exercises foster problem-solving skills, and all problems feature step-by-step solutions. 1997 edition. 368pp. 6 1/2 x 9 1/4. 0-486-47418-6

CHANCE, LUCK, AND STATISTICS, Horace C. Levinson. In simple, non-technical language, this volume explores the fundamentals governing chance and applies them to sports, government, and business. "Clear and lively ... remarkably accurate." – *Scientific Monthly.* 384pp. 5 3/8 x 8 1/2. 0-486-41997-5

FIFTY CHALLENGING PROBLEMS IN PROBABILITY WITH SOLUTIONS, Frederick Mosteller. Remarkable puzzlers, graded in difficulty, illustrate elementary and advanced aspects of probability. These problems were selected for originality, general interest, or because they demonstrate valuable techniques. Also includes detailed solutions. 88pp. 5 3/8 x 8 1/2. 0-486-65355-2

EXPERIMENTAL STATISTICS, Mary Gibbons Natrella. A handbook for those seeking engineering information and quantitative data for designing, developing, constructing, and testing equipment. Covers the planning of experiments, the analyzing of extreme-value data; and more. 1966 edition. Index. Includes 52 figures and 76 tables. 560pp. 8 3/8 x 11. 0-486-43937-2

STOCHASTIC MODELING: Analysis and Simulation, Barry L. Nelson. Coherent introduction to techniques also offers a guide to the mathematical, numerical, and simulation tools of systems analysis. Includes formulation of models, analysis, and interpretation of results. 1995 edition. 336pp. 6 1/8 x 9 1/4. 0-486-47770-3

INTRODUCTION TO BIOSTATISTICS: Second Edition, Robert R. Sokal and F. James Rohlf. Suitable for undergraduates with a minimal background in mathematics, this introduction ranges from descriptive statistics to fundamental distributions and the testing of hypotheses. Includes numerous worked-out problems and examples. 1987 edition. 384pp. 6 1/8 x 9 1/4. 0-486-46961-1

Mathematics-Geometry and Topology

PROBLEMS AND SOLUTIONS IN EUCLIDEAN GEOMETRY, M. N. Aref and William Wernick. Based on classical principles, this book is intended for a second course in Euclidean geometry and can be used as a refresher. More than 200 problems include hints and solutions. 1968 edition. 272pp. 5 3/8 x 8 1/2. 0-486-47720-7

TOPOLOGY OF 3-MANIFOLDS AND RELATED TOPICS, Edited by M. K. Fort, Jr. With a New Introduction by Daniel Silver. Summaries and full reports from a 1961 conference discuss decompositions and subsets of 3-space; n-manifolds; knot theory; the Poincaré conjecture; and periodic maps and isotopies. Familiarity with algebraic topology required. 1962 edition. 272pp. 6 1/8 x 9 1/4. 0-486-47753-3

POINT SET TOPOLOGY, Steven A. Gaal. Suitable for a complete course in topology, this text also functions as a self-contained treatment for independent study. Additional enrichment materials make it equally valuable as a reference. 1964 edition. 336pp. 5 3/8 x 8 1/2. 0-486-47222-1

INVITATION TO GEOMETRY, Z. A. Melzak. Intended for students of many different backgrounds with only a modest knowledge of mathematics, this text features self-contained chapters that can be adapted to several types of geometry courses. 1983 edition. 240pp. 5 3/8 x 8 1/2. 0-486-46626-4

TOPOLOGY AND GEOMETRY FOR PHYSICISTS, Charles Nash and Siddhartha Sen. Written by physicists for physics students, this text assumes no detailed background in topology or geometry. Topics include differential forms, homotopy, homology, cohomology, fiber bundles, connection and covariant derivatives, and Morse theory. 1983 edition. 320pp. 5 3/8 x 8 1/2. 0-486-47852-1

BEYOND GEOMETRY: Classic Papers from Riemann to Einstein, Edited with an Introduction and Notes by Peter Pesic. This is the only English-language collection of these 8 accessible essays. They trace seminal ideas about the foundations of geometry that led to Einstein's general theory of relativity. 224pp. 6 1/8 x 9 1/4. 0-486-45350-2

GEOMETRY FROM EUCLID TO KNOTS, Saul Stahl. This text provides a historical perspective on plane geometry and covers non-neutral Euclidean geometry, circles and regular polygons, projective geometry, symmetries, inversions, informal topology, and more. Includes 1,000 practice problems. Solutions available. 2003 edition. 480pp. 6 1/8 x 9 1/4. 0-486-47459-3

TOPOLOGICAL VECTOR SPACES, DISTRIBUTIONS AND KERNELS, François Trèves. Extending beyond the boundaries of Hilbert and Banach space theory, this text focuses on key aspects of functional analysis, particularly in regard to solving partial differential equations. 1967 edition. 592pp. 5 3/8 x 8 1/2. 0-486-45352-9

INTRODUCTION TO PROJECTIVE GEOMETRY, C. R. Wylie, Jr. This introductory volume offers strong reinforcement for its teachings, with detailed examples and numerous theorems, proofs, and exercises, plus complete answers to all odd-numbered end-of-chapter problems. 1970 edition. 576pp. 6 1/8 x 9 1/4. 0-486-46895-X

FOUNDATIONS OF GEOMETRY, C. R. Wylie, Jr. Geared toward students preparing to teach high school mathematics, this text explores the principles of Euclidean and non-Euclidean geometry and covers both generalities and specifics of the axiomatic method. 1964 edition. 352pp. 6 x 9. 0-486-47214-0

Mathematics-History

THE WORKS OF ARCHIMEDES, Archimedes. Translated by Sir Thomas Heath. Complete works of ancient geometer feature such topics as the famous problems of the ratio of the areas of a cylinder and an inscribed sphere; the properties of conoids, spheroids, and spirals; more. 326pp. 5 3/8 x 8 1/2. 0-486-42084-1

THE HISTORICAL ROOTS OF ELEMENTARY MATHEMATICS, Lucas N. H. Bunt, Phillip S. Jones, and Jack D. Bedient. Exciting, hands-on approach to understanding fundamental underpinnings of modern arithmetic, algebra, geometry and number systems examines their origins in early Egyptian, Babylonian, and Greek sources. 336pp. 5 3/8 x 8 1/2. 0-486-25563-8

THE THIRTEEN BOOKS OF EUCLID'S ELEMENTS, Euclid. Contains complete English text of all 13 books of the Elements plus critical apparatus analyzing each definition, postulate, and proposition in great detail. Covers textual and linguistic matters; mathematical analyses of Euclid's ideas; classical, medieval, Renaissance and modern commentators; refutations, supports, extrapolations, reinterpretations and historical notes. 995 figures. Total of 1,425pp. All books 5 3/8 x 8 1/2.
Vol. I: 443pp. 0-486-60088-2
Vol. II: 464pp. 0-486-60089-0
Vol. III: 546pp. 0-486-60090-4

A HISTORY OF GREEK MATHEMATICS, Sir Thomas Heath. This authoritative two-volume set that covers the essentials of mathematics and features every landmark innovation and every important figure, including Euclid, Apollonius, and others. 5 3/8 x 8 1/2. Vol. I: 461pp. 0-486-24073-8
Vol. II: 597pp. 0-486-24074-6

A MANUAL OF GREEK MATHEMATICS, Sir Thomas L. Heath. This concise but thorough history encompasses the enduring contributions of the ancient Greek mathematicians whose works form the basis of most modern mathematics. Discusses Pythagorean arithmetic, Plato, Euclid, more. 1931 edition. 576pp. 5 3/8 x 8 1/2.
0-486-43231-9

CHINESE MATHEMATICS IN THE THIRTEENTH CENTURY, Ulrich Libbrecht. An exploration of the 13th-century mathematician Ch'in, this fascinating book combines what is known of the mathematician's life with a history of his only extant work, the Shu-shu chiu-chang. 1973 edition. 592pp. 5 3/8 x 8 1/2.
0-486-44619-0

PHILOSOPHY OF MATHEMATICS AND DEDUCTIVE STRUCTURE IN EUCLID'S ELEMENTS, Ian Mueller. This text provides an understanding of the classical Greek conception of mathematics as expressed in Euclid's Elements. It focuses on philosophical, foundational, and logical questions and features helpful appendixes. 400pp. 6 1/2 x 9 1/4. 0-486-45300-6

BEYOND GEOMETRY: Classic Papers from Riemann to Einstein, Edited with an Introduction and Notes by Peter Pesic. This is the only English-language collection of these 8 accessible essays. They trace seminal ideas about the foundations of geometry that led to Einstein's general theory of relativity. 224pp. 6 1/8 x 9 1/4. 0-486-45350-2

HISTORY OF MATHEMATICS, David E. Smith. Two-volume history – from Egyptian papyri and medieval maps to modern graphs and diagrams. Non-technical chronological survey with thousands of biographical notes, critical evaluations, and contemporary opinions on over 1,100 mathematicians. 5 3/8 x 8 1/2.
Vol. I: 618pp. 0-486-20429-4
Vol. II: 736pp. 0-486-20430-8

Physics

THEORETICAL NUCLEAR PHYSICS, John M. Blatt and Victor F. Weisskopf. An uncommonly clear and cogent investigation and correlation of key aspects of theoretical nuclear physics by leading experts: the nucleus, nuclear forces, nuclear spectroscopy, two-, three- and four-body problems, nuclear reactions, beta-decay and nuclear shell structure. 896pp. 5 3/8 x 8 1/2. 0-486-66827-4

QUANTUM THEORY, David Bohm. This advanced undergraduate-level text presents the quantum theory in terms of qualitative and imaginative concepts, followed by specific applications worked out in mathematical detail. 655pp. 5 3/8 x 8 1/2. 0-486-65969-0

ATOMIC PHYSICS AND HUMAN KNOWLEDGE, Niels Bohr. Articles and speeches by the Nobel Prize–winning physicist, dating from 1934 to 1958, offer philosophical explorations of the relevance of atomic physics to many areas of human endeavor. 1961 edition. 112pp. 5 3/8 x 8 1/2. 0-486-47928-5

COSMOLOGY, Hermann Bondi. A co-developer of the steady-state theory explores his conception of the expanding universe. This historic book was among the first to present cosmology as a separate branch of physics. 1961 edition. 192pp. 5 3/8 x 8 1/2. 0-486-47483-6

LECTURES ON QUANTUM MECHANICS, Paul A. M. Dirac. Four concise, brilliant lectures on mathematical methods in quantum mechanics from Nobel Prize-winning quantum pioneer build on idea of visualizing quantum theory through the use of classical mechanics. 96pp. 5 3/8 x 8 1/2. 0-486-41713-1

THE PRINCIPLE OF RELATIVITY, Albert Einstein and Frances A. Davis. Eleven papers that forged the general and special theories of relativity include seven papers by Einstein, two by Lorentz, and one each by Minkowski and Weyl. 1923 edition. 240pp. 5 3/8 x 8 1/2. 0-486-60081-5

PHYSICS OF WAVES, William C. Elmore and Mark A. Heald. Ideal as a classroom text or for individual study, this unique one-volume overview of classical wave theory covers wave phenomena of acoustics, optics, electromagnetic radiations, and more. 477pp. 5 3/8 x 8 1/2. 0-486-64926-1

THERMODYNAMICS, Enrico Fermi. In this classic of modern science, the Nobel Laureate presents a clear treatment of systems, the First and Second Laws of Thermodynamics, entropy, thermodynamic potentials, and much more. Calculus required. 160pp. 5 3/8 x 8 1/2. 0-486-60361-X

QUANTUM THEORY OF MANY-PARTICLE SYSTEMS, Alexander L. Fetter and John Dirk Walecka. Self-contained treatment of nonrelativistic many-particle systems discusses both formalism and applications in terms of ground-state (zero-temperature) formalism, finite-temperature formalism, canonical transformations, and applications to physical systems. 1971 edition. 640pp. 5 3/8 x 8 1/2. 0-486-42827-3

QUANTUM MECHANICS AND PATH INTEGRALS: Emended Edition, Richard P. Feynman and Albert R. Hibbs. Emended by Daniel F. Styer. The Nobel Prize–winning physicist presents unique insights into his theory and its applications. Feynman starts with fundamentals and advances to the perturbation method, quantum electrodynamics, and statistical mechanics. 1965 edition, emended in 2005. 384pp. 6 1/8 x 9 1/4. 0-486-47722-3

Physics

INTRODUCTION TO MODERN OPTICS, Grant R. Fowles. A complete basic undergraduate course in modern optics for students in physics, technology, and engineering. The first half deals with classical physical optics; the second, quantum nature of light. Solutions. 336pp. 5 3/8 x 8 1/2. 0-486-65957-7

THE QUANTUM THEORY OF RADIATION: Third Edition, W. Heitler. The first comprehensive treatment of quantum physics in any language, this classic introduction to basic theory remains highly recommended and widely used, both as a text and as a reference. 1954 edition. 464pp. 5 3/8 x 8 1/2. 0-486-64558-4

QUANTUM FIELD THEORY, Claude Itzykson and Jean-Bernard Zuber. This comprehensive text begins with the standard quantization of electrodynamics and perturbative renormalization, advancing to functional methods, relativistic bound states, broken symmetries, nonabelian gauge fields, and asymptotic behavior. 1980 edition. 752pp. 6 1/2 x 9 1/4. 0-486-44568-2

FOUNDATIONS OF POTENTIAL THERY, Oliver D. Kellogg. Introduction to fundamentals of potential functions covers the force of gravity, fields of force, potentials, harmonic functions, electric images and Green's function, sequences of harmonic functions, fundamental existence theorems, and much more. 400pp. 5 3/8 x 8 1/2. 0-486-60144-7

FUNDAMENTALS OF MATHEMATICAL PHYSICS, Edgar A. Kraut. Indispensable for students of modern physics, this text provides the necessary background in mathematics to study the concepts of electromagnetic theory and quantum mechanics. 1967 edition. 480pp. 6 1/2 x 9 1/4. 0-486-45809-1

GEOMETRY AND LIGHT: The Science of Invisibility, Ulf Leonhardt and Thomas Philbin. Suitable for advanced undergraduate and graduate students of engineering, physics, and mathematics and scientific researchers of all types, this is the first authoritative text on invisibility and the science behind it. More than 100 full-color illustrations, plus exercises with solutions. 2010 edition. 288pp. 7 x 9 1/4. 0-486-47693-6

QUANTUM MECHANICS: New Approaches to Selected Topics, Harry J. Lipkin. Acclaimed as "excellent" (*Nature*) and "very original and refreshing" (*Physics Today*), these studies examine the Mössbauer effect, many-body quantum mechanics, scattering theory, Feynman diagrams, and relativistic quantum mechanics. 1973 edition. 480pp. 5 3/8 x 8 1/2. 0-486-45893-8

THEORY OF HEAT, James Clerk Maxwell. This classic sets forth the fundamentals of thermodynamics and kinetic theory simply enough to be understood by beginners, yet with enough subtlety to appeal to more advanced readers, too. 352pp. 5 3/8 x 8 1/2. 0-486-41735-2

QUANTUM MECHANICS, Albert Messiah. Subjects include formalism and its interpretation, analysis of simple systems, symmetries and invariance, methods of approximation, elements of relativistic quantum mechanics, much more. "Strongly recommended." – *American Journal of Physics.* 1152pp. 5 3/8 x 8 1/2. 0-486-40924-4

RELATIVISTIC QUANTUM FIELDS, Charles Nash. This graduate-level text contains techniques for performing calculations in quantum field theory. It focuses chiefly on the dimensional method and the renormalization group methods. Additional topics include functional integration and differentiation. 1978 edition. 240pp. 5 3/8 x 8 1/2. 0-486-47752-5

Physics

MATHEMATICAL TOOLS FOR PHYSICS, James Nearing. Encouraging students' development of intuition, this original work begins with a review of basic mathematics and advances to infinite series, complex algebra, differential equations, Fourier series, and more. 2010 edition. 496pp. 6 1/8 x 9 1/4. 0-486-48212-X

TREATISE ON THERMODYNAMICS, Max Planck. Great classic, still one of the best introductions to thermodynamics. Fundamentals, first and second principles of thermodynamics, applications to special states of equilibrium, more. Numerous worked examples. 1917 edition. 297pp. 5 3/8 x 8. 0-486-66371-X

AN INTRODUCTION TO RELATIVISTIC QUANTUM FIELD THEORY, Silvan S. Schweber. Complete, systematic, and self-contained, this text introduces modern quantum field theory. "Combines thorough knowledge with a high degree of didactic ability and a delightful style." – *Mathematical Reviews.* 1961 edition. 928pp. 5 3/8 x 8 1/2. 0-486-44228-4

THE ELECTROMAGNETIC FIELD, Albert Shadowitz. Comprehensive undergraduate text covers basics of electric and magnetic fields, building up to electromagnetic theory. Related topics include relativity theory. Over 900 problems, some with solutions. 1975 edition. 768pp. 5 5/8 x 8 1/4. 0-486-65660-8

THE PRINCIPLES OF STATISTICAL MECHANICS, Richard C. Tolman. Definitive treatise offers a concise exposition of classical statistical mechanics and a thorough elucidation of quantum statistical mechanics, plus applications of statistical mechanics to thermodynamic behavior. 1930 edition. 704pp. 5 5/8 x 8 1/4. 0-486-63896-0

INTRODUCTION TO THE PHYSICS OF FLUIDS AND SOLIDS, James S. Trefil. This interesting, informative survey by a well-known science author ranges from classical physics and geophysical topics, from the rings of Saturn and the rotation of the galaxy to underground nuclear tests. 1975 edition. 320pp. 5 3/8 x 8 1/2. 0-486-47437-2

STATISTICAL PHYSICS, Gregory H. Wannier. Classic text combines thermodynamics, statistical mechanics, and kinetic theory in one unified presentation. Topics include equilibrium statistics of special systems, kinetic theory, transport coefficients, and fluctuations. Problems with solutions. 1966 edition. 532pp. 5 3/8 x 8 1/2. 0-486-65401-X

SPACE, TIME, MATTER, Hermann Weyl. Excellent introduction probes deeply into Euclidean space, Riemann's space, Einstein's general relativity, gravitational waves and energy, and laws of conservation. "A classic of physics." – *British Journal for Philosophy and Science.* 330pp. 5 3/8 x 8 1/2. 0-486-60267-2

RANDOM VIBRATIONS: Theory and Practice, Paul H. Wirsching, Thomas L. Paez and Keith Ortiz. Comprehensive text and reference covers topics in probability, statistics, and random processes, plus methods for analyzing and controlling random vibrations. Suitable for graduate students and mechanical, structural, and aerospace engineers. 1995 edition. 464pp. 5 3/8 x 8 1/2. 0-486-45015-5

PHYSICS OF SHOCK WAVES AND HIGH-TEMPERATURE HYDRO DYNAMIC PHENOMENA, Ya B. Zel'dovich and Yu P. Raizer. Physical, chemical processes in gases at high temperatures are focus of outstanding text, which combines material from gas dynamics, shock-wave theory, thermodynamics and statistical physics, other fields. 284 illustrations. 1966–1967 edition. 944pp. 6 1/8 x 9 1/4. 0-486-42002-7